EQUINE
GENETICS
&
SELECTION
PROCEDURES

ii.

1978

Equine Research
PUBLICATIONS

15048 Beltway Drive Dallas, Texas 75240

WRITTEN BY:
the research staff of Equine Research Publications

ILLUSTRATED BY:
Patti R. Strauch, D.V.M.

VETERINARY RESEARCH EDITORS:
Lorraine W. Chalkley, M.S., D.V.M.
W. R. Cook, F.R.C.V.S., Ph.D.

EDITOR/PUBLISHER:
Don M. Wagoner

ACKNOWLEDGEMENT PAGE

Equine Research Publications wishes to express sincere appreciation to:

T.W. Martin, D.V.M. (Practitioner, Breeder)
W.C. McMullan, D.V.M., M.S. (Equine Clinician)
J.R. Joyce, D.V.M., M.S. (Equine Clinician)
D.V. Hanselka, D.V.M., M.S. (Equine Clinician)

for their technical contributions and invaluable assistance in the preparation of this book.

TABLE OF CONTENTS

INTRODUCTION . . . .**xv**

I. THE ORIGIN AND DEVELOPMENT OF THE HORSE

1. Evolution — The Dawn of Today's Horse 1
   Natural Selection 1
   Dawn Horse 2
   Primitive Forest Horse 4
   Primitive Plains Horse 6
   Early Migration 11
   Massive Migrations 14
   Ancestral Horse Types 17

2. Domestication and Breed Development 23
   Artificial Selection 31
   Breeds of Horses 35
     Light Horses 38
       The Arabian 38
       The Thoroughbred 40
       The Quarter Horse 42
       The Standardbred 44
       The Morgan 46
     The Color Breeds 47
       Parti-colored Horses 47
         Appaloosas 47
         Paints, Pintos and Morocco Spotted Horses 49
       Palominos 50
       Buckskins 51
     Gaited Horses 52
       American Saddlebreds 52
       Tennessee Walking Horses 53
       Fox Trotting Horses 54
       Paso Horses 54
     Draft Horses 55
     Pony Breeds 58

II. SELECTION

3. Important Selection Characteristics 63
   Selection Criteria 70
   Pedigree 71
      Close Ancestors 76
      Heritability 76
      Collateral Relatives 77
      Progeny Records 78
   Performance...79
      Objective vs. Subjective Records 79
      Reproductive Performance 80
   Conformation..82
      Head and Neck 85
      Body 89
      Legs 93
   Temperament 102
   Intelligence 102
   Trainability 103

4. Performance Selection Characteristics 105
   The Race Horse 105
      Pedigree 106
      Performance 108
      Conformation 109
      Elements of Speed 113
   The Harness Race Horse 116
   The Steeple Chaser 119
   The Hunter and Jumper 120
   Combined Training 125
   The Stock Horse...126
   The Endurance and Competitive Trail Horse 130
   The Draft Horse...133
   The Show Horse 137

III. APPLYING GENETICS TO SELECTION

5. Basic Genetic Definitions 141
   Chromosome 141
   Gene 142

Heredity 146
Environment 146

6. Breeding Aspects 147
Selecting a Stallion 147
   Reproductive Anatomy 148
   Reproductive Examination 149
   Reproductive Records 150
   Performance Records 151
   Siring Records 151
   Pedigree 153
   Stud Fee 153
   Theories on Selecting a Sire 154
      Nicks 154
      Broodmare Sires 154
      Families 155
Selecting a Broodmare 155
   Reproductive Anatomy 156
   Reproductive Examination 158
   Disposition 159
   Breeding Records 160
   Uterine Biopsy 160
   Pedigree 161
   Performance Records 161
   Theories on Selecting a Broodmare 162
      Maternal Influence 162
      Performance vs. Pedigree 163
Fertility 164

7. Inbreeding and Outbreeding 169
Culling 169
Inbreeding 171
   Disadvantages of Inbreeding 171
   Advantages of Inbreeding 174
   Closebreeding 175
   Linebreeding 177
Relationships 179
Inbreeding Coefficient 181
Outbreeding 186
   Outcrossing 187
   Linecrossing 188

Grading 189
Crossbreeding 189
Species Hybridization 190

8. **Environment and Heredity** 193
    Genetic Variation 194
    Environmental Variation 194
    Judging Genetic Potential 196

9. **Breed Improvement Through Applied Genetics** 199
    Selection Methods 200
        Number of Traits 200
        Tandem Method 201
        Minimum Culling Level 203
        Selection Index 205
    Selection Pressure 207
    Heritability Estimates 207
        Predicting Breeding Values 208
    Quantitative Inheritance 210
    Qualitative Inheritance 213
    Selection For a Trait 217
        Backcrossing 218
            Introducing a Dominant Trait 219
            Introducing a Recessive Trait 224
    Selection Against a Trait 227

10. **Controversial Theories of Selection** 231
    Galton's Law of Hereditary Influence 231
    Stamina Index 233
    Stamina of a Stallion's Progeny 234
    Vuillier Dosage System 234
    Bruce Lowe System 237
    Telegony 238

11. **Coat Color and Texture** 239
    The Mechanics of Coat Color 240
    Black and Liver: The B Locus 244
    Bay and Seal Brown: The A Locus 245
    Chestnut and Black: The E Locus 247
    Grey, Roan and Dominant White: The Epistatic Modifiers ..249

Grey: The G Locus 250
    Bloody Shoulder Markings 251
Roan: The R Locus 252
Dominant White: The W Locus 255
Diluted Coat Colors: The C and D Loci 256
    The C Locus 257
        Palomino and Cremello 257
        Diluted Seal Browns 259
        Buckskin/Dun: Dilution of Eumelanin 259
        Perlino 260
        Red Dun 260
    The D Locus 260
        Isabella 261
        Grulla 261
        Buckskin/Dun: Dilution of Eumelanin
            and Phaeomelanin 262
    Combined Effects of c^{cr} and D Dilution 262
Silver Dapple: The S Locus 263
Flaxen Mane and Tail 265
Pinto/Paint: Tobiano and Overo Spotting 266
    Tobiano: The T Locus 266
    Overo: The O Locus 267
Appaloosa Spotting 268
    Basic Appaloosa Gene 270
    White Blanket Gene 270
    Spotting Gene 271
    Leopard and Roaning Patterns 271
    Diminishing Contrast 273
White Markings: The Face and Legs 274
Colors That Always Breed True 295
Colors That Never Breed True 295
Colors That Breed True When Homozygous 295
Coat Texture 296
    Feathering 297
    Curly Hair 298
    Whorls (Trichoglyphs) 300
    Appaloosa Spots 301

12. **Heredity and Gaits** 303
    Walk 304
    Trot 307
    Canter 310

Backing 314
Running Walk 316
Rack 316
Fox Trot 316
Pace 317
Paso Gaits 317

13. Inherited Abnormalities 325
Conformation 329
Head and Neck 330
Parrot Mouth 330
Bulldog Mouth 331
Predisposition to Periodontitis .332
Big Head (Equine Osteomalacia) 333
Ewe Neck 334
Cresty Neck 334
Limbs 334
Abnormal Leg Set 334
Epiphysitis 336
Contracted Digital Flexor Tendons 337
Upward Fixation of the Patella .338
Lateral Luxation of the Patella .339
Contracted Heels 339
Flat Feet 340
Osteochondritis Dissecans 341
Hip Dysplasia 341
Stiff Joints 342
Fibular Enlargement 342
Polydactyly 342
Abrachia 343
Body 344
Umbilical Hernia 344
Lordosis (Sway Back) 345
Kyphosis (Roach Back) 348
Scoliosis (Lateral Curvature of the Spine) 348
Multiple Exostosis 348
The Eye 348
Iris 349
Heterochromia 349
Aniridia 350
Coloboma Iridis 351
Iris Cysts 351

Hyperplasia of the Corpora Nigra 352
Cornea..352
    Microcornea 352
    Congenital Keratopathy 352
    Dermoids 352
    Melanosis 353
Lens..353
    Cataracts 353
    Lens Luxation.................................355
    Periodic Ophthalmia 355
    Persistent Hyaloid Vessel 356
Retina 357
    Absence of the Retina 357
    Glaucoma 357
    Congenital Night Blindness 357
    Optic Nerve Hypoplasia........................358
Other Defects 359
    Microphthalmia 359
    Entropion and Ectropion 360
    Atresia of the Nasolacrimal Duct...................360
Skin 361
    Photosensitization 361
    Squamous Cell Carcinoma 362
    Melanoma 363
    Lack of Chestnuts and/or Ergots 364
    Dry Coat (Anhidrosis) 365
    Pinky Syndrome...............................366
    Variegated Leukotrichia 366
    Subcutaneous Hypoplasia 367
    Epitheliogenesis Imperfecta 367
    Linear Keratosis 367
    Atheroma (Sebaceous Cysts) of the False Nostril 367
    Mohammed's Thumbprint 367
    Mallenders and Sallenders 367
Reproductive System 368
    Cryptorchidism 368
    Scrotal Hernia 371
    Hypothyroidism 372
    Nymphomania 373
    Pneumovagina 374
Digestive System 376
    Cleft Palate 376

CONTENTS

Atresia Coli, Atresia Recti and Atresia Ani 377
Esophageal Defects 379
Shistosoma Reflexum 379
Nervous System 380
Temperament 381
Hereditary Ataxia 382
Wobbler Syndrome 382
Congenital Occipito-Atlanto-Axial
Malformation (OAAM) 386
Cerebellar Hypoplasia 387
Cerebellar Ataxia 388
Congenital Hydrocephalus 389
Idiopathic Epilepsy 390
Shivering 390
Respiratory System 391
Roaring 391
Heaves 394
Epistaxis 395
Pharyngeal Cysts 395
Circulatory System 395
Patent Ductus Arteriosus 397
Interventricular Septal Defects 398
Patent Foramen Ovale 398
Persistent Right Aortic Arch 399
Blood 401
Neonatal Isoerythrolysis 401
Combined Immunodeficiency Disease 404
Hemophilia 406

IV. CYTOGENETICS AND PROBABILITY

14. The Cell 413
Cell Structure 414
The Structure of Inheritance 415
DNA 419
RNA 422
Karyotypes 422
Mitosis 426
Stages of Mitosis 427
Interphase 427
Prophase 427

    Metaphase 428
    Anaphase 428
    Telophase 429
  Meiosis 429
    Spermatogenesis Part I 430
    Spermatogenesis Part II 434
    Oogenesis 436

15. Sex Chromosomes 439
  Sex Determination 439
  X-Linked Traits 440
  Y-Linked Traits 443
  Sex Influenced Traits 443
  Sex Limited Traits 443
  Sex Chromosome Abnormalities 444
    Small Hard Ovaries (Aneuploidy 63 XO)............444
    Intersex...445
      Male Pseudohermaphroditism 445
      Testicular Feminization 447

16. Complicating Factors of Inheritance 449
  Multiple Factors...................................449
  Multiple Alleles 449
  Modifying Genes 450
  Gene Interactions 450
    Dominance 450
      Complete Dominance 450
      Codominance 451
      Incomplete Dominance 451
      Overdominance 451
    Epistasis 451
    Penetrance and Expressivity 452
  Linkage and Independent Assortment 452
  Pleiotropic Effects 454
  Prepotency 454
  Combination of Traits 455

17. Mutations 457
  Structural Mutations 458
  Point Mutations 460
  Gross Mutations 462

Chromosomal Aberrations 462
    Nondisjunction 463
    Centric Fusion 463
Chromosomal Rearrangement 464
    Crossing Over 464
    Translocations 464
    Deletions, Inversions and Ring Chromosomes 465

18. Population Genetics 469
Elements of Progress 469
Laws of Probability 471
Gene Frequency 475
Inbreeding and Genotypic Frequency 480
Additive Genes 481
Population Mean 482
Variance and Standard Deviation 483
Errors in Measurement 485
Association of Traits 488

BIBLIOGRAPHY 493
APPENDIX 501
Paternity Tests 501
Relationships Between Blood Types
    and Performance Ability 505
Comparative Analysis of Plasma Protein
    in Various Breeds 505
Congenital Defects in Foals 507
Comparison Between Horse and Ass 510
Points of the Horse 512
Skeleton of the Horse 513
GLOSSARY 515
INDEX 531

INTRODUCTION

EQUINE GENETICS AND SELECTION PROCEDURES is today's most comprehensive and instructive study on all aspects of equine selection. This text has benefited from the efforts of many researchers and will hopefully stimulate further research in this interesting and vital area. Controversial and conflicting viewpoints have been thoroughly analyzed to present only the most accurate and up-to-date conclusions.

Evaluation of a horse's natural potential to perform or produce is essential within a successful operation. **EQUINE GENETICS AND SELECTION PROCEDURES** is designed to provide the horseman with an important guide to establishing an effective selection program — one that fits his own goals and requirements.

Another important objective of this text is to present both the horseman and the serious student with a reference to inherited traits. By carefully defining all scientific terms and by applying important genetic concepts to practical selection, this text helps to bridge the literature gap between the horseman and the geneticist. Genetic concepts normally regarded as difficult have been organized and presented in a clear concise manner, so that all horsemen can apply current data to their breeding and selection programs.

Perhaps the most important objective of **EQUINE GENETICS AND SELECTION PROCEDURES** is to provide a guide to understanding the procedures involved in identifying and selecting horses with the greatest genetic potential for a specific career — breeding, performance, etc. This enables the horseman to improve herd quality and avoid economic waste.

section I.

the origin and development of the horse

1

EVOLUTION: THE DAWN
OF TODAY'S HORSE

The development of the modern horse from Eohippus, a 12-inch fox-like mammal, takes place over a period of nearly 60 million years. Evolution illustrates how nature chooses individuals who can escape predators and, at the same time, adapt to drastic changes in the environment. It also explains why the horse developed his distinctive features (i.e., size, limb structure, eye placement, etc.).

As the history of the horse is traced, the reader should remember that, throughout evolution, many types of prehistoric horses developed. Natural selection (survival of the fittest) determined that only one pathway would continue, eventually producing the present day equine species. As the earliest form of equine selection, evolution provides a natural introduction for a study of equine genetics and selection procedures.

Natural Selection

To understand how the horse survived drastic environmental changes encountered over the years, the word mutation must be defined. Within each living cell, the units of inheritance, or genes, are carried upon long protein strands, or chromosomes. Each gene contains a unique message that is carefully transmitted to the protein production sites within the cell. (The building blocks for living tissue are made at these sites.) If a chromosome loses a section of information, or if a section becomes turned around or twisted, a genetic message may be altered. This results in a mutation, or sudden variation in protein formation. If a protein is altered, a related function or characteristic of the living body may also be altered. If mutations had never occurred, primitive species would have been identical; they would have survived or perished as a complete group. Mutations did occur, however, and provided variation that gave some animals better survival characteristics. Imagine, for example, that a group of primitive horses had eyes placed toward the front of the head. One strange mutant, with eyes located toward each side of his head, was born within this group. Because of his increased field of vision, the mutant could detect predators far more efficiently than his companions. His survival, and the survival of his offspring, were favored by this mutant trait. This simplified example is provided to show that mutations played an important role throughout evolution. (Refer to **"Mutations"** under **CYTOGENETICS AND PROBABIL-ITY**.)

The first chapter of the horse's evolution begins during the age of reptiles (Mesozoic Era, which began about 200 million years ago and lasted for an estimated 130 million years). According to many authorities, the first known ancestor of all present day mammals was THERIODANT, a small, nimble animal that inhabited the swamps of what is now North America. Although Theriodant's appearance contrasted sharply with that of the dominant reptile population, those differences eventually favored his progeny when the swamps disappeared and the forests turned to plains. Unlike the reptiles, Theriodant's legs were underneath his muscular body. This feature gave the prehistoric mammal agility, enabling him to dart in and out of the underbrush to escape his more cumbersome enemies. As time passed, even more changes took place to give Theriodant different survival characteristics, depending on the demands of his environment. Eventually, his progeny developed along various pathways to form the mammal dynasty of today's world.

Dawn Horse

Geologists divide the earth's history into four distinct eras: Precambrian, Paleozoic, Mesozoic, and Cenozoic. The end of each period (except the present Cenozoic Era, which began about 60 to 70 million years ago), was marked by profound geological changes and the development of new life forms. The end of the Mesozoic Era, for example, was marked by turbulent climatic conditions, earthquakes, and upheavals which remolded the earth's surface. Gradually, conditions stabilized; torrid jungles were replaced with semi-tropical temperatures, forests, and lagoons. The beginning of the Cenozoic Era marked the end of reptile dominion. As the swamps receded, the reptiles perished, leaving the small mammal population at the threshold of their future dynasty.

EOHIPPUS ("dawn horse"), the first known ancestor of the horse, was one of the first mammals present during the first Cenozoic, or Eocene, epoch. (The Cenozoic Era is divided into five distinct epochs: Eocene, Oligocene, Miocene, Pliocene, and Pleistocene.) The remains of at least 13 different Eohippus types have been discovered in the Eocene deposits of Wyoming, New Mexico, Colorado, Utah, and England (possibly also in France, Belgium, and Switzerland). These prehistoric skeletons show that Eohippus stood approximately 12 to 14 inches at the shoulder (variations in size are directly related to location and possibly related to environmental conditions within each area). The primitive horse was also characterized by an arched back, rounded body, slender legs, and weight-bearing foot pads. He had three toes on each hind foot and four toes on each front foot. Small splint bones (remnants of non-functional toes) on each foot suggests that Eohippus' ancestors had five toes on each foot. Some authorities believe that the end of each toe was encased in a small hoof-like toenail.

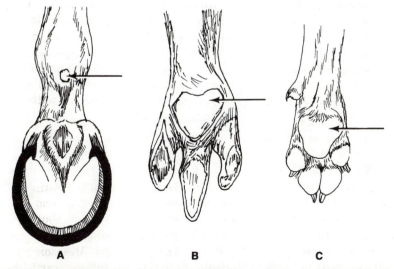

A   **B**   **C**

Bottom surface of the horse's hoof (A), tapir's foot (B), and dog's paw (C), comparing position and size of the horse's ergot (remnant of a primitive foot pad) with the foot pads of his distant relatives, the tapir and the dog (see arrows).

Although Eohippus might have been an easy victim of his early hostile environment, he survived and developed into several modified forms. Scientists have suggested reasons for his survival:

1) Small size indicates that Eohippus was a creature that necessarily had to avoid predators. He probably possessed quick reflexes and a tendency to panic, which helped him survive.

2) Speed and agility were probably the two most important survival features. Concealed by dense vegetation along the river banks, Eohippus ran quietly upon his soft foot pads like a dog, sometimes leaping through the bushes like a rabbit. Slender agile legs and flexible gripping toes enabled the prehistoric horse to maneuver rapidly over rough, uneven ground.

3) The primitive horse's food supply (soft forest leaves) was bountiful during this period. Eating and drinking without leaving the dense cover along the river may have been another important survival aspect.

Several Eohippus types are believed to have wandered from the American continent across the Bering Strait land passage and into parts of what is now Europe. Many back and forth migrations probably occurred at this time. For reasons unknown to man, the European ancestor of the horse did not survive. The North American varieties continued, changing slightly in form (i.e., Orohippus and Epihippus) and gradually adapting to environmental changes which marked the beginning of a new epoch.

Primitive Forest Horse

The Oligocene epoch began over 38 million years ago and was probably characterized by continued regression of swamps and expansion of forests and plains. At this point, the primitive horse is referred to as MESOHIPPUS. (Later Oligocene types are often referred to as Miohippus or "middle horse.") Mesohippus was about six to eight inches taller than his predecessors. This increase in size was an important survival characteristic, since it helped the primitive horse reach the soft forest leaves on which he fed.

Increased size may have served another important purpose. Unlike his ancestors, Mesohippus could not rely on dense vegetation for concealment and protection. For this reason, longer legs, and a consequent increase in speed, enhanced his survival. Mesohippus was probably beginning to place less weight upon his foot pads and more upon his toes. In his new form, the primitive horse had three toes on each foot. (One toe from each front foot had become remnant splint bones; later, these bones were replaced by remnants of the two outer toes which became the present day splint bones.) Although weight was distributed evenly upon each of the functional toes, the middle toe (which, millions of years later, evolved into the hoof) was relatively larger.

Mesohippus' skull was slightly larger than that of Eohippus, suggesting that some degree of intelligence aided his survival. Because a wider field of vision and rapid response to visual stimuli helped the horse detect and flee from his enemies, mutations which caused placement of the eyes back toward the sides of the head (lateral eye placement) were favored by natural selection. With lateral eye placement, two separate pictures are viewed at one time (monocular vision). Together, these pictures are conceived by the brain as flat, a feature which allows detection of even the slightest movement. (The horse is capable of focusing on a small area with both eyes and therefore has slight binocular vision.)

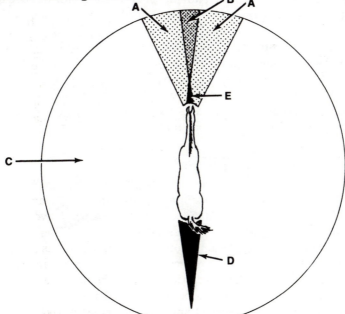

Lateral eye placement allows the horse to see either directly forward (areas A & B) or laterally and to the rear (area C). He cannot see both to the front and sides without slight eye movement. There is a small area (B) that the horse can view with both eyes at the same time (binocular vision). Black areas (D & E) indicate the horse's blind spots.

Primitive Plains Horse

The gradual geological changes which resulted in the Great Plains of North America occurred during the 16 million year period known as the Miocene epoch. Responding to changes in available food sources (forest leaves to tough pampas grass), the primitive horse developed several specialized structures and changed slightly in appearance. The name MERYCHIPPUS is used to describe the primitive horse of the Miocene period, the third epoch of the Cenozoic Era.

Merychippus was characterized by his ability to grasp, crop, and grind grass with modified teeth. The primitive short-crowned teeth of earlier species had been replaced by sharp incisors and grinding molars. To compensate for the wearing effects of the grinding motion, Merychippus had deep jaws with high-crowned teeth that were capable of continual eruption (i.e., in today's horse, four to five years of true growth is followed by gradual movement of the teeth from the jaw as their sockets become shorter with age).

Tushes (early canine teeth) and wolf teeth are remnants of primitive teeth, which dwindled during the horse's evolution. Because these teeth served no essential purpose, mutations which caused their absence were not selected against. Occasionally, vestiges of these primitive teeth will appear in today's horse. Tushes, which grow from both the upper and lower jaws, are usually limited to the males; they are not associated with any significant problems. Wolf teeth, on the other hand, erupt from the upper jaw (just in front of the first premolar) and are usually extracted as they often cause gum irritation, interference with the bit, and root fusion with the first premolars.

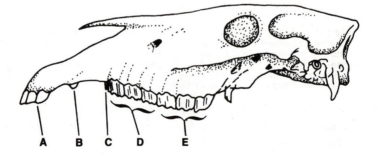

Canine teeth, or tushes (B), which grow from both the upper and lower jaws, do not usually cause gum irritation or interference with the bit, as do primitive wolf teeth (C), which grow only from the upper jaw. Incisors (A), premolars (D), and molars (E) are important to normal prehension, mastication, and digestion.

The new diet also had an effect on the early horse's digestive system. Although most of the nutrients were absorbed into the blood through the small intestine, the cecum and the large colon became somewhat more important when tough grass was introduced into the diet. Bacteria, located in these areas, aided the horse by digesting cellulose (fibrous plant tissue which is especially prominent in tough grasses). Eventually, the cecum became an enlarged fermentation vat, protected by the presence of an extra rib. Basically a wanderer, Merychippus ingested small amounts of food at frequent intervals. For this reason, a small stomach (still characteristic of today's horse) fit the primitive grazer's needs.

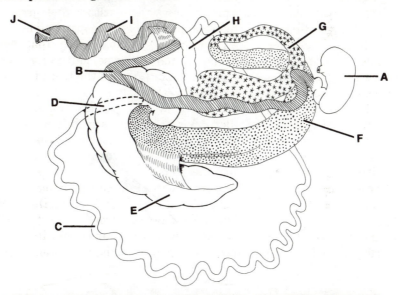

The horse's digestive tract is the result of dietary changes throughout evolution. The small stomach (A) is a reflection of the early horse's need to eat small amounts frequently as he migrated. The enlarged cecum (E) provided a larger area for bacterial digestion of tough grasses. (A) stomach, (B) duodenum, (C) jejunum, (D) ileum, (E) cecum (fermentation vat), (F) ventral colon, (G) dorsal colon, (H) transverse colon, (I) small colon, (J) rectum.

Like his ancestors, Merychippus' survival depended on his ability to detect and flee from predators. Vision, reflex, and speed were probably his three most important characteristics:

1) Although there is no muscular adjustment for focus in the horse's eye, he is capable of focusing on an object by raising or lowering his head. This movement allows light rays to strike at different places on the irregularly shaped retina until the sharpest possible

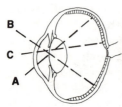

Cross-section of the equine eye. The irregular shape of the retina allows the focal length to vary with the position of the head. To see close objects, the horse raises his head (A) allowing light to fall on the upper retina. To see distant objects, the horse lowers his head (B), directing light towards the bottom of the retina. The horse holds his head level to focus on middle distances (C); light falls on the central part of the retina.

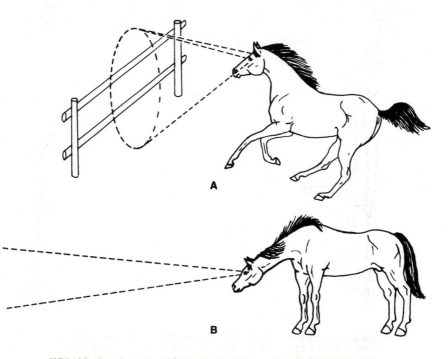

With his head carried high, the horse can focus on objects in close proximity (A). By lowering his head, the horse focuses on objects at a distance (B).

picture is formed. Because of his almost circular field of vision (lateral eye placement), and an ability to focus upon objects at a distance by lowering his head, the horse can keep a close watch on the distant horizon while grazing. This is an important survival feature which probably developed during, or just prior to, the Miocene epoch.

2) Quick reflexes and a tendency to panic were also important characteristics during the horse's evolution. The horse of today still

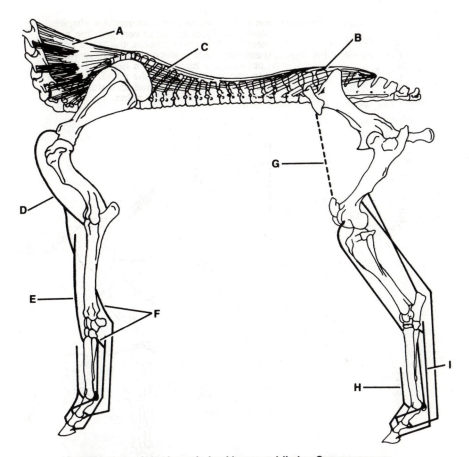

Lateral view of the horse's backbone and limbs. Suspensory apparatus allows the horse to lock his legs in a standing position while sleeping; the back is supported like a suspension bridge between the limbs. (A) Serratus cervicis muscle, (B) Longissimus dorsi muscle, (C) Spinalis dorsi muscle, (D) Biceps brachii muscle, (E) Lacertus fibrosus muscle, (F) Superior and inferior check ligaments, (G) Tensor fasciae latae muscle, (H) Suspensory ligament of the fetlock, (I) Volar annular ligament of the fetlock.

exhibits this sensitive behavior in that he is easily frightened by strange situations or sudden movements. The specialized anatomy which supports the horse's head and back, and prevents overflexion of the joints, also allows the horse to sleep while standing — the best possible position for sudden flight from danger. This stay apparatus is a complex system of ligaments, muscles, and fibrous bands, which support the body over three limbs much like a hammock or sling. One limb is cocked in a resting position.

3) Short bursts of speed were also important to the primitive horse's survival. When frightened by a predator, the herd ran at top speed; the slow, the aged, and the sick were more susceptible to attack. When a kill had been made, panic subsided and the herd stopped running.

Over the years, speed was enhanced by an overall increase in size (especially leg length) and the development of hinge-like leg joints which allowed efficient back and forward motion. Many authorities believe that, due to gradual drying of the terrain, the early horse's speed was increased by running on his toes. Merychippus, for example, had a specialized weight-bearing toe between two lesser, nonfunctional toes on each foot. As his size increased, protective hooves (modified toenails) were incorporated on each functional toe to act as shock absorbers. Without the cushioning effect of these hooves, the horse's slender legs could not have withstood the stress of support coupled with the concussion produced while running.

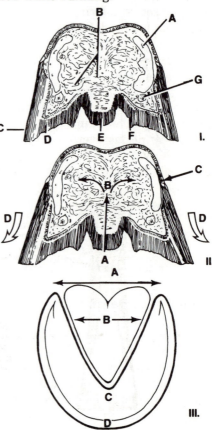

I. Cross section of the posterior third of the hoof. The lateral cartilage (A), digital cushion (B), hoof wall (C), sole (D), frog (E), bars (F) and wing of the coffin bone (G), prior to concussion.

II. Cross section of the hoof after concussion. Shock is dissipated through the frog (A) and into the digital cushion (B), causing the lateral cartilages to flatten (C) and the hoof wall to expand slightly (D).

III. Viewing the bottom of the hoof, expansion of the heel (A) and frog (B) is an important shock absorbing mechanism. The hoof wall (D) absorbs much of the stress and should be strong with slight flexibility. If the sole (C) contacts rough ground, bruising may result.

Early Migration

The arrival of the Pliocene epoch was accompanied by dry climate and sparse vegetation. The primitive horse adapted to these changes by wandering, constantly in search of water and grass. Finding protection in numbers, he traveled in herds. (Today's horse still expresses a basic herd instinct.) Characteristics such as speed, vision, reflex, and ability to utilize available food, were constantly selected for (i.e., survival of the fittest).

Within this period, the equine ancestor is referred to as PLIOHIPPUS. Migration, isolation, and environmental pressures probably contributed to the gradual development of many Pliohippus types. Although scientists cannot give a detailed description of each, they speculate that some varieties resembled the Tarpan, or perhaps Przewalski's horse, since these species have not been subjected to hundreds of years of artificial (man-made) selection. Other Pliohippus types might have resembled the zebra, with its short, muscular neck and protective coloring.

Unlike his predecessors, Pliohippus had only one toe on each foot. The non-functional toes (two on each foot) had atrophied into remnant splint bones (which are still present in the modern horse), leaving one superior weight-bearing toe and highly specialized toenail (hoof) on each foot. Skeletal remains show that equine anatomical proportions (i.e., size of the head compared to limb and body size) have changed little since Pliohippus roamed the North American plains over five million years ago. The modern horse, for example, varies in size and type, but his basic skeletal structure is quite similar to that of his Pliocene ancestors.

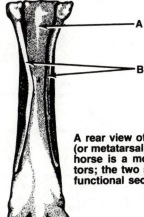

A rear view of the cannon area showing the metacarpal (or metatarsal) bones (A & B) illustrates how the modern horse is a modified reflection of his three-toed ancestors; the two splint bones (B) are remnants of the once functional second and fourth toes.

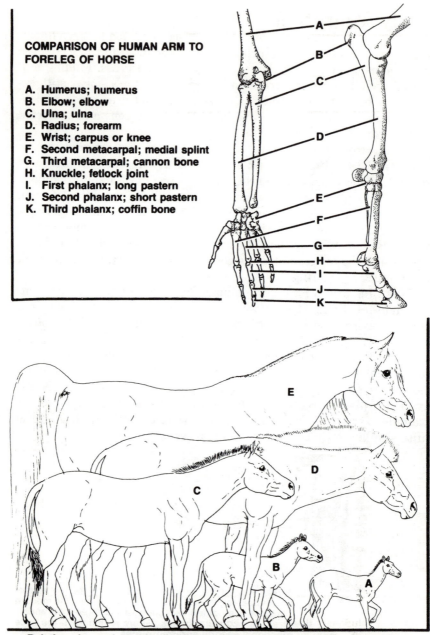

COMPARISON OF HUMAN ARM TO FORELEG OF HORSE

A. Humerus; humerus
B. Elbow; elbow
C. Ulna; ulna
D. Radius; forearm
E. Wrist; carpus or knee
F. Second metacarpal; medial splint
G. Third metacarpal; cannon bone
H. Knuckle; fetlock joint
I. First phalanx; long pastern
J. Second phalanx; short pastern
K. Third phalanx; coffin bone

Relative size and structural differences between four primitive ancestors of *Equus caballus*. (A) Eohippus, (B) Mesohippus, (C) Merychippus, (D) Pliohippus. Except for the size differences between Pliohippus and *Equus caballus* (E), skeletal structures are quite similar between the two types.

Evolutionary flow chart for Equus caballus. Note the limb development and numerous dead-end pathways.

Massive Migrations

Pliohippus was the product of persistent modification throughout severe changes in climate, terrain, and available food sources. Nearly 57 million years had lapsed since the days of Eohippus; the early equine species had become well established on the American continent by late Pliocene time. A new epoch (Pleistocene), commonly referred to as the Ice Age, brought alternate periods of cold and warmth, and massive migration of the American mammals. Ecological systems were interrupted by the movement of giant sheets of ice, as the polar ice caps expanded. Authorities believe that the ice descended and regressed several times during this period, leaving scars (e.g., huge lakes) upon the land.

Several Pliohippus forms traveled to South America and almost survived the glacial period. Other types crossed the Bering Strait land passage and migrated to various parts of Asia, Europe, and Africa, where they evolved for thousands of years, eventually forming the equine family of today. Several reverse migrations probably occurred before rising waters (melting ice) covered the intercontinental land passage. Although many Pliohippus types remained and many migrated back from the Old World, the equine population on the American continent suddenly diminished. (The so-called "wild horses," or mustangs, of America were actually descendants of Spanish stock, introduced to the continent during the 16th century.)

Why the horse suddenly became extinct in America is a question which still puzzles paleontologists (scientists who study ancient fossils, bones, and rocks). Because it coincided with glacial regression, extinction cannot be attributed to the cold. The survival of companion

San Diego Zoo Photo by Ron Garrison

Hartman's Mountain Zebra (Equus hippotigris zebra hartmannae)

Damaraland Zebra (Equus quaggoides burchelli antiquorum)

Grevy's Zebra (Equus dolichohippus grevyi)

grazers, such as the American Bison, indicates that food shortage was not a contributing factor. An interesting theory suggests that the horse and distant relations, such as the rhinoceros, were victims of a viral, bacterial, or fungal epidemic.

Fortunately, the early equine forms adapted well to the various climates of the Old World. Migrating into Africa, one Pliohippus type (Plesihippus) gradually developed into several distinct zebra types. Three species exist today: Equus grevyi, Equus burchelli, and Equus zebra. Present day zoologists believe that the domestic (African) ass

San Diego Zoo Photo

San Diego Zoo Photo

ABOVE: Somali Wild Ass (Equus asinus somaliensis)
RIGHT: Transcaspian Kulan (Equus hemionus kulan)
BELOW: Persian Onager (Equus hemionus onager)

San Diego Zoo Photo by Ron Garrison

originated in North Africa. The early species might have resembled today's Nubian and Somali Wild Asses. Some authorities believe that a distant relative of Pliohippus (Neohipparion) founded the hemionid, or Asiatic ass, species (i.e., kiang, onager, kulan).

Many theories, some of them sharply contrasting, have been provided to explain the beginning of Equus caballus, or the true horse. The following discussion will examine the relatively recent history of the horse.

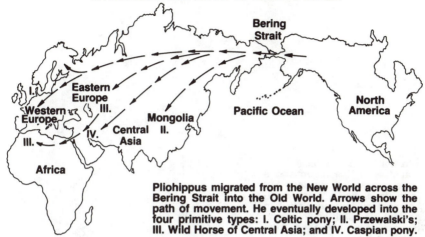

Pliohippus migrated from the New World across the Bering Strait into the Old World. Arrows show the path of movement. He eventually developed into the four primitive types: I. Celtic pony; II. Przewalski's; III. Wild Horse of Central Asia; and IV. Caspian pony.

Ancestral Horse Types

As Pliohippus crossed the Bering Strait and migrated across what is now the Eurasian continent, he encountered various climatic and geographical conditions which encouraged his evolution into the ass, the zebra, and four basic types of horses. Each type developed characteristics needed to survive in the environment encountered: in areas with temperature and altitude extremes; on moist, rocky coastland; flat, treeless plains; barren flatlands; and scorching deserts.

One of these four appeared in northwestern Europe during the Paleolithic Period (or the Old Stone Age) as early as 1,000,000 B.C. That the horse had reached Switzerland by 50,000 B.C. is almost certain: one of the earliest known engravings of the horse was uncovered at Schaffhausen, Switzerland, and has been dated to the end of the Paleolithic's Magdalenian Period. The find, called the "commando baton" by archaeologists, shows two wild horses engraved on a reindeer horn. Primitive cave drawings in France (dated to about 23,000 B.C.) also show distinct horse figures: a heavy, draft type horse is drawn next to a smaller, more refined pony type.

A small pony, standing about 12-2 hands and closely resembling today's Exmoor, was probably the first type to develop in Europe. He had a dark bay (or brown) water-resistant coat to protect him from the moist environment. He lived in the mountainous coastal areas of western Europe, which demanded endurance and sure-footedness, both qualities which he developed. This early horse gradually evolved into two types: the Celtic pony and the primitive heavy horse of the European forests. The Celtic lived along harsh coastal regions where there was little food. These conditions resulted in his development as

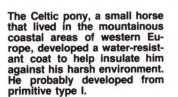

The Celtic pony, a small horse that lived in the mountainous coastal areas of western Europe, developed a water-resistant coat to help insulate him against his harsh environment. He probably developed from primitive type I.

The European forest horse (primitive heavy horse) resembled the Celtic pony with his large barrel and relatively short legs. But the heavy horse grew to massive proportions in his forest environment.

a small, short-legged animal. The heavy draft horse, on the other hand, thrived in lush forests, where he grew to massive proportions but retained many of his early characteristics: short legs (relative to body size), a large barrel and a heavy coat. This primitive giant was used to develop the European Great Horse of the Middle Ages, and eventually gave rise to the French-Belgian Ardennes breed known since Julius Caesar's time. Skeletal remains found in Britain and dated to the Pleistocene geological epoch support the theory that both types existed before domestication.

courtesy of San Diego Zoo   Photo by Ron Garrison

Przewalski's horse (Equus caballus przewalski) is thought to be a direct descendant of the primitive type that inhabited Mongolia and northern Eurasia. Note the coarse head, the absence of a forelock and the erect mane.

A second type of horse, small and stocky like the original European pony, remained in northern Eurasia and made its home on the Mongolian steppes (level, treeless plains) and in China's Hwang Ho valley. This pony had a coarser head than his European counterpart and, unlike the European pony, he retained Pliohippus' primitive black dorsal stripe. He probably had a yellowish coat color with a dark, stiff mane and no forelock. His thick, shaggy winter coat protected him from the cold frosts that blanketed Mongolia's steppes and mountain ranges. So, while the western European pony developed moisture resistance, the northern pony became resistant to cold. Today's Przewalski's horse, found in the Altai Mountain region of Mongolia and southern Sinkiang, closely resembles this type. He is thought to be a direct descendant of this Asiatic wild horse.

The Przewalski's type (primitive type II) was small and stocky like the Celtic pony. He lived mainly on the steppes of Mongolia, and developed cold resistance to protect himself from frequent frosts.

Another type developed from the stock that migrated toward Central and Southern Asia and settled in present day Iran. This horse stood about 15 hands, and is thought to be the largest of the primitive types. He had a long neck and head, with a small forehead and a Roman nose. He was probably rather large-boned, with long legs, large ears and a long back. To further emphasize his lack of refinement, he was slab-sided, with a sparse mane and tail, and a low tail set. Although unattractive by modern standards, the horse was slender and swift and, because of his environment, learned to survive in arid conditions.

This type of horse lived in Turkestan (now part of the Soviet Union east of the Caspian Sea) and in Iran and Afghanistan, which lay to the south. Drawings and skeletal remains indicate that he migrated westward across southern Iran about 3000 B.C., and that he probably continued toward Syria. Under man's selective breeding, this tall, rangy type contributed to the development of the Numidian. The Numidian was used as a war horse and entered Egypt and northern Africa with invading armies. Many Numidians reached the Barbary Coast (now including Algeria, Morocco, Tunisia and other North African states). There, he was used in a selective breeding program by the Algerians to develop the Barb, the native horse of North Africa. (Some of the early Barbs may have crossed an ancient land bridge from North Africa into Spain. The Barb almost certainly served as foundation stock for Spain's Andalusian breed.)

The fourth type of primeval horse, sometimes referred to as the Caspian pony, stood about 12 hands, and had fine bones, light legs, a

The Wild Horse of Central Asia (type III) was probably the tallest of the primitive horses — he was rather large-boned, but slender and swift, and he adapted to his environment by developing drought-resistance. The Caspian pony (type IV) was probably the most refined of the primitive types. He developed stamina and heat-resistance.

high-set tail, silky mane and tail, and a slightly concave face. He may have lived as far north as the steppes between the Black and Caspian Seas, but generally was most populous throughout Mesopotamia (the historic region roughly analogous to present day Iraq) and in the drier regions of Syria, Canaan and North Africa. This horse developed stamina and heat-resistance necessary to survive in his desert environment. He developed a small, concave head, large nostrils, and a relatively short neck. These heat-resistant qualities, combined with his small, graceful build, make him the most likely forerunner of today's Arabian. Some scientists believe that the Tarpan, a primitive breed now virtually extinct, but once abundant in Central Asia, was related to this primeval horse type. The Tarpan probably developed from a cross between the Caspian pony and the Celtic pony or the Przewalski's-type horse. Although the original Tarpan no longer exists, a breed of horses resembling that type has been created through crosses between Shetlands, Welshes, Icelandics, and other pony breeds.

Crosses between wild types were common and, as the years passed, more distinct groups formed. Environment played an important role in the development of these geographic breeding groups but, after domestication, man's influence became predominant. By the 15th century, man's selective breeding had shaped the horse into numerous types suited for specific purposes. Many of our modern breeds developed from these early types.

Photo by Everett A. Thrall

This modern Tarpan gelding resembles the original wild Tarpans with his dark points, zebra markings and dorsal stripe.

Zoological Classification of the Modern Horse

Kingdom:	Animalia (all animals)
Subkingdom:	Vertebrata (animals with internal skeletons)
Phylum:	Chordata (animals that have backbones)
Class:	Mammalia (animals that produce milk for their young)
Cohort:	Ferungulata (hoofed and carnivorous mammals)
Order:	Perissodactyla (odd-toed mammals that have hooves; includes the horse, the rhino and the tapir)
Family:	Equidae (members of the Order Perissodactyla excluding the rhino and the tapir)
Genus:	Equus (horses, asses, hemionids and zebras)
Species:	Equus caballus (the horse; includes the domestic horse, Przewalski's horse, the Tarpan, etc.)
Subspecies:	Equus caballus caballus (the domestic horse)

American Tapir. This three-toed mammal of South and Central America is a distant relative of the horse. His size and shape reflect that of the earliest horse, Eohippus. Like Eohippus, the tapir inhabits dense jungles, taking refuge along river banks.

2

DOMESTICATION & BREED DEVELOPMENT

The final chapter of the horse's evolution involves his domestication. Man used the horse throughout history to help him progress from a hunter-gatherer to a builder of cities. When the evolving horse migrated from the New World to the Old, Stone Age man himself was undergoing evolution. He lived in cave-like dwellings and existed by hunting wild animals and gathering fruits. Although he tamed the dog for companionship, he preyed upon the horse and other animals. Archaeologists have uncovered evidence in Solutre' in the Rhone Valley of France that indicates that the wild horse was the early Solutreans' chief quarry. One ancient campsite in southern France (dated to about 23,000 B.C.) was surrounded by the cracked bones of as many as 100,000 horses.

As man passed into the New Stone Age, West Europeans and mountain dwellers continued to rely on the horse for food but, in more fertile parts of the world, dramatic changes were occurring. These changes precipitated the domestication of the horse. In Mesopotamia, an area that lay between the Tigris and Euphrates rivers, man was taking the first steps toward civilization. Many areas may have nurtured primitive agriculture, but Mesopotamia is generally regarded as the cradle of civilization. It was here that man found the soil and climate ideal for cultivation. Cultivation led to the development of agricultural villages. This restricted the formerly nomadic man to a smaller hunting area, and he soon depleted herds that had provided his meat. Domestication occurred at this point so that man

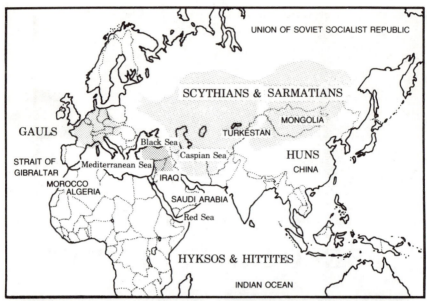

The horse was first domesticated in three areas: Central Asia (in Turkestan and Mesopotamia (now Iraq); China (in the Hwang Ho valley and on the Mongolian steppes); and northwestern Europe (primarily in Sweden and France). The Scythians and Sarmatians, whose range is shown above, were among the first to domesticate the horse. The Chinese and Gauls also tamed the horse by 1800 B.C., as did several nomadic Asian tribes (Hyksos, Hittites and Huns).

could have a dependable food supply. In time, man learned to use animals to ease his labor.

The ass, onager and camel were domesticated before the horse, and probably well before 3000 B.C. Carvings in the remains of the Royal Cemetery of Ur (one of the three major cities of Sumeria, an area in southern Mesopotamia) show onagers drawing chariots. Also, some are shown being ridden with a nose ring and a strap around their jaws. Man soon recognized the horse as a substantial improvement over the onager and the donkey: he was both faster and more trainable. Many scientists believe that, by 2000 B.C., the city of Ur had made several adaptations in house design to accommodate increasing pack horse and donkey traffic. Houses in Ur were rounded at the corners to allow easy passage of heavily loaded animals, and the steps of some were designed for equine use.

Still, the most conclusive evidence that the Mesopotamians had domesticated the horse was uncovered in 1935 in the highlands of southwestern Persia, along with artifacts known to be used by the Sumerians at that time. The engraved stone, called seal 105F, clearly

courtesy of Cheyenne Mt. Zoological Park, Colorado Springs, Co. Photo by Bob McIntyre

The onager (Equus hemionus onager) was ridden and driven in Central Asia well before the Asians domesticated the horse.

pictures horse heads arranged in four horizontal rows. This is be-lieved to be the oldest known pedigree of animal breeding. It is of special interest to equine geneticists and breeders because it indicates that men in Elam (to the east of Sumeria) were selectively breeding horses as long as 5000 years ago. The seal shows horses with three mane types (erect, hanging and maneless) and three head profile types (convex, straight and concave). Possibly the mane types were designations used to separate mares, stallions and foals. On the other hand, they might have distinguished wild from tame horses. The three profile types probably signified different ancestry. These types exist today, most notably in horses of Andalusian, Tarpan and Arabian descent.

Additional evidence indicates that the Sumerians and Elamites had domesticated the horse by this time. This evidence supports the theory that the seal represents a pedigree and not a mere enumera-tion of wild animals. Also in Elam, in the layers of earth between the ancient capitol cities of Susa I and Susa II, scientists have uncovered

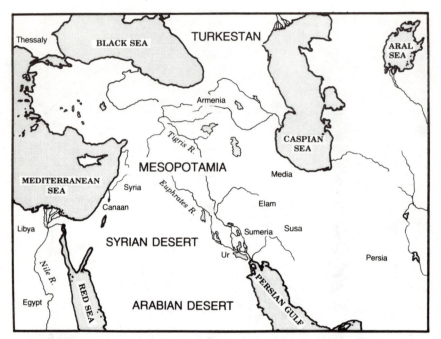

Mesopotamia and Turkestan were sites of early domestication. There, man tamed primitive types III and IV — the Wild Horse of Central Asia and the Caspian pony. The Hyksos and Hittites entered Egypt across the Sinai Peninsula and introduced the two-wheeled chariot.

a fragment of bone engraved with the figures of a horse and rider. The bone was found in a layer which has been dated to about 2800 B.C. To the west, archaeologists working in the upper Euphrates Valley uncovered drawings of a hitched horse wearing a snaffle-type bit, which they believe were drawn about 3000 B.C. Also, an amber statuette of a horse found in the tomb of the Queen of Ur dates to 4000 B.C. These finds suggest that civilized man had domesticated the horse in Central Asia by 2800 B.C.

But Mesopotamia has not yielded the only evidence of early domestication. The horse might also have been tamed at this time in northwestern Europe, where Stone Age man first hunted him, then decided he would be useful in hunting other quarry. ("Hors" is an Anglo-Saxon root meaning "swiftness.") Old Stone Age European man certainly knew the horse: primitive drawings of horses have been found in caves throughout France and in Britain. The first evidence of the horse's domestication appeared in Jarrestad, Sweden. The Jarrestad rock drawings have been dated to the Bronze Age, a period of specialization in metalurgy which culminated in the

Artist's charcoal sketch of horse drawing found on the wall of a French cave. Note the animal's short legs, round barrel and thick neck.

aforementioned Sumerian and Mesopotamian civilizations. The drawings might pre-date domestication in Mesopotamia. They show six mounted men, with four of the six using reins to control their mounts. Two types of horses are distinguishable — a Tarpan type (probably found in the ancestry of Europe's Gotland pony) and the Przewalski's type (which probably gave rise to the Fjord pony). Further evidence that Europeans had domesticated the horse by this time is provided by some engravings found on the stone casing of a grave in Kivik, Sweden. The engravings, which show horses and charioteers, probably were drawn about 2300 B.C., which would approximate the domestication of the horse in Mesopotamia.

The Kivik engravings represent the earliest known depictions of the chariot. The chariot's existence in northern Europe was affirmed when the remains of an ancient sun chariot and horse were unearthed in Seeland, Denmark. Scientists estimate that they probably were used about 1500 B.C. The chariot, however, might have been devised concurrently in the Hsia region of China. Although the Far East was generally slower to domesticate animals than other regions, the agricultural development of the fertile Hwang Ho valley was not far behind that of Mesopotamia. The development of agricultural villages probably prompted the domestication of the horse in China also. Remains of horse-drawn chariots have been uncovered in the Hwang Ho valley and dated to about 1640 B.C., and written records indicate that two- and four-horse chariot burials for Chinese war lords were quite common at this time. To the north, remains of horses of the Przewalski type have been found along with tools used by Paleolithic Chinese man. The Chinese and Mongolians (their neighbors to the north) probably originally envisioned the horse as a useful source of meat and milk, then found him suited for work and warfare.

Domestication may have followed a similar pattern in Turkestan, a region now occupying portions of the southern Soviet Union between the Black and Caspian Seas and China. There, on the steppes north of the seas, Indo-European tribes began domestication of what they had come to regard as a primary food source. Thousands of horses ranged wild on Turkestan's fertile grasslands, and it was a simple matter for the pastoralists to corral these horses. The Turkmenes raised horses much as modern cattlemen raise cattle: they existed off mare's milk and the meat of each year's colt crop. Only fillies were allowed to grow to maturity. Once grown, the mares in season were tethered on the steppes and, hopefully, bred by wild stallions. The mares then conceived to provide sustenance for another year. This was one of the earliest forms of artificial selection: for milk-producing mares and colts for food. However, as civilizations grew to the south, the Turkmenes realized that periodic raids on villages there could provide them with treasure as well as food and drink. So horses ceased to be their food source, and, instead, were taught to pull sleds and carry riders. The Turkmenes began to value the horse for his speed and strength.

One of the Turkmene tribes, the Sarmatians, valued speed so highly that they began breeding race horses in Ashkhabad (northern Persia) as early as 1000 B.C. These breeders selected for size and weight-carrying ability as well: they needed large, sturdy horses for battle, and fast, strong horses for polo and racing. Breeders throughout Armenia and in the Persian city of Media developed the Nisean strain of horses, which served as the foundation stock for the present day Turkmene breed.

The Sarmatians' numbers eventually dwindled, until they were replaced by the Scythians (800-400 B.C.), a band of nomadic conquerors who ranged across Eurasia from the Danube to the Chinese border. The Scythians, considered barbaric by the Romans, sustained themselves primarily off the milk and flesh of a large herd of horses. But they needed their horses for pillaging and plundering as well as for sustenance, so they came to value speed over bulk. The Scythians were excellent horsemen and trained their mounts to exist on the barley they carried with them during extended raids. They staged many such raids on Mesopotamian cities, and in turn were attacked frequently by strong southern troops. (During one attack in 522 B.C., Persia's King Darius I led more than 30,000 mounted cavalrymen.) The cultures, due to geographic proximity, mingled considerably, so it is difficult to assign specific developments to either. However, the Scythians probably invented the first ringed snaffle bit, and they were the first to teach their horses to stretch out and lower their backs or to kneel for easier mounting. Drawings and paintings of kneeling horses have been found on statues and ceramics uncovered in Scythian tombs. (The Scythians achieved some distinction as trainers. In Spain, in about 90 B.C., Scythian-trained mounts carried Iberian warriors into battle. The cavalrymen rode two-to-a-horse during the initial charge. Then, when the fighting grew heavy, each horse would kneel and allow one soldier to dismount to fight on foot.)

To the south, the Persians heightened mounted warfare with the invention of body armor and the lance. The Persians also invented hobbles about 700 B.C.

By about 1800 B.C., Pliohippus' descendants had been domesticated in Persia, Iraq, China and northern Europe, and probably were being tamed in the southwestern parts of Asia. Evidence that the horse had traveled as far as Kultepe in Asia Minor appeared about 2200 B.C. (He probably was the tall, lanky type native to Turkestan and Mesopotamia, continuing his migration toward the west.) The Canaanites, who lived between Jordan and the Dead and Mediterranean Seas in Israel, probably domesticated the horse about this time. Domestication had certainly taken place in Israel by 1680 B.C., when the Hyksos and Hittites (groups of Canaanites and Amorites) entered Egypt. There, they conquered the Pharoah and introduced the driving horse and the two-wheeled chariot. The Egyptians embraced the horse warmly: they idolized and immortalized him in paintings, writing and sculpture. Statuettes and monuments of horses and riders abounded during this period, and one statue of a mounted horseman shows that Egyptians were riding their horses by 1580 B.C. Literature from this period includes the Kikkulu text, written in Egypt in 1360 B.C. The text was one of the first to present detailed instructions on training. Egyptians also sacrificed horses and slaves: bits carved from bone and ivory are among the remnants of sacrificial burials found in Egyptian tombs. According to the Bible, the

Egyptians were accomplished equestrians by the time of the Jewish Exodus.

The Egyptians needed courageous horses to help them hunt lion, wild ox and ostrich. Their horses came from the short-coupled, heat-resistant stock that inhabited Canaan and Syria — stock that exhibited considerable strength, quality and refinement. The Libyans reportedly imported some of these quality horses from Egypt in 1600 B.C., possibly preferring the more refined Egyptian type to their coarser native steeds. These Egyptian imports, when crossed with the native horses, produced a horse that was hardy and slender. The Algerians bred a particularly pure strain called Numidians, and rode them using only a crop for control. These horses quickly spread to the Barbary Coast, where they became known as Barbs.

Domestication and equestrian skill spread throughout southern Europe by way of Egypt, as Egyptian colonists moved into new territory with their stock. Greece was largely unpeopled in 1500 B.C., but Egyptians had settled in the Thessaly region where they continued to breed riding and driving horses. A clay figurine from Mycenae dated 1300 B.C. indicates that riding had indeed reached Greece. By 648 B.C., chariot and flat racing were so well established that they were included in that year's Olympiad.

As Thessaly and other regions developed into prominent horse breeding districts, a man named Xenophon, the father of modern horsemanship, began plying his theories. It was Xenophon who first asserted, "no foot, no horse," and who first recommended handling a horse from both sides to prevent him from becoming one-sided. Xenophon borrowed the ancient Persian belief that a horse being purchased should be ridden and checked for soundness, then placed back in his stall and later re-examined. So, as long ago as 1000 B.C., horsemen were checking for warm and cold lamenesses!

To the west of Greece, in southern Italy, the Sybarites had developed considerably advanced methods of training by 800 B.C. The Sybarites were a group of former Greek and Egyptian traders known for their luxuriant lifestyle. They often lay in bed until noon, when they would arise and spend the entire day grooming and exercising their horses. The Sybarites probably invented dressage: they taught their horses to perform intricate dances and maneuvers to music. Rome became an equestrian center during the sixth century B.C., but never achieved distinction for its breeding programs. Romans, like all other early horsemen, rode without saddles. (A bowl painting dated 500 B.C. and found among Roman ruins shows saddleless, stirrupless riders.)

The Romans did use ringed snaffles, which probably were invented by the Scythians and introduced to the Italian Etruscans between 700 and 600 B.C. The curb bit appeared in Rome about 55 B.C., but it probably was invented hundreds of years earlier by the Gauls in northern Europe and carried back to Rome after Caesar's conquest of England. (The Gauls might have needed the more powerful bit to control their heavier-necked native horses.)

So domestication proceeded across Eurasia, following the primitive horse's paths across the continent and taking place wherever man needed an equine helpmate. Steadily, the numbers of tame horses grew and the numbers of wild animals declined, until man no longer found it feasible to select his horses from wild herds. Instead, he began breeding his stock to obtain the qualities he desired: beauty, speed, courage, strength, endurance and tractability.

Artificial Selection

The primitive horse was accustomed to roaming free, but he adjusted to man's control. Although early drawings depict chariot horses and chargers breaking away from their riders and drivers, the horse seemed to accept human control with relative ease. It is theorized that his compliance may have stemmed from his basic desire for companionship and instinct to follow an established leader. During his domestication, the horse was removed from his natural habitat and placed under restraint. With man as his sole provider and protector, he learned to respond to commands to earn his master's favor. Xenophon, the Greek, wrote that the horse should learn to associate man with things that give him pleasure: food, rest, relief from flies, etc. Xenophon became one of the first recorded advocates of the equine pleasure principle and training by reward rather than punishment. He believed that the horse was a gift from God that should not be abused.

In contrast to nature, the earliest breeders selected for temperament and learning ability: if a horse refused to obey, he was liberated. He returned to the wild to produce more disobedient horses, but as wild herds were crowded out of fertile grassland areas his chances of survival declined. The more agreeable animals remained in the comparative safety of captivity, where they were bred to other docile animals to produce good-tempered foals.

Wild horses band together for protection from predators and for assistance in finding food and shelter: they rely on the herd leader for their survival. During the domestication process, the horse learned to transfer much of this reliance to man, who provided nourishment, protection and companionship. In return for his assistance, man employed the horse: he taught him to overcome some of his natural fears to hunt lions, tigers and other wildlife. With a strong degree of selection for disposition and trainability, other traits important in the wild became less valuable to the horse: good hearing, the ability to attack with the teeth and forelegs, the ability to effect quick escape. Man continued to select for speed, but his methods were more subjective than nature's. In the wild, horses that lacked the requisite speed were overtaken by predators and eliminated from the herd. In captivity, early horsemen devised other tests. One such test for speed was conducted by locking horses in a pen and withholding their water for several days. When all the horses were sufficiently thirsty, the horsemen released the herd and pointed them in the direction of a known water source. The horses that arrived first were the prime candidates for breeding and training.

The Scythians practiced more barbaric selection methods. They needed fast, hardy horses for their raids, so they began their selection by galloping their horses heavily for several days. Afterward, they reduced their food intake. At the end of two weeks, they staked the animals in the open and threw ice water on them. Then, the horses were galloped for several more days, this time carrying more than

These Przewalski's horses, who live in a protected yet natural environment, behave much as their ancestors did more than 5000 years ago. Speed, alertness and power were traits that helped the wild horse survive. After domestication, man selected for other traits as well — disposition, trainability and compliance.

courtesy of San Diego Zoo Photo by F.D. Schmidt

300 lbs. of weight. As with natural selection, only the fittest survived to reproduce.

The horse learned to respond to man's wishes, but in the process many of his primitive responses were dulled. The flight instinct was perhaps most severely affected, because the horse found that he could not escape man's constraints. Confinement was a new experience for the wandering grazer, and many modern day stable vices indicate that the horse has adjusted to it rather poorly. Vices such as cribbing, weaving and bolting feed stem from boredom: they are vices a horse would not develop in the wild. Strict confinement certainly conflicts with the horse's desire to travel, developed over millions of years during his treks across several continents.

Man's modification of the horse's diet presented several problems. The primitive horse, both as a browser and a grazer, fed throughout the day and had no need for a large stomach capacity. Consequently,

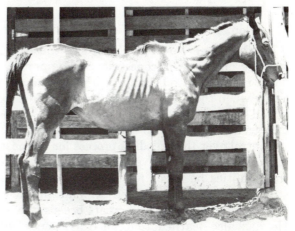

Windsucking, like cribbing and many other stable vices, often stems from boredom.

he developed a stomach that was small in relation to his size and total daily feed intake. Domestication led to a diet rich in grain supplements and limited to a few feedings per day. The introduction of large quantities of rich foods into his small stomach subjected the horse to digestive and stress-related disorders such as colic and founder.

Man-made environmental pressure caused orthopedic problems. The horse's new responsibilities, which included carrying riders and pulling heavy loads, caused added stress to fall on his highly specialized legs and hooves. No one is exactly sure when man began to appreciate the importance of the horse's legs and hooves, but prevention of lameness has been researched and emphasized for hundreds of years. The Persians had invented horseshoes as early as 500 B.C. The shoes, which really were pieces of cloth wrapped around the hoof and fastened above the fetlock, provided some protection from abrasion and aided footing on ice and rocks.

The Persians were among the first horsemen to realize the importance of hoof care. Their first horseshoes were pieces of cloth wrapped around the hoof and fastened above the fetlock.

Man's influence over the horse also caused significant reductions in fertility. In the wild, conception rates approach 100% — mares cycle more regularly, and stallions are able to detect estrus more effectively when they run free with their bands of mares. Captivity has an adverse effect on the breeding process. Some of the problems it creates are discussed in **"Breeding Aspects."**

Control over the breeding process for the past three thousand years has enabled man to develop specific groups and types of horses with the qualities he needs for particular purposes. He has learned that stringent selection of those animals allowed to produce the next generation can eventually create breeds of beauty and/or usefulness. He has also learned that he can tamper with nature and create imbalances that would not have existed in the wild, by selecting strongly for certain traits that are accompanied by undesirable traits. This could be the reason for the relatively high occurrence of certain inherited abnormalities such as Combined Immunodeficiency Disease (CID) in the Arabian breed. Perhaps the coding gene for the disease is transmitted along with some desirable characteristic that has been selected for since the breed's development. Sometimes man errs by selecting for what is beautiful but not functional. In many breeds, selection for small, dainty hooves has led to stress-related problems such as navicular disease.

Also, horsemen developing breeds and types have discovered that selection for specific performance traits may preclude versatility. A breed that has achieved distinction for working cattle might not be suitable for dressage, steeplechasing or draft work. Similarly, breeds developed for versatility might never excel at functions requiring specific skills or traits. Morgans, for example, work well as cow ponies, pleasure driving and riding horses, show horses and hunters or jumpers. As a group, however, their performance in each category could not match that of a breed developed for one specific job — most Morgans could not compete with a Quarter Horse cutting horse or a Thoroughbred hunter.

Breeds of Horses

A breed of horses is a group with distinctive characteristics that are transmitted to their offspring. These characteristics may include conformation, color, performance, intelligence and disposition. Together, they form a composite type that constitutes a specific breed. The promoters of some breeds stress certain skills and abilities: Standardbreds, for instance, are bred almost exclusively for speed in harness over a mile. Other breeds are touted for their versatility: Arabians, Quarter Horses and Morgans are examples of excellent all-around performers.

Breeds develop through a combination of natural crosses and artificial selection. Initially, environment played an important part in the

The Shetland pony and the draft horse descended from the primitive horse of northwestern Europe. The Shetland traces to the primitive horse through the Celtic pony, and the draft horse through the heavy forest horse. This photograph, entitled "Impudence and Dignity," was taken at Iowa State University in 1906. Although information about the draft horse and Shetland foal is sketchy, many contend that the draft horse is Jalop, a Percheron stallion owned by the college and one of the breed's greatest sires.

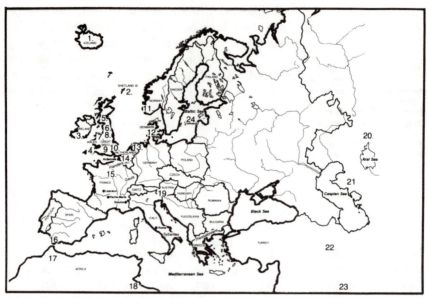

1. Icelandic 2. Shetland 3. Connemara 4. Exmoor 5. Clydesdale 6. Fell 7. Welsh 8. Cleveland Bay
9. Shire 10. Suffolk 11. Fjord 12. Jutland 13. Friesian 14. Belgian 15. Percheron 16. Andalusian
17. Barb 18. Numidian 19. Lippizaner 20. Przewalksi's 21. Tarpan 22. Turkmene 23. Arab 24. Gotland

Principal areas of development of some breeds.

development of horse types: extremes in climate, topography and vegetation necessitated adaptations that affected ultimate size, conformation and abilities. Adaptations of the original four types of horses resulted in many breeding groups, each uniquely suited to its particular environment. In western Europe, some of the Celtic pony stock migrated to the Shetland Islands north of Scotland, where they developed into the diminutive, hardy Shetland pony. The original European horses moved into Britain, Wales and Scotland; each herd developed characteristics to help the horses survive in their new environment. Some stock moved to Norway and developed into the Fjord pony, while others migrated to the Swedish island of Gotland and evolved into the Gotland pony.

 Formal breeds developed when man selectively bred these early types to fix certain desirable characteristics. Often, an existing type suited man's purposes, and few outcrossings were necessary to develop a functional breed. Most of the ponies mentioned remain relatively pure today: they are almost direct descendants of one early type of horse. Some breeds have been developed through centuries of outcrossings with numerous other breeds. Establishing type can be especially difficult when breeders select for a trait that cannot be

This Suffolk Punch mare, Read's Vicky, like all members of the breed is a chestnut. Breeders have fixed the chestnut color in these draft horses, much as breeders of the Cleveland Bay have fixed the bay coat color in their horses.

courtesy of American Suffolk Horse Association

fixed; that is, when the genes that code for the trait cannot be passed uniformly from one generation to the next. Breeders of the Suffolk Punch, an English draft horse breed, have fixed the chestnut coat color in their horses. Palomino breeders, however, have been unable to develop a strain of true-breeding palominos. (For an explanation of the inheritance of palomino coloration, refer to **"Coat Color and Texture."**)

During the developmental process both desirable and undesirable characteristics occasionally become fixed in the breeding population. This is especially true of breeds that trace to only a few foundation stallions who have passed on their total genetic characteristics. The process that fixes these genes can be demonstrated through a hypothetical example: A stallion who sires foals with unusually attractive heads may be used to develop a breed with that particular trait. He is bred to 100 mares in a season and the first foal crop has the desired head conformation. However, most are also calf-kneed. Later foal crops reveal the same defect. After several decades of breeding successive generations, a breed of horses with beautiful heads is created but, because so many of the foundation horses were also calf-kneed, that trait also has become a breed characteristic.

One sire can have a tremendous influence on a breed; in fact, many breeds have developed from a single founder. Standardbreds, Tennessee Walking Horses, Saddlebreds and Morgans each have been strongly influenced by one prepotent stallion in their breed.

There are more than 300 breeds of horses today, with at least 34 light horse, 7 pony and 6 draft horse breeds represented in the United States. The most popular and influential breeds are discussed in this section.

LIGHT HORSES

Light horses — unlike draft horses or ponies — are usually lean-legged, athletic animals built for speed and agility. Most modern light horse breeds developed from horses of the third and fourth primeval types: the lanky Andalusian ancestor and the more refined Caspian pony. These types produced horses suitable for driving, racing, pleasure riding, exhibition work and, occasionally, light packing. Light horses still are used for these purposes, as well as showing, hunting, distance riding and general ranch work. They usually stand between 14-2 and 17 hands at the withers and weigh between 900-1400 lbs. Many light horse breeds developed in the United States when settlers in different regions were faced with certain needs. These breeds may still excel at the purpose for which they were bred. This section will mention the most common uses of each light horse breed, but the reader should remember that most are versatile and probably have been employed in many activities. The histories that accompany each breed description were derived from many reliable sources, but the reader should also remember that most sources are in disagreement about exact names, places and dates.

The Arabian

The Arabian is considered the oldest breed of light horse and has had a profound influence on the development of most other light horse breeds. It has been selectively bred for more than 1500 years in the Arabian Peninsula, yet the breed is not entirely native to Arabia. The Arabian probably evolved from the Syrian stock that developed into the Barb (through the Algerian Numidian) and the Turk (through the Nisean). The Arabian's progenitors may have been transported from Egypt to Arabia in about 200 A.D., with improvement of the stock beginning about 632 A.D. when the prophet Mohammed returned from his raid on Mecca with five Egyptian-bred horses.

The Egyptians are credited with the first efforts toward improvement of the breed when they selected for speed, but the Bedouins (desert Arabs) initiated the most dramatic improvements. They valued mares highly and syndicated their best, believing that the female line was dominant over the sire. Arabs may have been the first to practice artificial insemination: tribesmen stole semen from their enemies' best stallions and used it to inseminate their mares.

Most modern hot-blooded breeds trace to the Arabian; even the cold-blooded Percherons claim some Arabian ancestors. The first

Champion Ansata El Sherif (*Ansata Ibn Halima x *Ansata Bint Bukra) exemplifies many of the characteristics Arabian breeders seek in a modern Arabian stallion. He is owned by Dr. and Mrs. C.E. Hardin of Santa Ynez, California.

major importation of Arabians into the United States was in 1906 when Homer Davenport imported 25 desert purebreds. Recently, American breeders have imported both Egyptian and Polish-bred Arabians to supplement their stock.

Arabians usually stand between 14-1 and 15-1 hands, and average 800-1100 lbs. in weight. They are distinguished by several conformation characteristics, including small, triangular heads, dished faces and wide-set eyes. An Arabian's forehead is broad; his arched neck is moderately long and set high on the shoulder.

The Arabian's back is short and straight with a long, flat croup. Some fanciers attribute the Arabian's short back to a reduced number of lumbar vertebrae. But studies show that, like most domestic horses, the Arabian normally has six lumbar vertebrae. Some Arabians have one fewer thoracic vertebrae (17 instead of 18) but this seldom affects back length.

*Bask (Witraz x Balalajka), Legion of Merit Champion, exhibits excellent Arabian conformation. He was imported from Poland in 1963 by Lasma Arabians, Scottsdale, Arizona.

The first Arabian registry in America organized in 1908. Prior to that date, American Arabians were registered with the Jockey Club. Arabians may be grey, bay, chestnut, black and occasionally roan. White markings on the face and legs are common. Arabians are used primarily for pleasure and show, although they are also raced and used as endurance and stock horses. The breed is noted for docility and stamina.

The Thoroughbred

The Thoroughbred originated in England when native English mares (many of them Connemaras, Cleveland Bays and Galloways) were crossed with Barbs imported during the days of Julius Caesar and again during the Crusades. King Henry II staged England's first horse race in 1174 and made racing a national passion. Between 1660 and 1685, King Charles II imported Barb mares from Spain, Turkey and Italy, but they failed to improve the traits desired by English breeders. The English light horses were still small and light-boned in 1685, and English breeders wanted stronger backs, increased speed and stamina. They began importing stallions, primarily Barbs, Turks and Arabians, to improve the stock. Of the 174 stallions imported before 1750, three led to the formation of the foundation Thoroughbred lines: the Byerley Turk, the Darley Arabian and the Godolphin Barb.

The Byerly Turk, the first of the three to enter England (1689), bred successfully with the small English mares. His great-great-grandson, Herod, founded one of the three foundation lines. The Darley Arabian, brought to England in 1706, was foaled in Syria in 1700. His great-great-grandson, Eclipse, founded another foundation line. Eclipse, a superior race horse unbeaten in 26 starts, was the first great Thoroughbred to emerge after the breed became recognized. Nearly 90% of all Thoroughbreds trace to Eclipse. The Godolphin Barb, probably born on Morocco's Barbary Coast, was the last of the three to be imported. His origins are obscure (he may have been a cart horse in Paris before his importation in the 1720 s), but his career at stud is well-documented. His grandson, Matchem, became another foundation sire.

The Oriental sires, particularly the three mentioned above, almost immediately began siring winning horses. The Darley Arabian sired the first great English race horse, Flying Childers, and also his full brother Bartlett's Childers. Although Bartlett's Childers was never raced or trained because he was a bleeder, he sired many notable race horses and became the paternal grandsire of Eclipse. After the days of

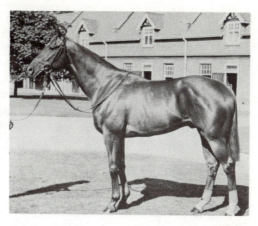

Crown Prince is generally considered to represent the most desirable features of conformation for the Thoroughbred horse.

Eclipse the breed improved so much that Charles Darwin, father of the theory of evolution, wrote, "By a process of careful selection and by careful training, the whole body of English race horses have come to surpass in fleetness and size the parent stock."

While the English were developing their horses, the Americans were also improving their Thoroughbreds. Bull Rock, a 21-year-old son of the Darley Arabian, arrived in Virginia in 1730, and the development of the American Thoroughbred began. England's first listing of pedigreed Thoroughbreds, *THE INTRODUCTION TO A GENERAL STUD BOOK,* was published by James Weatherby, Jr. in 1791. Two years later, Volume I of the *General Stud Book* was published. The first volume of the *American Stud Book* (now maintained by the Jockey Club) appeared in 1873.

America's first Thoroughbred races were held in the 1730s, but tracks did not begin catering exclusively to the developing breed until 1863. Kentucky, a border state during the Civil War, became the nation's leading breeding center for Thoroughbreds because it was spared much of the destruction that ravaged the rest of the rural South.

Thoroughbreds, developed primarily for speed at intermediate distances (from ¾ to 1½ miles), excel at the extended gallop. They normally range in height from 15 to 17 hands. Weight for horses in racing condition ranges from about 900-1150 lbs. Their conformation varies somewhat because they have been bred for speed rather than uniformity of appearance. General characteristics include long, smooth muscling, with powerful hindquarters. Most good Thoroughbreds have short backs, deep chests, and length in the forearm and gaskin.

Polo, a mounted game played at almost a continuous gallop, places strong demands on both horse and rider.

A Thoroughbred may be black, bay, brown, chestnut, or grey. Few white Thoroughbreds have been born. Occasionally, one is registered as a roan, but it is usually a grey. (All grey Thoroughbreds trace to Alcock's Arabian, one of the Arabians brought to England during the 1600s.)

The Thoroughbred has served as foundation stock for a number of other breeds (including the Standardbred) and has influenced breeds such as the American Saddlebred, Morgan and Quarter Horse. Thoroughbreds have achieved most notoriety as race horses, but they are also used extensively as hunters, jumpers, polo ponies, endurance horses and saddle horses.

The Quarter Horse

The Quarter Horse originated in the United States, and may even have developed as a type before the Thoroughbred was established in England. Quarter Horses had been bred for running short distances years before the breed association was formed in 1940. Early settlers in the East raced their horses on streets that seldom were longer than a quarter mile. They came to value the horses' short bursts of speed and ability to navigate rugged courses hewn out of the wilderness. Janus, a Godolphin Barb grandson who was imported in 1752, influenced the breed by siring many good quarter running horses. Janus was known for his speed at four miles, yet he proved most successful as a sire of horses with speed over short distances. He also sired many famous Thoroughbred stallions and mares.

As settlers moved westward, their short distance running horses were crossed with mares of Spanish descent. The resulting horses retained their speed and gained increased endurance and stamina. One of these horses, Steeldust, was foaled in Illinois and moved to Texas

This Quarter Horse typifies the balanced Quarter Horse conformation and muscle definition sought by many modern breeders.

in 1846. He achieved great fame in match races as one of the fastest horses of his time over the quarter mile. His influence was so great that, until about 1938, horses of the Quarter type were often called "steeldusts."

These horses also became known for their adaptability to ranch life. The Quarter Horse has been recognized as the ideal cow horse for years. The same attributes that gave him speed and power for short-course racing qualified him as a stock horse: a muscular, sturdy body, tremendous drive and courage. Quarter Horses were powerful enough to pull against roped steers, and fast enough to outrun the speediest calf.

In recent years, increased infusions of Thoroughbred blood have added greater speed and staying power to the running Quarter Horse. The Thoroughbred blood also is increasing the Quarter Horse's average height, which ranges from about 14-3 to 16 hands. Average

Many of today's running Quarter Horses have strong Thoroughbred ancestry. This introduction of Thoroughbred blood has contributed to increased height, weight and staying ability within the Quarter racing type.

weight is about 950 to 1300 lbs. Quarter Horses are used as pleasure, show, stock and race horses. There are more Quarter Horses registered in the world than horses of any other breed.

The Standardbred

The Standardbred is another breed of United States origin. Once known as the American Trotting Horse, the Standardbred was derived from a mixture of Thoroughbred, Norfolk Trotter (predecessor of the Hackney), Barb, Arabian, Morgan and Canadian and Narragansett Pacer horses. Thoroughbred top crosses on native trotting and pacing mares produced many of the fastest early horses that were more rugged than their Thoroughbred ancestors. The breed developed primarily in the North, where better roads created a demand for better driving horses. Races between fast trotters and pacers became commonplace on New England streets. These horses often spent the week behind the plow or cart but, on weekends, raced down dirt roads pulling sulky-type vehicles. (Post-Revolutionary War anti-British sentiment led many colonists to consider flat racing un-American: they wanted a sport uniquely their own, and quickly adopted harness racing.)

The Standardbred traces to Messenger, a grey Thoroughbred stallion imported from England in 1788. Messenger, who himself descended from all three foundation sires (Herod, Matchem and Eclipse) and traced to Flying Childers in the direct sire line, compiled a good, but unremarkable, racing record. In America, he was bred to native riding mares and Thoroughbreds. An inclination to trot soon began appearing in his offspring. Then Hambletonian 10, a great-grandson of Messenger, was foaled in Sugar Loaf, N.Y. in 1859, and the breed took a definite shape. Hambletonian became the father of the Standardbred breed: he sired 1,321 foals in his 21 years at stud. Four of his sons founded the Standardbred foundation sire lines. Hambletonian's influence was astounding: more than 99% of the trotters and pacers racing in the country today trace to him, making him the undisputed father of the breed.

The designation "Standardbred" was derived from the original performance requirements for registration — if a horse could meet the performance standard, he gained admittance to the registry. Originally, a horse had to cover a mile in 2:30 at the trot, or 2:25 at the pace. (Genetic factors, training and shoeing help determine whether a horse will trot or pace.) The standard was revised as the breed improved; it now stands at 2:20 at the trot for young horses and 2:15 at the trot for older horses. Registration on performance alone has not

courtesy of the United States Trotting Association

Bret Hanover, an outstanding Standardbred pacer, as a four-year-old paced a mile in 1:53 3/5. He is a son of Adios.

been granted since 1933. Generally, both sire and dam must be registered Standardbreds for a horse to be eligible. (Exceptions are sometimes granted but only when certain stringent requirements are met.) The forerunner of the current *Standardbred Stud Book* was published in 1871. Since then, records have shown dramatic increases in speed because of breed, track and equipment improvements. The performance standard has aided selection for speed: horses are usually evaluated on their speed over a mile, a fairly objective measurement. Their records are less subject to interpretation than those of horses that race under the saddle. The Standardbred's supremacy

The Standardbred stallion, Adios, demonstrates superior conformation for his breed. His performance as a sire of speed is impressive — few harness horses have ever negotiated a mile in 1:55 or faster, yet most of those horses are sons of Adios.

courtesy of the United States Trotting Association

in the harness racing world is virtually unchallenged: it is one of the few livestock breeds that the United States has consistently exported rather than imported.

Standardbreds usually are less leggy, smaller and sturdier than Thoroughbreds. Their average height ranges from 14.2 to 16.2 hands, with 850-1150 lbs. as the average weight. Colors include bay, brown, chestnut, black and grey. Standardbreds are used primarily for harness racing, and often in roadster classes.

The Morgan

The Morgan is the only horse breed named after a single horse, Justin Morgan. Every horse in the breed is descended from this stallion. The horse was named after his owner, Justin Morgan, who lived in Vermont in the late 1700s. He obtained the young horse in payment of a debt. The colt grew into a magnificent horse: he was a dark bay with a stocky build, standing 14 hands tall and weighing about 950 lbs. Justin Morgan probably was of Arabian and Thoroughbred descent: he reportedly was by a Thoroughbred stallion named True Briton and out of a daughter of Diamond. The horse was compact, spirited and extremely prepotent: he and his offspring were known for their ability at driving, racing, pleasure riding and farm work. Until the development of the Standardbred breed in the mid-1800s, they were desirable road horses. Morgan blood influenced the Standardbred and has been instrumental in the development of the American Saddlebred, Tennessee Walking Horse and Quarter Horse breeds. The registry for Morgan horses was started in 1894, and was taken over by the Morgan Horse Club in 1930.

Morgans average 14-1 to 15-1 hands at the withers, and weigh from 900-1200 lbs. All solid colors, with the exception of white,

courtesy of Windswept Place, Allen, Texas

Funquest Wampor, a national champion English pleasure Morgan gelding.

appear in the breed. Morgans are characterized by a straight or slightly dished face; prominent eyes set far apart; small, wide-set ears; a small muzzle and a prominent jaw. Morgans also have a crested neck, great depth and slope to the shoulder and a short back. The Morgan has a high-set tail, and gives the overall impression of power in a compressed frame. The breed is noted for its easy-keeping qualities, and its gentleness and hardiness. Morgans are used primarily for pleasure, show and endurance.

Sources of other light horse breeds are scattered throughout the world, including many gaited horse and color breeds. In addition, man has developed dozens of draft horse breeds. A representative few of each class are discussed in this section.

THE COLOR BREEDS

Parti-Colored Horses

Spotted horses were first depicted on cave walls in Lascaux and Peche-Merle, France, about 20,000 years ago, and numerous artifacts indicate that they abounded in Persia and Egypt. The Appaloosa traces through the Andalusian to early Spanish and West European spotted horses. Horses colored in an overo or tobiano pattern (pintos) also have been in existence for thousands of years. (Refer to **"Coat Color and Texture."**) In this section, parti-colored horses will be divided into two types: 1) the Appaloosa, and 2) the Paint, Pinto and Morocco Spotted Horses.

APPALOOSAS

Egyptian art of the first millenium B.C. shows that the Appaloosa type of spotted horse apparently was held in great esteem. Chinese art depicts the Appaloosa type circa 500 B.C., and the Persians began selectively breeding them a few hundred years later. The Chinese may have been the first to attach a mystical importance to these horses: in 102 B.C. Emperor Wu sent an expedition to Turkestan for 30 "heavenly horses," or Appaloosas, for his private stock. The horses achieved even greater importance in China during the T'ang dynasty, 618-906 A.D. By the 18th century they were being bred in England, France, Sweden, Holland and Denmark (home of the Knabstrup spotted breed, similar to the Appaloosa). They have been bred in Spain (where the American stock originated) since 100 A.D.

The Appaloosa entered the United States during the 17th century, with the early Spanish explorers. The Nez Perce Indian tribe of the northwestern United States trained these horses and developed them for their own use. The tribe lived in Washington, Oregon and Idaho, establishing their villages along the flat, grassy Palouse River valley — hence the name "Appaloosa." They selectively bred these horses from 1730 until 1877, when extensive wars disrupted their tribal life. In 1938, the Appaloosa Horse Club formed to promote the breed.

The Appaloosa's markings include white sclera around the eye, parti-colored or mottled skin (particularly on the muzzle and genitalia) and striped hooves. The mane and tail are sometimes sparse, although this is seen less frequently as the breed becomes more refined. Appaloosa registration requirements state that to be eligible a horse must stand more than 14 hands tall, and may not be of pony, draft, overo or tobiano ancestry. There are two basic Appaloosa spotting patterns: the blanket pattern, in which a white blanket, with or without colored spots, covers the hips of a colored horse; and the leopard pattern, a white horse with colored spots. There are also roan-type appaloosa patterns and variations of the two listed above. Appaloosas may be of any color except grey, which the association now prohibits because of the effect of the greying gene on the color pattern. (Refer to **"Coat Color and Texture."**)

The Nez Perce gelded inferior stallions and selected their best mares to produce a combination war horse, race horse, hunting horse and long-distance mount. Today, the Appaloosa possesses the speed, stamina and hardiness that adapted him for those purposes.

Appaloosa blanket, one of several color patterns found within the Appaloosa breed.

PAINTS, PINTOS AND MOROCCO SPOTTED HORSES

In addition to the characteristic Appaloosa coloration, horses may be spotted in a tobiano or overo pattern. (Refer to **"Coat Color and Texture"** for complete color descriptions.) These horses are called Paint, Pinto or Morocco Spotted Horses, depending on the registry involved. These horses have a colorful history: one of the earliest Egyptian representations of a horse was a painting of a pinto dating to 1500 B.C. But the three breed associations that register these horses are of relatively recent United States origin. Horses eligible for registration include colored horses with natural white markings. The amount of white each registry requires varies from three spots above the knees to 10% or more of the body area above the knees and hocks, not including the face. Characteristics sometimes associated with the spotting include an apron or bald face, white on the jaws, blue or white (glass) eyes, pink skin under white hair and a parti-colored mane and tail. (Blue eyes are acceptable because they are common to horses of spotted breeding — especially when the white areas extend over the eye.)

The American Paint Horse Association formed in 1965 primarily to register horses of Quarter Horse or Thoroughbred descent: they register quarter-type or stock-type animals taller than 14 hands. The Pinto Horse Association, founded in 1956, also registers stock type horses of Quarter Horse and Thoroughbred breeding. Additionally, they accept pleasure-type animals with Arabian bloodlines, and saddle-type horses that trace to the Saddlebred. The Morocco Spotted Horse Cooperative Association of America, formed in 1935, began by

This overo Paint exhibits several breed characteristics: extensive white spotting, a bald face and a parti-colored mane and tail.

registering crosses of piebald (black and white) and skewbald (any other color and white) English Hackneys and French Coach horses with Morocco Barb horses. Arabian blood has also been added to the breed. The registry excludes horses of draft, pony or Appaloosa breeding, but registers gaited horses and horses of Tennessee Walker, Thoroughbred, Arabian, Hackney and Morgan descent.

Palominos

The Palomino was first recognized as a breed in the United States. The first registry formed in 1936, but palomino horses existed long before that time. The color was mentioned as early as 800 B.C., and a tapestry worked in France during the 11th century shows horses of the palomino color. In Spain, palominos were considered "The Horse of the Queen," and the country's best specimens were reserved for the royal family and for high-ranking military officials. (Commoners were discouraged from riding them.) The Spaniards, who had obtained their golden horses from Arabia and Morocco, brought them to the New World in 1519.

Palomino horses in the United States may be registered with either the Palomino Horse Association (PHA, formed in 1936) or the Palomino Horse Breeders of America (PHBA, formed a few years later). While the latter accepts only dark-skinned horses for registration, their color requirements are similar. A Palomino should have a light mane and tail, and his body color should be golden. (A deep, rich gold is ideal, but the registries accept horses several shades lighter or darker than the ideal.) The mane and tail may be white, ivory or silver, but may contain no more than 15% black or chestnut colored

This Palomino has an ivory mane and tail, and the rich gold color that breeders seek. He exhibits Quarter Horse conformation.

The coat pattern for registered buckskins or duns is characterized by a deep gold to light yellow body color with dark points (mane, tail and legs).

hair. Generally, the term "palomino" refers to a color and not a specific type. The PHBA requires that both parents be registered with their organization, or that one be a registered Palomino and the other a registered American Saddlebred, Quarter Horse, Morgan, Tennessee Walker, Arabian or Thoroughbred. The horse resulting from such matings should have characteristics of his parent breeds as well as the palomino color. Ponies and draft horses are excluded.

Buckskins

Buckskin, dun and grulla horses are collectively registered by two organizations: the American Buckskin Registry Association, Inc. and the International Buckskin Horse Association. Both have organized since 1963. Although the buckskin color is found in a number of breeds, these associations believe that the color traces to the Spanish Sorraia breed and the Norwegian Dun. Both are true-breeding for their colors. These horses probably were crossed with Barb and Arab mares in Spain, and some of the descendants were among the horses brought to America by the Spaniards.

Buckskin and dun horses are defined by the registries as being basically yellow-bodied, with dark points (mane, tail and legs). (Refer to **"Coat Color and Texture"** for color definitions.) These horses may have a dorsal stripe, shoulder stripes and/or zebra-like leg markings. Grulla is a smoky blue or mouse color with dark points, a dorsal stripe and often zebra-like markings on the legs, withers and ears. A red dun is a chestnut dun with darker points and the dorsal stripe. (Refer to **"Coat Color and Texture."**)

courtesy of American Saddle Horse Breeders' Association
Photo by John R. Horst

Society Rex (Kalarama Rex x Spoonbill) is an outstanding Saddlebred stallion.

GAITED HORSES

Gaited horses originally were developed to provide comfortable rides over long distances. As more people started riding for pleasure, the popularity of the smooth-striding gaited horses increased. Gaited breeds include the American Saddlebred, the Tennessee Walking Horse, the Fox Trotting Horse and the paso breeds.

American Saddlebreds

The American Saddlebred was first bred in the southern and east-central United States. It is descended from Thoroughbred, Canadian and Narragansett Pacer, Standardbred, American Trotter, Morgan, Arabian and Galloway stock. While the eastern United States' road system made harness horse travel practical, settlers needed good riding horses further inland. Southerners also wanted quiet horses with good ground-covering strides. In the mid-1700s, colonists were becoming increasingly disenchanted with everything English — including posting, an English convention. They wanted horses they could sit comfortably at the trot.

While a number of sires contributed to the breed's success, one clearly dominated. His name was Denmark. He was a consistent, but not outstanding, four-mile Thoroughbred race horse. He achieved little fame on the track, but became known as an outstanding sire of saddle horses. Denmark is considered the breed's foundation sire: more than half the Saddlebreds in the first stud book (published in 1891) trace to him in the male line. The other important Saddlebred line descended from Mambrino Chief, foaled in 1844, a great-grandson of the imported Messenger. (Mambrino Chief also founded a trotting line of Standardbreds. Messenger was the horse who gave rise to the Standardbred breed.) In Saddlebreds, Denmarks are known for their exquisite finish and easy riding qualities, and the Chiefs, for their brilliant action at the trot.

The American Saddlebred stands about 15-16 hands and weighs about 1000-1200 lbs. He has a long neck rising from the shoulder at a high angle, giving him high head carriage. Saddlebreds have sloping shoulders and pasterns and level croups. They may be bay, brown, black, chestnut, grey or roan, or even pinto or palomino. White markings are common. Depending on training, a Saddlebred can be either three- or five-gaited. If a horse is five-gaited, one of his gaits is a "slow" gait (such as an amble, fox trot, slow pace or running walk) and the other is the rack, a flashy, four-beat nonpacing movement.

American Saddlebreds are used extensively as show animals, and some are used as pleasure horses.

Tennessee Walking Horses

The Tennessee Walking Horse was developed from the Thoroughbred, Saddlebred, Standardbred, Morgan and Canadian and Narragansett Pacer breeds. It originated in the South more than 100 years ago when plantation owners needed a utility horse, one suitable for riding, driving and tilling the fields. A Standardbred named Allan F-1 served as the breed's primary foundation sire. This stallion (foaled in Kentucky in 1886) showed an inclination to pace early in his training. His owner, who wanted trotters, decided to sell him. Allan found a new home in Tennessee, where his owner liked his running walk and bred him to his best mare. The resulting offspring, the stallion Roan Allan F-38, embodied the ideal walking horse charac-

The Tennessee Walking Horse is known for his comfortable gaits and stylish execution of the running walk.

teristics. Nearly all Tennessee Walkers have descended from Allan through his son Roan Allan.

Tennessee Walking Horses are larger and stouter than their Standardbred ancestors and have excellent dispositions. They have shorter necks and lower head carriage, as well as sloping croups. Many walkers are slightly sickle-hocked and have long, sloping pasterns. They come in roan, grey and all solid colors, with an abundance of blacks and bays.

The characteristic running walk is a four-beat gait in which the hind foot oversteps the front foot, usually by 10 or 15 inches. The walk is accompanied by a rhythmic nodding of the head.

Tennessee Walking Horses for years were evaluated on the basis of their stamina and comfortable riding qualities. Today, they are ridden primarily for pleasure and show.

Fox Trotting Horses

The fox trotting horse developed from crosses between Arabians, Morgans, Saddlebreds, Standardbreds and Tennessee Walking Horses. The breed originated in the Ozark Hills of Arkansas and Missouri, where farmers needed horses that could cover long distances at a comfortable gait. Two registries currently enroll fox trotting horses, with the gait being the primary registration requirement. The Missouri Fox Trotting Horse Breed Association was formed in 1948, and the American Fox Trotting Horse Association organized in 1970.

Fox trotting horses stand about 14-2 to 15-3 hands and weigh approximately 950-1200 lbs. Colors include palomino, black, chestnut (sorrel) or roan, with chestnut being the most common.

The fox trot is a four-beat, broken diagonal gait in which the front foot strikes the ground before the diagonal hind foot. The hind foot then slides into place to softly contact the ground. Fox trotting horses are used primarily for pleasure riding and trail riding.

Paso Horses

Paso horses descended from Andalusian, Barb and Friesian horses introduced by the Spanish into South America and the Caribbean. They first reached the New World in 1493 on Columbus' second voyage. The influence of the smooth-striding Andalusian on the paso breeds is evident; also, some horsemen believe that the Narragansett Pacer influenced the developing breeds during the 1600s. Peruvian Paso refers only to those horses from Peru or directly descended from Peruvian stock. Paso Fino includes horses of Puerto Rican, Columbian, Cuban and Dominican Republic extraction.

The Peruvian Paso is noted for its inherent smooth fast gait (broken pace) which is characterized by "termino" (the outward movement of the forelimb during extension).

Pasos range in weight from 700-1200 lbs., and in height from 13-15 hands. (Height is increasing in United States-bred horses because of better nutrition. In addition, height among Peruvian Pasos is more standardized than in Paso Finos.) These horses have low-set tails and sloping croups. The cannon bones of the hindlegs may slope forward, making them slightly sickle-hocked. Pasos tend to have crested necks and the Andalusian convex nose.

The basic paso gait is a smooth, lateral four-beat movement similar to a broken pace. The horse moves both lateral legs together, but the hind foot touches the ground a fraction of a second ahead of the front foot. This makes the gait less jarring than a true pace. The paso gait has five forms, ranging from the slow "paso fino" gait to the "paso largo," which may exceed the speed of the canter.

Paso horses are normally docile and easily trained. They are used for pleasure, show, parade and distance riding.

DRAFT HORSES

Draft horses were once common in the United States but, since mechanized agriculture, their numbers have decreased. There has been a resurgence of interest in the animals for exhibition and show purposes during recent years. Draft horses stand about 16-18 hands and weigh about 1400-2200 lbs. They tend to be proportionately wider than light horses, with straighter shoulders and massive necks. Draft horses have round, heavily muscled hindquarters and large hoofs. Because of their great size and strength, a gentle disposition is of primary importance.

Development of Draft Breeds

The ancestors of today's light horse probably lived on grassy steppes and plains, but the forerunner of today's draft horse was a forest dweller. The large forest type originally may have developed from the same stock as the Celtic pony. He is usually classified as Equus caballus germanicus; fossil remains indicate that, although he was considered massive by early Europeans, he probably stood no more than 15 hands tall. This early draft horse resembled the stereotyped "bulldog" Quarter Horse in build: he had strong hindquarters and ample bone. The forest horse's conformation differed markedly from that of desert and steppe horses. Unlike his fleet-footed light horse relatives, he radiated power and solidarity. He needed strength because his habitat, with its heavy forestation and uneven terrain, made escape from his enemies difficult. Where the steppe horse fled, the forest horse often had to stand and fight. Consequently, he developed a powerful build: a blocky body supported by relatively short legs and large feet.

During the first century of the Christian era, many of these wild forest horses were captured and trained to pull heavy carts and chariots. Their bulk and more upright foreleg conformation made them more efficient than light horses in pulling heavy weights at slow speeds. Man began breeding these native cold-bloods to develop the Old Flemish draft horse type. These horses helped form several draft breeds. The Ardennes, one of the first, originated along the French-Belgian border, and today is considered the breed closest to the primitive heavy forest horse. The Belgian, a popular modern breed, probably formed from crosses between the wild black Flemish horses and the developing Ardennes. Of the United States' five most popular draft breeds (Percherons, Clydesdales, Belgians, Shires and Suffolks) the Belgian most nearly resembles the black forest horse in build: Belgians are very compact with heavily crested necks.

The Ardennes also contributed to the development of the Percheron, a breed that originated in the La Perche region of France. Percherons developed from crosses between the native French Ardennes, and Spanish Barbs and Arabs that were imported after the 8th century. Further crosses during the 18th century added Danish, English and Belgian blood to the developing breed. Then, in 1820, two imported grey Arabian stallions were bred extensively to Percheron mares. These stallions contributed to the Percheron's refinement and are responsible for nearly half of all Percherons being grey. (Most of the others are black, although bays, browns, chestnuts and roans do occur.) Because Percherons have attractive, snappy trots, they were often preferred as heavy coach horses.

Other draft breeds developed from wild black horses that roamed the French and Belgian lowlands. (These horses were collectively referred to as "the wild black horse of Flanders.") Medieval man selectively bred these sturdy wild horses to develop and ideal war horse. The result was a type unsurpassed in strength and weight-carrying ability — the European "Great Horse." ("Great Horse" is another collective noun used to describe all members of this early type.) The Great Horse was bred to carry armor-clad knights (often weighing more than 400 lbs.). The horse also wore protective armor. What he lacked in speed and agility, he made up for in sheer power: he was slow and steady and made the ideal charger. Most modern draft breeds developed from the Great Horse. Several, such as the black Dutch Friesian and the French Brabant, probably trace directly to him. Other draft breeds were created through crosses between Oriental stock and the early developing native draft breeds, such as the Ardennes, Friesian, Brabant and Old English Black Horse.

The Shire, a modern breed, traces directly to the Great Horse through the Old English Black Horse, a type developed during the Renaissance. The Shire was named for his place of origin, the lowland Shire countries of east central England. During the breed's early days of development, the horses were especially popular with Yorkshire farmers, who valued their docility. Many American farmers believed that the Shire's excessive size made him impractical for agricultural work: studies show that the smaller the horse, the stronger it is in relation to its weight. Optimum weight for a typical 16 hand draft horse has been calculated at about 1540 lbs. Shires, the largest of the drafters, typically stand about 17 hands and weigh as much as 2400 lbs. Bay, brown and black are the most common Shire colors, but greys and roans occasionally are seen.

Shire blood helped develop the Clydesdale, a breed that originated in Scotland along the Clyde River valley. Crosses between the native Scotch variety of Great Horse and the Old English Black Horse produced the first Clydesdales. Then, in the 18th century, breeders imported a black Flemish stallion and numerous Shire stallions to contribute to the breed's hardiness. Finally, outcrosses with the Cleveland Bay added style and action to the Clydesdale's size. Clydesdales today are usually bay or brown; a few exhibit the ancestral black color, and white markings are common. Like Shires, Clydesdales have "feathering" (long, silky hair on the lower legs).

Clydesdale team in harness.

The Suffolk, a native English draft horse, differs from most other draft breeds in that it has little or no Flemish blood. Many authorities believe the Suffolk descended from the horses that once inhabited Solutre in the Rhone Valley of France. The Suffolk's comparatively small size and chestnut coat lend credence to the theory: Solutrean horses were smaller than the forest horses, and they gave rise to the predominantly chestnut German Schleswig and Danish Jutland breeds. The Suffolk probably developed from crosses between the English Great Horse and the smaller native Solutrean stock. Because he is more compact and more efficient at feed conversion than other draft breeds, the Suffolk traditionally has been popular as a farm work horse in England. In the United States, however, some farmers thought him too small to cross with American light horses to produce a drafter they considered large enough to work in the fields. This was unrealistic, because the Suffolk is a game worker and an economical keeper. His compact, "punched up" appearance encouraged the use of his common name, the Suffolk Punch. Although the breed has existed since 1586, all Suffolks today trace to a single prepotent horse, the Crisp stallion of Ufford, foaled in 1786.

PONY BREEDS

Ponies stand less than 14-2 hands (58 inches) at the withers when fully mature. They are used for a variety of activities and differ widely in conformation according to their primary purpose. Depending on the breed, ponies may resemble either light or draft horses in build.

Most pony breeds originated in areas where harsh climate and scanty vegetation had a dwarfing effect on animal size. Most descended from the original European Celtic pony, and a great many breeds developed in England. The Fell pony, a typical breed, is native to the Cumberland and Westmoreland areas of England. For several centuries, these ponies were used to carry lead from inland mines to ports on the northeast coast. The Exmoor, the oldest of the pony breeds, existed on the wild moors of Devon and Somerset in prehistoric times. He responded to his rugged environment by developing extreme hardiness and, for that reason, has been used as foundation stock for many other pony breeds. The Welsh pony, which originally lived in Wales, possessed the Exmoor's hardiness and combined it with strength, agility and sure-footedness. During the 1800s native Welsh ponies were crossed with Arabians, Hackneys and Thoroughbreds, which lent the breed added refinement and speed. The Welsh often is crossed with larger saddle horses to produce intermediate-sized riding horses or small hunters. Purebred Welsh ponies can be classified as one of three principal types: the Welsh mountain pony (known for his natural jumping ability), the Welsh pony (a slightly larger riding pony) and the Welsh Cob (the largest of the three types, with both jumping ability and stamina).

The Shetland, one of the smallest ponies, evolved on the inclement Shetland Islands. Its isolation prevented outcrossing with other breeds and, combined with scanty food supplies, led to the pony's diminutive size. Centuries ago, Shetlands were transported to Scotland and England for work in the coal mines. The ponies possess amazing strength for their size: they derive most of their power from short backs, deep girths and well-sprung ribs. Two principal types exist: the native British type, which has a short, round-barreled draft horse build, and the more slender, refined American version.

Native English ponies generally are rounder or draftier: they resemble their Celtic pony ancestors. Other pony breeds have been developed through liberal infusions of light horse blood. The Irish Connemara resulted from crosses between the native Celtic pony, the Barb and Arab stock. The Pony of the Americas (POA) developed from Black Hand No. 1, its foundation sire, a cross between an Appaloosa mare and a Shetland pony stallion. Black Hand inspired the development of the intermediate-sized United States breed. POAs today contain Quarter Horse, Arabian and Galiceno blood.

Miniature horses stand less than 34 inches at the shoulders when fully mature. They are registered by the American Miniature Horse Registry, a division of the Shetland Pony Club. The tiny horses are rare, but are becoming somewhat popular as family pets. They have been bred down from full-size horses and ponies.

section II.

selection

3

IMPORTANT SELECTION
CHARACTERISTICS

The purpose that a horse will serve largely determines the criteria used in its selection. Some will be chosen for a single use while others will serve in a number of capacities. Matching the horse to the function is obviously the strongest consideration.

Selection, like ownership, is affected by numerous external factors, such as available time, special interests, money and facilities. Economic considerations are often of great importance to the horseman. To save himself time and bother, a horseman should begin by establishing an approximate price range for the horse he wants.

When the horseman decides how much money he can invest, he should envision the "ideal" horse for his purposes. This "ideal" will serve as the horseman's goal and, in his attempt to achieve this goal, he should examine ideal breed descriptions and records of horses that have proven to be outstanding. He should then set standards for characteristics such as the following:

1. breed
2. pedigree
3. performance
4. conformation
5. soundness
6. blemishes
7. size
8. color
9. sex
10. reproductive performance
11. disposition
12. intelligence
13. trainability
14. training
15. age
16. condition
17. health
18. suitability to facilities

Finding a horse that meets the ideal in each of these categories is not a realistic goal. (Thus, the old saying, "There is no such thing as a perfect horse.") The horseman will have to compromise by setting minimum acceptable standards, perhaps by assigning point values for each desirable characteristic. He must decide how far from the ideal the horse can be and still fulfill the purpose(s) for which he is intended. This applies to consideration of individual traits (how straight must the legs be?) and to comparisons between traits (would excellent conformation make a horse with a lackluster pedigree more acceptable?). This type of selection — assigning point values and determining minimum scores — is called the Selection Index method and is covered in more detail later. Weighing these standards enables a horseman to exercise his judgment and arrive at an index, after balancing many factors according to his needs and preferences. Such an index allows the horseman to make exceptions for horses that are superior in one respect and inferior in another. To establish guidelines and prepare an index, the horseman should consider certain traits in detail:

1. **Breed:** The term breed refers to a group of horses with a particular function, form and/or color that have been carefully bred until their desirable characteristics have become fixed.

2. **Pedigree:** The pedigree — which shows a horse's line of descent — can be a valuable tool in predicting an animal's worth and performance potential. It is only useful when treated as an indicator and not as a guarantee of ability. A horseman who evaluates the merits of individual horses and compares their pedigrees is using the pedigree wisely. But one who insists on obtaining the get of a particular sire or dam is limiting his choice to a very small field: he is making a certain pedigree his goal and may overlook faults that otherwise would make the horse unacceptable. (Refer to **"Pedigree"** later in this chapter under **SELECTION CRITERIA**.)

3. **Performance:** The horseman should decide upon an acceptable level of proven performance in a mature animal. For a younger horse, he must decide how much potential it should exhibit to match that level. This is difficult since performance records are subject to both hereditary and environmental influences. Some good horses are handled improperly or campaigned lightly and have not had an opportunity to realize their full potential. On the other hand, poorer quality animals shown to the best advantage by skilled trainers may compile

impressive records. (Refer to **"Performance"** under **SELECTION CRITERIA** later in this chapter for an expanded discussion on this subject.)

4. **Conformation:** Emphasis on correct conformation will make selection for performance easier since a horse must be conformed properly to perform efficiently. A horse should be balanced and have good basic structure with no conformational defects that could lead to debilitating unsoundnesses. (Refer to **"Conformation."**) In show horses, overall excellence of conformation might be a primary consideration. But in performance horses, specific good traits may outweigh the lack of perfection in other traits. (Strong hindquarters and straight legs in a racing Quarter Horse might outweigh a somewhat shortish neck.) There are also conformation faults that should be avoided when selecting a horse for breeding purposes because those faults can be passed on to its progeny. (Refer to **"Conformation"** later in this chapter and under **INHERITED ABNORMALITIES**.)

5. **Soundness:** Many conformational defects can impair a horse's action and ability to function normally, and can often result in unsoundness. Many horses used in strenuous work or competitive situations do not remain sound throughout life. The horseman should decide what potential for continued soundness he requires relative to the functions the animal will be expected to perform. Older horses with proven performance records may have acquired some unsoundness that prevents them from being shown or raced, but, if the unsoundness is not the result of a heritable fault, they can still be useful as breeding stock. Many horses will be suitable for uses less strenuous than for which they were originally intended. (A hunter may become a child's pleasure mount, etc.) The type and degree of unsoundness relative to intended use will determine whether the horse is a worthwhile investment. The horseman must decide what type of unsoundness (if any) or potential unsoundness he can accept.

6. **Blemishes:** The intended primary use of a horse will dictate the type of blemishes that are acceptable. Some blemishes may be unattractive, but will have no effect on a horse's performance. Minor blemishes might be tolerable in a performance or breeding animal, but unacceptable in a halter horse. The horseman should set his own minimum standards: would firing marks or wire scars impair a horse's effectiveness or draw criticism from show judges or the buying public? These blemishes may be important under those circumstances, but will have no effect on the quality of the animal's offspring.

7. **Size:** Size requirements often stem from personal preferences, but can also be important in determining how a horse will perform. A race horse with determination and a will to run can overcome deficiencies in size and achieve victories over much larger horses, but generally, sizeable horses that are long in the forearm and gaskin make the best race horses. (Length in the forearm and gaskin leads to long strides.) In hunting competition, large hunters often have an advantage over smaller horses because hunt courses are designed for horses with fairly long strides. The horseman must decide if size will be a limiting factor. Would he reject a yearling with outstanding conformation because he is a few inches shorter than average yearlings of his breed? While such a lack of size might handicap a race horse, jumping horse or hunter, it might actually be preferred in a pleasure horse. The horseman should also remember that horses mature at different rates: a colt that is comparatively small as a yearling might be larger than horses of his same age as a two-year-old.

8. **Color:** Among the most emotional of all selection factors are color choices. Whether or not horsemen believe that color and performance are genetically correlated, most have strong likes and dislikes for certain colors and markings. Many discriminate against horses with white feet. Their prejudice may appear arbitrary, but it has a sound basis in fact: horses with white socks or stockings will have light-colored hooves (assuming that the white extends to the coronary band) and white hooves may have a thinner, more fragile wall than those of their darker counterparts. Furthermore, white markings, both on the face and other parts of the body, have pink skin under the coat that is more subject to sunburn, photosensitization and plant allergies. (Refer to **"Coat Color and Texture."**)

Arbitrary or not, color preferences have led to the formation of such breed registries as Paint, Pinto, Palomino, Albino and Appaloosa. Color is of great importance in those instances because of registry requirements. In addition, some breeds have characteristic colors even though color preference was not the basis for breed formation. These include the Lippizaner (grey), Suffolk (chestnut) and Cleveland Bay breeds. Color and markings in all breeds, while regulated somewhat by registries, are subject to occasional fads.

9. **Sex:** A horse's sex will often influence its temperment and performance. Stallions are naturally somewhat more aggressive, and mares are often said to be less consistent performers than geldings. A

mare or stallion might be selected for showing and working qualities early in life and later, if the pedigree is right, be a valued addition to a breeding program. But a horseman interested solely in performance has the option of choosing a gelding. They often make quieter mounts than mares or stallions because their dispositions are not subject to hormonal influences. They present fewer management problems and often sell at lower prices than breeding animals of comparable quality.

10. **Reproductive performance:** Reproductive ability in horses is difficult to assess. One of the most accurate indicators, progeny records, can generally be obtained from the major breed associations. A horse's own offspring provide the only reliable measure to its success as a producer. But for those that have no mature progeny, reproductive records of close relatives, and a reproductive examination of the horse by a qualified veterinarian, are the best indicators. (A breeder should also consider a stallion's prepotency, fertility, and average foal crop evaluation, as well as a mare's fertility and mothering ability.)

11. **Disposition:** An even temperament and an eagerness to perform are important qualities in all types of horses. Quiet, tractable horses minimize management problems: they are less likely to hurt themselves or the people and horses around them than excitable or ill-tempered horses. Disposition seems to be quite heritable, but is also

This Przewalski's mare will influence the temperament of her growing foal. A good disposition and mothering ability are important considerations in broodmare selection.

San Diego Zoo Photo

determined by environmental influences: a foal learns many behavioral responses from his dam, and may mimic her reaction to people, unfamiliar situations, etc. This makes disposition of broodmares more important than some horsemen realize, as gentle, quiet mares are more likely to raise foals with good dispositions than those that are ill-tempered.

12. **Intelligence:** There are no easy IQ tests for horses, but their behavior and certain points of their conformation provide valuable clues to their intelligence. A broad forehead indicates sufficient space for a large brain; large, alert eyes and erect ears are also indicative of intelligence. An alert horse, interested in his surroundings, usually learns more quickly than one that is disinterested.

13. **Trainability:** This refers to certain physical and mental characteristics that make a horse suitable for a particular job. It is a quality especially difficult to assess in young horses that have had little handling or training. The horse provides clues to his particular predisposition toward trainability: a horse that is responsive to a human voice, that handles quietly from the ground and that shows interest in his surroundings, is probably more trainable than a stubborn or willful animal. To be easily trained for a particular purpose, a horse must have certain inherited instincts or characteristics. Whatever the type of horse, he must have the physical structure that enables him to perform well in his specialty. A large, sluggish or awkward horse would fail as a cow horse since he could not be quick enough or work close enough to the ground. Similarly, a horse without natural animation and high action would fail as a fine harness, formal driving or park horse.

14. **Training:** The job that a horse is expected to perform helps determine selection. How well trained the horse needs to be to perform that job depends to a great extent on the ability and expertise of the proposed new handler. A seasoned and talented horseman might prefer that the horse be green or totally untrained. A novice might require a highly trained animal. This is a very subjective matter in the overall area of selection. The horseman should weigh the relative importance of the horse's training against his particular needs to determine the one most suitable for him.

15. **Age:** A horse's age will influence his market and resale value. Weanling race prospects are often acquired for substantially less than they would bring as yearlings. Stakes-quality Quarter Horse and Thoroughbred race horses usually perform their principal work as

two- and three-year-olds, although some race much longer. Similarly, Standardbred trotters and pacers are raced primarily when young, but may still be performing in their early teens. Hunters, jumpers and distance horses are seldom used heavily until they are five years of age or older. Barrel racing horses often turn in their best performances while in their teens. Breeding stock that receive proper care may produce over a span of 20 years or so. Some genetically coded traits will be accentuated with age and/or use, such as greying, swayed backs, and breakdown due to faulty conformation. Age might also affect a horse's keeping qualities, but horses receiving the proper care generally remain functional into their twenties.

16. **Condition:** A horse in poor condition should not be discriminated against if his condition has been caused by mismanagement. The ribs may show on a horse that has been neglected, and his coat may be poor, but proper nutrition, exercise and grooming may restore him to peak condition. On the other hand, poor condition may indicate a chronic poor doer — a horse that will not improve adequately even with the best of care. Condition in a young horse is important because a young horse that has been fed a poor diet may take years to recover from certain deficiencies. An over-fed, over-weight young horse might also have problems later in life if his developing leg bones and joints and his relatively small hooves are stressed excessively. Condition in mature horses can affect fertility — an older mare in poor condition might be difficult to breed, and a stallion in poor condition might have a lowered sperm count. A sleek, shiny coat alone does not always indicate good condition. A rough-coated horse of average or moderate body weight might be healthier than one that is fat and has a shiny coat, especially in winter.

17. **Health:** A knowledgeable horseman or trainer can usually assess the health and basic soundness of most horses. But even experienced horsemen normally secure veterinary assistance when there are questions regarding old injuries, insurance or breeding soundness, or when buying a horse at great expense. In addition to the vital functions (temperature, pulse, respiration, gut sounds, etc.) the horse should be checked for respiratory and circulatory problems, and should be checked for loss of vision, conformation defects, lameness, blemishes, reproductive defects, vices and diseases. Horses with predispositions toward certain conditions should be examined for signs of their onset: grey horses should be inspected for melanoma, overweight horses with small hooves should be checked for signs of navicular disease, etc.

18. **Suitability to facilities:** Some horses may require more attention than others, depending on their ages and uses. (This is not meant to imply that any horse should be neglected.) Some broodmares may do well on pasture with only a run-in shed for protection from the elements. Others, such as former race mares "letting down" from the track, or high-strung maiden mares or poor conceivers, may need more personalized attention. A horseman should decide how much attention and care he is willing and able to provide. It is wise to heed the adage: "Don't buy a horse that has received better care than you are willing to give it."

Like any good businessman, a horseman looks for the best horse possible at the most reasonable price. The standard-setting process just described will help make this goal more obtainable. Some people waste their time trying to find an outstanding horse at an unrealistically low price. Most real bargains result from finding an outstanding horse — one with excellent pedigree, performance and conformation — at a good price, rather than an inferior horse at any price. Admittedly, some inexpensive horses that were unimpressive for one reason or another in the sale ring have gone on to become superior performance horses. These, however, are the exception rather than the rule. For every unlikely, inexpensive horse that develops into a great one, thousands like them never come close to succeeding. Of those that do succeed, many are disappointments as breeding stock because their ability is due to a lucky combination of genes, a combination they probably cannot pass on to their offspring. (A horse may pass individual good genes to his offspring, but cannot pass his particular *combination* of genes.) There is an element of luck involved in determining a horse's potential and actual performance, but the horseman can shift the odds to his favor by selecting wisely on the basis of certain pre-defined standards.

By having defined standards, the horseman will greatly improve his chances of finding the horse most suitable to him and to his purposes. This enables him to narrow the field and target his efforts.

Selection Criteria

When a horseman has set his goals and established minimum standards for desirable traits, he should make a more detailed study of pedigree, performance and conformation. These should be balanced according to the intended use or uses of the animal since their importance will vary according to the job the horse will perform. For a gelding, actual performance will be far more important than pedigree. In a prospective broodmare, however, pedigree is a vital consideration. Because she may carry genes for many beneficial traits from her ancestors, those traits could be passed on to her offspring, making pedigree a more likely consideration than actual performance in predicting her ability as a dam.

Pedigree

Successful horsemen attach varying degrees of importance to the pedigree. Many argue that, because inheritance is governed by chance, the presence of excellent ancestors in the pedigree does not guarantee that the offspring will receive their favorable genotypes. They point out that it is *possible* for an animal to have an outstanding great-grandsire and possess none of his genes. However, the *probability* is that approximately 12% of the animal's genes will come from that outstanding stallion.

Some horsemen believe that a pedigree serves only as a tenuous guide to an individual's genetic make-up, since favorable combinations of genes may not pass from parent to offspring. Most horsemen, however, rely heavily on the pedigree as an indicator of hidden strengths and weaknesses. They may search for an animal that has been linebred to a particular sire, because they believe that this type of inbreeding can duplicate desirable genes. (Inbreeding can also duplicate undesirable genes.) Other horsemen may look for a horse that resulted from a cross between two particular families, because mating two unrelated horses may produce a foal superior to either parent (due to hybrid vigor). These horses are sometimes said to have "nicked," meaning that their genotypes complemented one another to improve their offspring. Certain family crosses are generally thought to produce nicks: some Thoroughbred breeders favor crosses between Bold Ruler and mares that are daughters of Princequillo. Some Arabian breeders favored crosses between the Polish stallion *Bask and inbred *Raffles mares to produce show horses.

This reliance upon pedigrees developed centuries ago when Arabian horse dealers memorized and recited their horses' ancestry to prospective buyers. These verbal recitations of pedigrees were valued as a legitimate basis for selection, but led to exaggerated claims by unscrupulous dealers. The sellers knew that prominent ancestors would bolster a horse's market value and that the buyer would probably be unable to verify the lineage.

Written pedigrees arose to authenticate ancestry and insure fairness in these buyer-seller exchanges. Today, pedigree information can be found in stud books maintained by individual breed registries. Un-

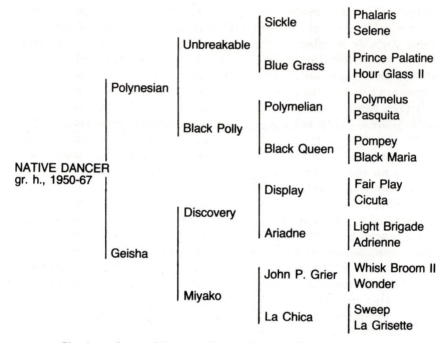

Simple pedigree of the grey Thoroughbred stallion Native Dancer.

		Sickle	Phalaris / Selene
	Unbreakable		
		Blue Grass	Prince Palatine / Hour Glass II
Polynesian			
		Polymelian	Polymelus / Pasquita
	Black Polly		
		Black Queen	Pompey / Black Maria
NATIVE DANCER gr. h., 1950-67			
		Display	Fair Play / Cicuta
	Discovery		
		Ariadne	Light Brigade / Adrienne
Geisha			
		John P. Grier	Whisk Broom II / Wonder
	Miyako		
		La Chica	Sweep / La Grisette

MEADOW LANDS
p, 3, 1:59 2/5

Sire
ADIOS, p, TT1:57 1/2
Sire of 75 in 2:00
including Bret Hanover,
p, 4, TT1:53 3/5

   Hal Dale
   p, 2:02 1/4
      Abbedale, p, 2:01 1/4
      Margaret Hal, p, 2:19 1/2h

   Adioo Volo,
   p, 3, 2:05h
      Adioo Guy, p, 2:00 3/4
      Sigrid Volo, p, 2:04

Dam
MAGGIE COUNSEL
Dam of 6 in 2:00 including
Meadow Rice, p, 3, TT1:58 1/5
  by The Widower
Meadow Chuck, p, 3, 1:58 4/5
  by Adios
Meadow Lands, p, 3, 1:59 2/5
  by Adios
Meadow Bucky, p, 3, TT1:59 1/5
  by Adios
Meadow Maid, p, 3, 1:59 2/5
  by Adios
Meadow Gold, p, 3, 1:59 4/5
  by Adios

   Chief Counsel,
   p, 3, TT1:57 3/4
      Volomite, 3, 2:03 1/4
      Margaret Spangler,
      p, 2:02 1/4

   La Reine,
   2, 2:27 1/4
      Peter Scott 2:05
      La Roya, 3, 2:08 3/4

Annotated pedigree of the Standardbred stallion Meadow Lands. Meadow Lands' sire, Adios, is the leading sire of speed in the breed's history. His dam, Maggie Counsel, produced more horses that covered one mile in 2:00 than any other broodmare in history.

fortunately, many modern pedigrees are still not much more helpful than those early oral recitations: they contain the names of ancestors for a variable number of generations (frequently three to five), but reveal no information about performance or collateral relatives. A simple pedigree is only helpful when the performance records of the ancestors are known, or if the horseman has had an opportunity to examine the ancestors himself. Names alone will not help predict performance of the offspring, and without performance records, the horseman must rely more on the individual's conformation, performance and produce records. The most useful pedigree is one that is annotated (i.e., it contains performance records of ancestors which can help predict usefulness and prepotency.)

Performance records can be important because they provide keys to the strengths and weaknesses of animals in a horse's pedigree. Although good bloodlines are vital indicators, insistence upon owning only those horses which have the "perfect" pedigree unnecessarily limits the horseman. Horses with undistinguished pedigrees, but good performance records, could prove successful as breeding stock, although they are the exception rather than the rule. (Also unusual are horses with excellent performance records, backed by fine pedigrees, who fail to pass their superiority to their offspring. Citation, who won the Triple Crown in 1948, was one of the best racing Thoroughbreds on record, yet he proved a disappointment at stud.)

courtesy of The Blood-horse

Citation, 1948 Triple Crown winner, was an outstanding race horse but achieved little distinction as a sire.

In genetics, the probability that a certain trait will be passed from parents to offspring is called a heritability estimate. Some traits, such as racing and jumping ability, are more heritable than others (e.g., pulling power). Racing performance in Thoroughbreds is thought to be quite heritable (i.e., 30-40% genetics vs. 60-70% environmental influence). Because of this heritability, the average earning indexes of the sire and/or dam seem to be positively correlated with those of the progeny. (Refer to **"Breeding Aspects."**)

Because an annotated pedigree contains records that indicate the ability of ancestors, it can help predict the potential ability of the offspring. A horseman could theoretically predict the probability of a trait being passed on to a horse's offspring, providing that he knows which ancestor possessed the trait, and to what degree he possessed it. For example: A breeder examines the pedigree of an outstanding trotting mare and finds that many of her ancestors and more than three-quarters of her collateral relatives were good trotting horses. He checks her progeny records and finds that three of her fillies, and both of her colts, have developed into good trotting horses. He knows that trotting speed is heritable and that his mare possesses it. Therefore, he can conclude that, when properly mated, she will probably pass this trait on to her next offspring.

Even when produce and performance records are available, it should be remembered that they are not always exact. Furthermore, environmental factors (training, nutrition, handling, opportunities) can have a striking effect on actual performance. A stallion in a certain horse's pedigree may have a poor race record, but he could have been a late bloomer retired early from competition because of injury.

Racing records tend to be quite objective and comparable between horses. Other records, which rely upon subjective judging, frequently are not. A horse with a seemingly outstanding show record, for example, may not live up to expectations. He may have been exhibited almost exclusively at small shows with limited competition, or shown only in one locality, or under judges known to have a preference for his particular type. Styles in show horses change, and a size, body style or bloodline that won in the show ring five years ago may be less popular today.

Certain bloodlines and families may experience periodic surges in popularity which can cause prices to skyrocket. Although the horseman must look to the future to update his ideal, he must also be able to hold steady through fads when planning for sales which may be two or more years in the future.

In selecting breeding stock, a pedigree that contains information on collateral relatives (those that share common ancestors, such as cousins, brothers, sisters, etc.) can help reveal the presence of favorable recessive genes whose effects are masked by less favorable dominants. A pedigree study may reveal that an above-average horse resulted from a lucky genetic combination if his collateral relatives were only average or poor performers. On the other hand, a horse of average ability might have consistently outstanding collateral relatives. He will usually be a better breeding prospect since his many good genes may be masked by dominant genes or unfavorable gene interactions.

To make the best use of a pedigree to predict a horse's potential, the horseman should consider the following:

1) close ancestors
2) heritability of desired traits
3) collateral relatives
4) progeny records
5) degree of inbreeding

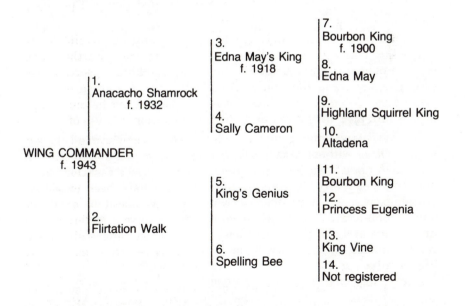

Pedigree of the Saddlebred stallion Wing Commander. In the absence of inbreeding, the most important ancestors in a horse's pedigree are the first 14: parents, grandparents and great-grandparents.

CLOSE ANCESTORS

The most important ancestors in a pedigree are generally the first 14: parents, grandparents and great-grandparents. Information on relatives past the third generation is less significant unless the horse is heavily inbred to a particular animal. This would increase the chances of that animal's genes being found in its descendants. By definition, all breeds are inbred, with some being more inbred than others. Thus it is highly probable that there are common ancestors in the seventh generation which should not be ignored. Because *all* of a horse's inheritance is filtered through his parents, they are the most important ancestors in his pedigree. More distant relatives are useful for the clues they provide to the parents' genotypes. A good rule of thumb for the horseman is to examine the horses in the first two generations (parents and grandparents). If none have any merit, he can usually disregard more distant ancestors: even if they are superior, many of their good genes obviously were not passed to their foals.

When successful horses are found close-up, within one or two generations, the pedigree is strong. Ancestors with excellent records are desirable, but their importance is diminished as they become further removed from the horse in question. That is why a horse with several better than average relatives within one or two generations has a stronger pedigree than one with outstanding great-grandparents and a number of mediocre immediate relatives. A good pedigree is one in which most of the close ancestors and relatives are strong in the desired traits. Such consistency in the pedigree implies a high degree of heritability of the desired characteristics.

HERITABILITY

Heritability estimates must also be considered. A stable of illustrious ancestors will not help the offspring if their strong points are only marginally heritable. A solid pedigree shows ancestors and collateral relatives frequently possessing traits that are highly heritable (speed, jumping ability, intelligence, desired mature size, etc.). The pedigree weakens as the frequency of these traits decreases. Traits that are only moderately heritable (pulling power, temperament, etc.) may show up consistently in a pedigree but even so, the pedigree is not an effective tool for measuring those traits in offspring which are more influenced by environment than genetic transmission. Certain qualitative traits (e.g., coat color, blood groups, and sex), have a heritability estimate of 100% since they are due entirely to genetic factors. Quantitative traits, such as speed, jumping ability, and intelligence are multi-genetic factors influenced by environment and are considered highly heritable if they fall in the 30-40% range. (For more information and for heritability estimates in horses, refer to **"Breed Improvement Through Applied Genetics."**)

COLLATERAL RELATIVES

A study of a horse's collateral relatives can help predict future performance in addition to revealing desirable recessive genes that are masked by dominants. Full brothers and sisters might especially resemble the horse because, on the average, they contain about 50% of the same genes. (They share 50% rather than 100% because the parental chromosomes split during meiosis. Chance determines which half passes from each parent to the offspring.) Even full brothers and sisters may contain completely different sets and combinations of genes. This explains why a full brother to a good race horse may be a disappointment. The chances of duplicating the genes of a good race horse by duplicating the mating are poorer than one might expect. The opportunity for variation is such that an equine brother stands no better chance of being a "twin" than do two sons in a human family.

A horse also has about 25% of his genes in common with his aunts and uncles. They are full brothers and sisters of his parents, and are therefore 50% related to his parents. Each grandparent shares about 25% of its genes in common with its second-generation descendant (the same amount shared by the individual with his aunts, uncles and half-siblings). This is because 50% of its genes were passed on to its immediate offspring which, in turn, passed on 50% of its genes to the foal. Therefore, the grandparents' genes are diluted to 25% in the grandson or granddaughter.

These collateral relatives are of equal value in making predictions because each one is likely to share 25% of the same genes as the prospect. Half-siblings are most easily evaluated because they are usually more accessible and more numerous than aunts, uncles, and the four grandparents. Obviously, if the horseman's survey contains the records of 10 half-brothers and 12 half-sisters, his predictions for the potential of the horse in question will probably be more accurate than if he had based them solely on the records of the grandparents.

PROGENY RECORDS

Progeny records have proven especially important in determining an animal's worth. Breeders often say that a pedigree tells what an animal *should* be, his actual performance and appearance tell what he *seems* to be, and his progeny tell what he actually *is*. This is true to a certain extent because a horse's progeny can reveal its genotype.

A generous crop of foals produced by one stallion and different mares could help a breeder determine whether or not the stallion carried certain desired genes, particularly if the genotypes of the mares are known. This type of progeny testing is not possible with mares because of the time involved.

Strength in a pedigree, indicated by ancestors and relatives with good performance records and the ability to transmit favorable genotypes, is important. Yet it by no means limits a horse to a certain standard or insures excellence. Actual performance may be the best guide to determine a horse's suitability for a particular job. (Refer to **"Inbreeding and Outbreeding"** for a discussion of how the degree of inbreeding may affect an individual.)

Performance

This is probably one of the most important factors in selection for mature horses and, to some extent, those that are young. Breed registries often maintain performance records. A horseman considering a prospect might check his performance record to see if he maintained a consistently high level of performance over a considerable period of time and under a variety of conditions. A consistently good horse will perform more effectively than one that turns in mediocre performances interspersed with occasional moments of brilliance. Selecting for a consistent performance record (like a consistent pedigree) makes choosing a horse less of a gamble and more of a science.

Comparing performance records can establish major differences in capabilities, but such a comparison should be guarded. Some records will reflect environmental influences more than a horse's actual ability. Others will be affected by subjective judging. The horseman can help equalize records by considering the many factors that affect performance (training, handling, nutrition, conditioning, age, sex, etc.) as well as the number and nature of opportunities a horse has had to prove himself. In evaluating a race horse, for example, the horseman should consider how many times and at what tracks the horse has raced. What was the caliber of competition and the average purse sizes? A horse with many starts and only a few wins could still be a wise choice if he has consistently placed immediately behind outstanding horses.

OBJECTIVE VS. SUBJECTIVE RECORDS

Certain records, such as speed measurements in race horses, are relatively objective. Unless disqualified, the first horse to cross the finish line wins. Many horses bred for racing never even reach the track because they perform poorly during training. Those which do reach the races must perform well, since a horse that consistently places out of the money will soon be retired. So much emphasis is placed on performance for the Standardbred breed in America that early registration was based on a minimum acceptable speed. Only those horses that could trot the mile in less than 2:30, or pace the mile in less than 2:25, were registered as Standardbreds. Other fairly objective evaluations of race horse performance include the Average Earnings Index (AEI) for Thoroughbreds and the Speed Index system (SI) used in both Thoroughbred and Quarter Horse racing.

Some abilities can be objectively assessed besides speed. These include pulling power in draft horses, jumping ability in hunters and jumpers, agility in stock horses, and quickness and dexterity in polo ponies. A horse strong in such natural abilities will display them consistently. When he performs, he will reveal his talent or special instincts, and a horseman can rate his performance by watching him work. This is not always possible with horses selected for more intangible qualities such as grace, beauty, action, style, etc. These are subject to individual tastes and preferences. An animal one horseman considers a good show prospect might place consistently lower than horses he considers inferior. Because of this variation in personal tastes, show records cannot always be accurate guides. Since judging is somewhat subjective, there is no insurance of uniformity between judges, classes and shows. Still, the horse that consistently places well in large classes and under many different judges is probably a good prospect.

Performances that are rated subjectively, such as harness, riding, and conformation classes, are especially difficult to compare. The judges often have somewhat divergent interpretations of the "ideal." Furthermore, the "ideal" show horse often undergoes changes through changes in fad and fancy. Comparisons should attempt to adjust for such fads. Did a Quarter Horse stallion prospect earn his show points years ago when the heavily muscled, "bulldog type" was popular? These comparisons should also account for environmental differences, such as size, nature and prestige of competitions, regional variations, intensity of show schedules, etc. Two show horses may have the same number of blue ribbons, but one may have been campaigned heavily to attain what the other won in fewer shows.

A horse's sex also has a bearing on the importance of performance records in the selection process. A mare from a good family might make an excellent broodmare prospect, even though she lacks strong performance credentials. (Lack of opportunity or injury may be factors in her background.) On the other hand, a gelding's performance alone should be a strong measure for his acceptability for the job he is expected to do.

REPRODUCTIVE PERFORMANCE

This quality in breeding animals is difficult to evaluate, primarily because progeny testing, the only true test of worth, is time-consuming and expensive. However, there are other guides to good prospects including pedigree, conformation, actual performance, veterinary records and the reproductive examination.

Obviously a good broodmare prospect is one with impressive performance credentials backed by a strong pedigree. A mare with a winning show or race career will normally produce better offspring than a poor performer. She will usually need some time to "let down" and adjust to a more sedentary life before being used as a broodmare. A horseman who insists on an immediate producer should look for a more settled mare, usually one at least four years old, who has already produced a foal.

If the mare lacks a proven record of performance, the pedigree requirements should be more stringent. The pedigree should show ancestors and collateral relatives that have proven themselves on the track or in the show ring. A broodmare prospect should have good conformation and be free from heritable defects such as roaring, parrot mouth, ewe-neck, bone spavin, etc. Animals with those defects should be selected against so that breed and species improvement will continue. Many horsemen believe that only the top 40 or 50% of all mares, and less than 5 to 10% of all colts, are of sufficient quality to be maintained as breeding stock.

Using only the best stallion prospects is critical to a breeding program. Although an inferior mare could average no more than one poor foal a year, an inferior stallion could sire dozens of disappointments each season. A stallion selected for breeding should have excellent conformation; culling for heritable defects should be even more stringent than it is for mares. A stallion who has had ample opportunity to prove himself as a sire and has produced only mediocre foals probably should be culled. Good disposition in a stallion is an asset: certain stallions are noted for producing bad-tempered offspring, while others seem to pass on quiet, docile temperaments. Because the stallion usually has little or no environmental influence on the foal, his contribution to temperament is generally thought to be genetic.

Outstanding sires usually have a pedigree containing an abundance of outstanding individuals with superior performances. Yet several horses with good pedigrees, but no performance records, have proven successful at stud. Otherwise good horses that have not performed because of injury or acquired disease (i.e., non-heritable) should not be overlooked as stud prospects. A horse named Bartlett's Childers was considered in his time to be practically worthless as a race horse. Because he was a bleeder, he was never trained. But he became a leading sire and his great-grandson, Eclipse, was one of Thoroughbred racing's greatest horses. Eclipse became the progenitor of the direct male line that has produced horses such as Whirlaway, Native Dancer, Bold Ruler, and Secretariat. A modern example in the Thoroughbred is Hoist the Flag. In 1971, as a three-year-old, he shattered a pastern and was fitted with an artificial foot. He earned less than $80,000 in his career. But by 1977, he had already produced 11 stakes winners and, in 1978, eight of his yearlings averaged nearly a quarter of a million dollars each at the Keeneland sale.

Fertility in prospective breeding animals should not be assumed; it should be carefully tested. A horseman buying a broodmare should obtain a complete foaling history and give special attention to the date of the last foaling, the number of years skipped, and any complications. A veterinarian should palpate the mare rectally to check the ovaries and reproductive tract, and look for signs of infection or transmissible venereal disease. A speculum examination, followed by culture and/or uterine biopsy, may also be recommended. Mares with a history of irregular heat cycles, or who have abnormal vaginal discharges, or fractured pelvises, or thick, cresty necks and a masculine appearance, will probably have breeding problems.

Stallions should also be checked for fertility. A qualified technician should perform a semen evaluation to determine numbers of viable sperm, motility, morphology, pH, etc. Stallions over 15, or those that have been heavily bred, should be more carefully inspected since old age and overuse can lead to infertility. Certain venereal diseases, infections, and even a high fever, can lead to temporary sterility. In some cases, the damage is irreversible.

A stallion's prepotency and value as a breeding animal is easier to determine than a mare's since he can sire many foals each year. A stallion used for breeding as a three-year-old will have yearlings or older horses on the ground by the time he enters his breeding prime at five or six years of age. The offspring can be evaluated for conformation and other important traits. In race horses, for instance, the offspring may race as two-year-olds and have an opportunity to display their ability. An accurate progeny evaluation would include offspring from many different mares born over a number of years, and could not be performed until the stallion had sired at least four or five foal crops. (Refer to **"Breeding Aspects."**)

Conformation

The horse's overall bodily shape or form is referred to as conformation. It has also been described as the relationship of form to function. Conformation will ultimately determine how a horse moves and withstands impact-related stress. A casual visual inspection can be misleading because many conformational defects that can lead to unsoundness are not obvious. A horse may appear perfect at first glance, but if some conformational defect prevents him from performing up to the standards required by the function he is to perform, he does not possess desirable conformation. That is why many horsemen rely on a veterinarian to conduct a soundness examination. A veterinarian can pinpoint blemishes versus unsoundness, and detect conformational traits that might lead to problems.

Conformation is basically the result of many heritable traits, although environmental factors help shape the horse's body. Because a foal will probably possess traits from both parents, a horseman planning a mating or buying a foal should be certain that both sire and dam have desirable conformation. Compatibility of conformational traits is especially important in planned breedings. A stallion strong in a particular trait (i.e., a broad, roomy chest) might partially compensate for a mare weak in that trait (a narrow chest). Judicious mate selection can occasionally result in foals superior to either parent. A mare with an excellent pedigree and an impressive record, but who toes out slightly, could be bred to a stallion with faultless forelimb conformation. This may correct or reduce the fault in the offspring. Selective breeding of this type may correct, or partially correct, certain undesirable traits. When more than one undesirable trait is involved, the offspring could develop more conformational problems than either parent. A mare with an especially attractive head and sway back may be crossed with a stallion possessing a straight, short back but a rather plain head. Instead of producing a foal with both *good* traits, the mating might result in a sway-backed horse with a plain head. (For a more complete explanation on selection for or against a trait in a breeding herd, refer to **"Breed Improvement Through Applied Genetics."**)

Although concepts of perfect conformation vary among breeds, all breed registries agree that the overall quality and balance of a horse's build should be symmetrical and proportional to its size. There are slight differences in this, but researchers have found that the musculoskeletal systems of outstanding performance horses of all breeds are surprisingly similar. Good horses are generally balanced: their bodies

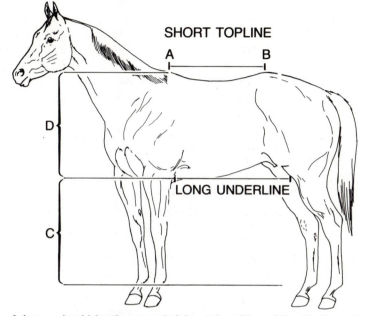

A horse should be the same height at the withers (A) as he is at the croup (B). His front legs (C) should be about the same length as the depth of his body (D). His topline (measured from A to B) should be short, and his underline long.

should be divisible into equal thirds — from about the point of the shoulder to the point of the withers, the point of the withers to the point of the hip, and the point of the hip to the point of the buttocks.

Although there are no absolutes, most horsemen agree on other basics of conformation: a good horse should be the same height at his withers as he is at the highest point of his hips. His front legs (measured from the fetlock upward) should be about the same length as the depth of his body, measured from underline to withers. He should have a long neck (for grace and flexibility), a long underline (for freedom of movement), and a short topline (for strength). He should have a long shoulder that slopes forward and downward from the withers to the point of the shoulder. This slope can be measured: if a line were drawn parallel to the ground from the rump to the neck, the shoulder should meet this line at about a 45-degree angle. This angle allows the shoulder muscles increased forward movement, resulting in a longer stride, reduced concussion, and a smoother ride. The pastern and the hoof wall should slope toward the ground at about the same angle as the shoulder to help the leg absorb shock. This slope can differ from the ideal 45 degrees and still be effective in shock absorp-

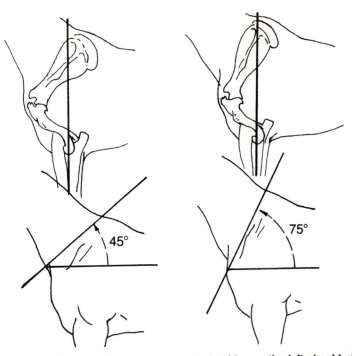

Angle of the shoulder. The well-angulated shoulder on the left should allow smoother action than the upright shoulder on the right. As the shoulder approaches a straight line, stress upon the limb increases due to jolting concussion upon joints with limited flexibility.

tion. It is the angled stacking of the leg bones that allows the joints to absorb much of the shock of impact. The result is a much smoother gait than bones stacked end-on-end could provide.

There are no horses with "perfect" conformation, and the degree of departure from perfection will vary. A horse may not have ideal conformation, but he is frequently able to compensate for his defects and remain serviceable. Sometimes compensatory devices are built-in. For example, a horse with rather straight shoulders (which would absorb little impact and make for a rougher gait) might have long pasterns. These absorb more shock than short pasterns and, therefore, provide an acceptably smoother ride. Another compensatory device might be seen in a horse with a wide chest. He might turn his feet inward to become pigeon-toed. This helps him balance his forequarter weight more evenly than if he stood wide (and placed excessive weight on the inside wall of his hooves). The stance also helps him keep a foot under each structural corner of his body. This horse will develop a stride to help compensate for being pigeon-toed: he will probably paddle (swing his forelegs in outward arcs), to keep his legs from interfering with one another as he moves forward.

A horse can have conformational defects and still be acceptable for certain purposes. Therefore, guidelines are not inflexible. They should serve only to aid the potential buyer, not absolutely dictate his choice of a horse. He should conduct a thorough examination of the horse from all angles, both up close and at a distance. After assessing conformation from each angle, he should judge the animal in motion, particularly at the walk and trot. This overall picture is important: a horse with a slight conformational defect should not be excluded from consideration if he is able to overcome his deficiency and perform acceptably. A horse should be judged by his ability to perform desired movements and functions, particularly if he is being selected for purposes other than breeding.

The following section discusses ideal traits, but makes only occasional mention of preferences among breeds. (A description of many undesirable conformational traits may be found in **"Inherited Abnormalities."**) For simplicity, this discussion is divided into three parts: the head and neck, the body, and the legs.

HEAD AND NECK

The head and neck affect the horse's movement and locomotion. They operate on the simple principle of a weight, lever and fulcrum to balance and propel the body. The head acts as a weight, which can raise or lower one end of the lever — the neck and backbone. This lever rotates about a fixed point, called the fulcrum, which in this case is the point formed by the attachment of the neck to the shoulder below and to the withers above. The lever, because it moves around the fulcrum, can raise heavy objects at the opposite end (i.e., the hindquarters). A galloping horse graphically demonstrates this principle: he uses his head as a weight, and his neck as a lever, to achieve greater speed and upward propulsion of the hindquarters. Basically, while the horse's hindlegs are pushing his body forward and his legs are extended in the air, his head and neck are raised. When his forelegs impact, the horse swings his head downward, pulling the neck

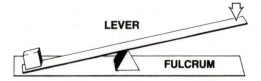

A lever pivots about a fixed point—the fulcrum. Weight applied to one end of the lever will raise a weight at the opposite end of the lever.

The lines on the photograph show how the head, neck and backbone function as a weight and lever at the gallop. The horse on the left has raised his head to collect his hind legs for another powerful thrust. The horse on the right has lowered his head so that his neck is nearly level with his backbone: this helps his weight roll over his forelegs.

down to shift his weight forward over his forelegs and raise his hind-quarters. With his hindquarters raised, he can collect his hindlegs for another powerful thrust, and swing his head upward again. This shifts his weight to the hindlegs, the forelegs are off the ground, and the process repeats itself. Because of its function as a weight on a lever, there is an interaction between head size and weight, neck length, and the ability to elevate the hindquarters.

In addition to function, the head is important for aesthetic reasons. It is the focal point of the horse's body and often reflects his temperament and intelligence. An attentive, inquisitive horse will usually have erect, alert ears, while an ill-tempered animal will frequently pin his ears back.

The shape of the head should follow breed type and be proportional to the body to aid in locomotion. The face should present a fairly straight profile when viewed from the side, although this feature varies among breeds. Arabian fanciers usually prefer horses with dished faces. They consider this trait a sign of refinement and good breeding even though other breeders may select against it. The dished profile is thought to be due to recessive genes. Many excellent draft horses have Roman noses, a trait that is undesirable in most breeds because it interferes with vision and may indicate coarseness.

A front view of the head should reveal a broad forehead in proportion to the size of the poll and the nostrils. The cranium (bones that encase the brain) should be large enough to provide adequate brain space. The face should taper to a small muzzle with large thin-walled nostrils capable of handling large quantities of oxygen.

A horse's eyes should be large, bright, alert, and widely spaced on the sides of the head to offer a proper range of vision. Hunters,

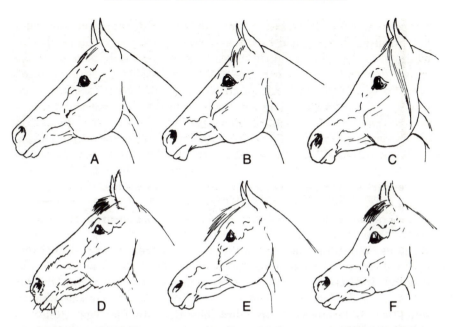

Profile types include:(A) Thoroughbred type, clean-cut with straight profile, fine ears; (B) Quarter Horse type, similar to Thoroughbred but with somewhat thicker jaw, making muzzle appear narrower, face often shorter than in Thoroughbred; (C) Arabian type, prominent forehead, face dished below eye, deep jowl, small muzzle, often low-set eye; (D) Coarse head, Roman-nosed with arched nasal bone, small eye; (E) "Camel" head (elk-nosed), with protruding lower lip, concavity in middle of face; (F) Pony type head, with face dished between eyes as well as along nose.

An attractive head, with small, erect ears; large eyes; a broad forehead; large nostrils and a face that tapers into a narrow muzzle.

jumpers, and race horses are exceptions to this general rule. Jumping and running horses need excellent forward vision and can see better when their eyes are slightly closer together. Most horsemen prefer warm, dark brown eyes because blue or glass-eyed horses tend to be more sensitive to light. Although white sclera (the area outside the iris, known as "the white of the eye" in humans) are not usually preferred, they are a breed characteristic of the Appaloosa. Good vision is vital: visual problems often lead to behavioral disturbances. Horses with good dispositions have been known to have abrupt changes in temperament when they begin to have problems with their vision.

The ears should be set well on a horse's head to give him the correct facial appearance. They should be light, alert, and not too extreme in size (ears too large or too small can make a face appear disproportionately large or small). Mares tend to have slightly larger ears than stallions. Whatever their size, a horse's ears are extremely sensitive: sharp, loud vocal commands can startle and upset a horse. Recent research indicates that a horse uses his sense of hearing to a greater extent than his senses of touch and sight. He may even be able to sense sound by picking up vibrations through his legs. (Refer to **"Evolution"** for more detailed discussions on both vision and hearing.)

The horse's jaws should be large and strong, and meet evenly without an overbite (parrot mouth) or underbite (bulldog mouth). If the incisors do not meet evenly, this could mean that the same is true with the molars. The molars must be correctly aligned to grind food properly and to produce an even wear on tooth enamel. A horse uses his lips to grasp food, so they should be firm and muscular. (For a more detailed explanation of a horse's teeth and prehensile lips, refer to **"Evolution."** Abnormalities of these structures are described in **"Inherited Abnormalities."**)

An attractive, graceful neck improves a horse's appearance and, as previously mentioned, figures prominently in his movement. Breed associations have various preferences in neck length and shoulder attachment, but there are some general characteristics that are desirable. Generally, the neck should be long and slender with a straight underline, a slightly convex topline, and a fine throatlatch. Length gives the horse maneuverability, improved balance and good carriage. Some width in the lower neck is necessary to safely encase the internal structures (windpipe, jugulars, etc.) However, a short, thick, "pony" neck and a thick throatlatch are undesirable because it limits the horse's ability to flex at the poll and inhibits lateral flexion. Ideally, the neck should have enough length and refinement to allow the horse to respond readily to the bit, and to collect and balance himself for difficult maneuvers.

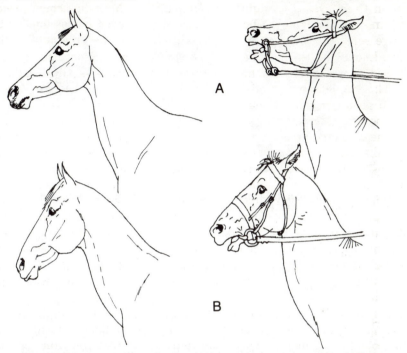

Two undesirable types of neck. Both of these horses tend to throw their necks upward when pressure is applied to the bit. (A) The ewe-necked horse has a light, concave neck with no flexion at the poll. (B) The horse with a close-coupled, "upside-down" neck cannot flex at the poll either — the heavy arch under his neck restricts collection.

BODY

The chest cavity, or thorax, on a good horse will appear deep and wide to accommodate the expanding lungs and heart. A spacious rib cage protects the heart and lungs within, and the liver and stomach behind it. Many of the muscles associated with the neck and legs are attached to it. The shape of the ribs partly determines the limitations of the chest. The ribs should be well arched and sufficiently spaced to allow room for lung expansion. Well curved ribs which project backward at a fairly large angle, and which have large spaces between them, allow the horse to have a long, deep underline and heartgirth (for power), and a short, straight back (for strength). A "slab-sided" horse (one with short, flat, straight, or upright ribs) has little room for development of heart and lung capacity. This will limit his per-

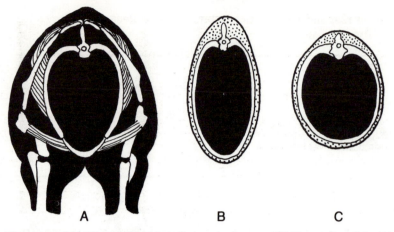

Thorax conformation. (A) Ideal, well-sprung rib cage. (B) Narrow barrel found on "slab-sided" horse. (C) Excessively wide barrel.

formance potential. When viewed from the front, the barrel should be plainly visible on either side of the horse.

The withers, located at the base of the neck, are a prominent and important part of a horse's structure. They should be well-defined and muscular, and slope smoothly into the back. Their definition will help anchor a saddle, while their muscularity will provide padding. High, sloping withers are often accompanied by long, sloping shoulders and longer muscles. These allow greater extension of the forelimbs and freer movement.

Large muscles, which extend from the withers down to the elbow, attach to the scapula (shoulder blade) to form the shoulder. Ideally, the shoulder is long and sloping, or what is termed "well laid back." The angle of its slope should approximate that of the pastern. The shoulder's main function is to help move the forelegs: since shoulder muscles extend the forearm, their length and slope will determine the length of the forearm swing. Straight shoulders restrict swing and stride, resulting in a choppy ride. On the other hand, long, sloping shoulders allow greater extension of the forelegs and lengthen the stride. A horse with a long, smooth stride contacts the ground less frequently than a straight-shouldered horse. This reduces impact-related stress on its hooves.

The back extends from the withers to the loins and supports much of the horse's weight. It also acts as a "suspension bridge" for the ribs, muscles and other structures. Because it must support this weight (and often that of a rider), a short, straight back is preferred. A sagging back (sway back) does not provide adequate support and an

High withers and straight shoulder conformation are exemplified by this horse.

arched back (roachback) is both unsightly and inefficient. For greatest strength, the back should be fairly short (with a long underline) and wide (an indication of adequate muscling). A longer back may give a more comfortable ride, but it is more likely to tire and sag under stress.

Certain breed enthusiasts contend that their horse's backs are shorter because the spinal columns have fewer vertebrae. Actually, the number of vertebrae varies more between individuals than between breeds. The backs of Arabian or Morgan horses may appear shorter than those of other breeds, but they generally possess the same number of vertebrae. The spinal column contains between 51 and 57 vertebrae with the most frequent variation occurring in the tail. These vertebrae are divided into those in the neck (7 cervical); the withers region and back (usually 18 thoracic, but can vary from 17 to 19); loin (6 lumbar); and hip and croup (5 fused sacral vertebrae). The caudal (tail) vertebrae may vary in number from 14 to 21. Arabians may differ more frequently than other breeds: their spines sometimes contain one less thoracic or lumbar (back) vertebrae and they frequently have fewer caudal (tail) vertebrae. Even so, the length of each vertebrae, not the total number, is what determines the length of the back. To illustrate this point: the horse has seven vertebrae in his neck, but so does the giraffe.

The flanks (on the lower body between the ribs and hips) should give the horse's middle a balanced appearance. On a well-fitted race horse with good muscle tone in his abdomen, the flanks may be cut up high. (A tight middle on other hard-working performance horses might indicate athletic condition in those horses, also.) High flanks should be discriminated against in a horse that does not have good muscle — or one with a weak loin. Horses that are excessively "tucked up" like a Greyhound have no strength for fast turns or maneuvers. (Such horses are called "hound-gutted" or "wasp-waisted.") Flank movements indicate a horse's wind capacity and breathing rate: these should be slow and regular (about 12 to 14 per minute) without panting or jerkiness. The presence of a heave line indicates that the horse has heaves, a respiratory disease that impairs breathing. (Refer to **"Respiratory System"** under **INHERITED ABNORMALITIES.**)

The loin, which lies above the flank, is the muscular portion of the horse's topline that extends from the last rib to the hips. Because it supports the lumbar veterbrae and transfers power from the hindlegs forward, the loin (or coupling) should be short and well-developed. A horse that is "well-coupled" in the loin has sufficient muscling to tense his spinal column to raise and propel the front portion of his body. These muscles should be strong and supple.

The hindquarters are the horse's main source of propulsion and should be long, well-muscled and powerful. The hips should be level with their points well-defined but not excessively prominent. (Prominent hip bones indicate poor condition and general lack of strength.) The croup, which extends from the loin to the base of the tail, should be long, uniformly wide and adequately muscled. Different breed enthusiasts prefer different croups. Although a steeply sloped croup usually restricts a horse to a shorter stride, draft owners will accept some sloping of this kind because their horses must take short, powerful steps. Arabian breeders prefer a flatter, nearly horizontal croup which gives a horse a longer, more flowing stride. A slope to the hindquarters is more frequently encountered in horses that perform the lateral gaits (e.g., pasos and some pacing Standardbreds).

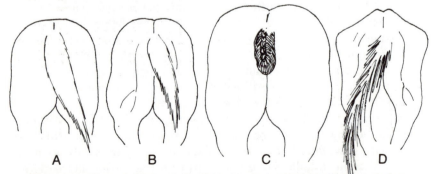

When viewed from behind, a horse's hindquarters should have a rounded appearance (A). Conformation faults include the following: (B) Pear-shaped hindquarters that are not adequately wide through the hips. (C) Double rump (desirable in a draft horse). (D) Rafter hips.

LEGS

The forelimbs carry about 65% of the total body weight. Since they support this weight, propel the body, and absorb the shock, they are susceptible to injury if not properly conformed. Proper sloping and angling of the bones help the legs absorb concussion. This is a major reason that horsemen look for long, sloping shoulders. The shoulder forms the main point of attachment of the forelimbs to the rest of the body. It is also attached to the vertebral column by sheets of muscles (not bones). A longer shoulder provides more points of attachment, resulting in greater support.

The angles at which the shoulder meets the arm (humerus), and the arms meet the forearm, help cushion impact because they allow give-and-take at the joints. The legs must be straight, as crooked legs may cause weight to be unevenly distributed. This will result in excessive pressure on a certain bone or joint, or excessive pull on a muscle, tendon or ligament. To determine straightness from a front view of the horse, an imaginary line dropped from the point of the shoulder to the center of the foot should bisect the knee, cannon and ankle. Deviation will cause stress to fall on the out-of-line area and result in problems such as splints, sidebones, windpuffs, etc.

A

Ideally, it should be possible to draw a straight line down the horse's front leg, through the middle of the knee, cannon, pastern, and foot. A horse with straight legs (A) will move in a straight line, while a horse that toes in (B) will paddle or move the foot through an outward arc before placing it on the ground. The horse that toes out (C) frequently strikes the opposite leg with the moving forefoot as it swings through an inward arc.

B

C

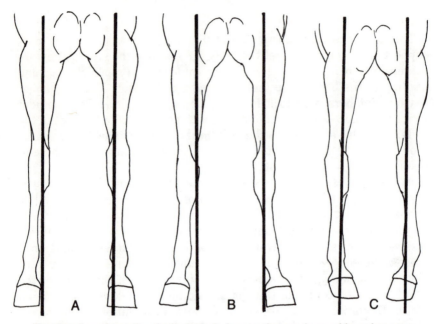

Forehand conformation faults include horses that are base-wide or base-narrow. A base-wide horse has a greater distance between the feet than between the forearms (A). This condition may be accompanied by toeing-out (B) or toeing-in (C).

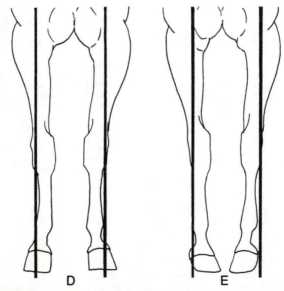

A base-narrow horse has less distance between the feet than between the forearms, and may toe-out (D) or toe-in (E).

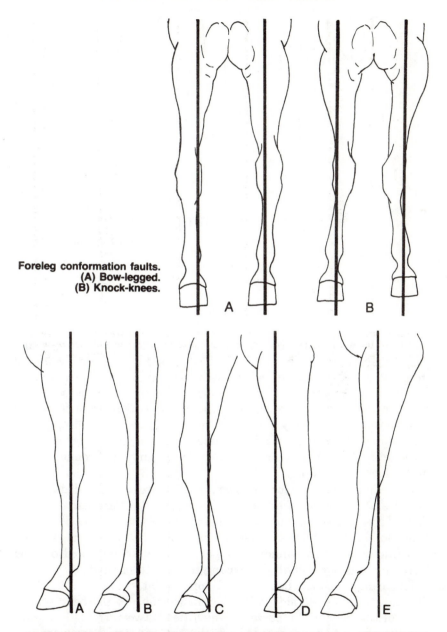

**Foreleg conformation faults.
(A) Bow-legged.
(B) Knock-knees.**

When viewed from the side, a straight line should bisect the leg and touch the heel of the foot (A). In a calf-kneed horse, this line will be too close to the front of the knee and too far back of the heel (B). In a buck-kneed horse, the line will fall behind the center of the knee (C). The foreleg may also be camped under (D) or camped out (E).

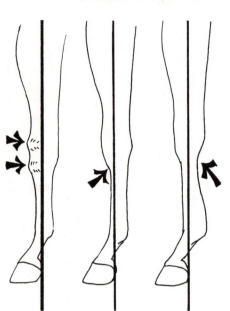

Foreleg faults also include
"open knee" (left), cut out
under the knee (center), and
tied in behind the knee (right).

When viewed from the side, the foreleg should be straight and perpendicular to the ground. The elbow should turn slightly outward (to point the toes forward and insure a straight, free stride) and blend into a long, smoothly muscled forearm. The knee (carpus) joins the forearm to the cannon and should be straight (both from front and side views) and squarely placed on the leg. A knee that is deep, thick, and wide will provide good cushioning for the leg, but an excessively large or swollen knee could indicate an old injury, poor conformation, and a lack of flexibility.

While horsemen like long forearms (which are accompanied by long muscles), they prefer short, flat cannon bones that are narrow when viewed from the front and wide when viewed from the side. This indicates solidarity which will reduce susceptibility to splints, breaks, and fractures. Short cannon bones are usually accompanied by long forearms, correspondingly increasing the length of stride.

The cannon bone should blend into a flat, wide fetlock. Some distance from side-to-side will allow more width of the bearing surfaces of the bones. The pastern, which lies between the fetlock joint and the coronary band, should be fairly long and slope downward in a continuous line with the hoof wall. The pastern slope should match the shoulder slope, with the ideal being about a 45 degree angle. Horses with short, straight pasterns tend to have rough, choppy

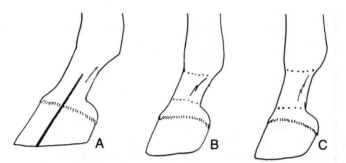

Pastern conformation. (A) Normal foot with an approximately 45° angle. (B) Short, upright pastern. (C) Long, upright pastern with a large angle.

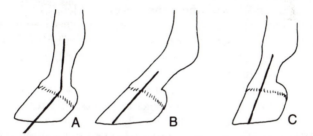

Trimming affects the axis of the foot and pastern. (A) In this foot with a long, upright pastern, the axis is broken due to improper trimming. Note that although the hoof slopes at a 45° angle, the pastern is very upright. (B) Long, sloping pastern that has been correctly trimmed shows an unbroken axis. (C) A correctly trimmed upright pastern with an unbroken axis.

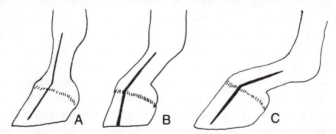

Broken axis of the foot and pastern. (A) long toe-low heel. (B) short toe-high heel (C) long pastern with excessive slope ("coon-footed").

gaits, predisposing them to navicular disease because of increased stress on the hoof from the leg bone. Horsemen may also be critical of horses with excessively long pasterns. Although they have a smooth gait, the tendons and ligaments of these horses are subject to more strain and may break down more easily from increased stress on the sesamoids.

A horse's hindlegs should be straight, strong, and well-muscled to aid in the forward propulsion of the body. The thigh bone (femur) should be short and sloping, and turned slightly outward at its point of origin. This will angle the horse's hocks almost imperceptibly inward, enabling him to collect himself more easily and work with his hocks closer together.

The stifle (which corresponds to the human knee) and the gaskin (or second thigh, which lies between the stifle and the hock) are integral parts of a horse's power plant and should be well-muscled. A long, muscular gaskin provides greater distance between the hock and the hip. The longer muscles exert more pull, lengthen the stride, and increase hindleg swing action.

The angle of the hock will affect the power and length of the stride to a certain extent. An almost straight hock, accompanied by a long, straight gaskin and short cannon bone, give a running horse his lengthy stride. They allow full extension of his hindlegs and, consequently, full use of the power in his hindquarters.

A stock horse usually works better with more angulation in the hock region. This enables him to execute pivots, sliding stops, and rollbacks more easily because he can "get his hocks under him." Tennessee Walking Horses and paso horses tend to be slightly sickle-hocked. Breeders believe they can cover ground more comfortably than a horse with straight legs. Still, sickle hocks are usually undesirable. Generally, horses with this characteristic are not as fast as horses with straighter legs. They also stand under too far, put excessive strain on their rear ligaments, and are more likely to develop a curb (a lump in the back of the lower part of the hock caused by spraining a small ligament).

Generally, the hock should be slightly angled when viewed from the side, large and wide, and should rest on correspondingly broad cannons. Straightness can be assessed from a side view by dropping an imaginary line from the point of the buttock perpendicular to the ground. On a horse standing with his cannons vertical, the line should touch the back tendon of the cannon throughout its length. The line will fall somewhere behind the tendon if the hind leg is too straight. On a leg that is too angled, the line will fall through, or in front of, the cannon.

When viewed from behind, the hocks should appear straight. Since a horse must collect and balance himself, and "work off his hocks" while in motion, they will encounter the least interference when they move in parallel planes with the cannons and fetlocks. Cow hocks (the points of the hocks curve inward) are usually viewed with suspi-

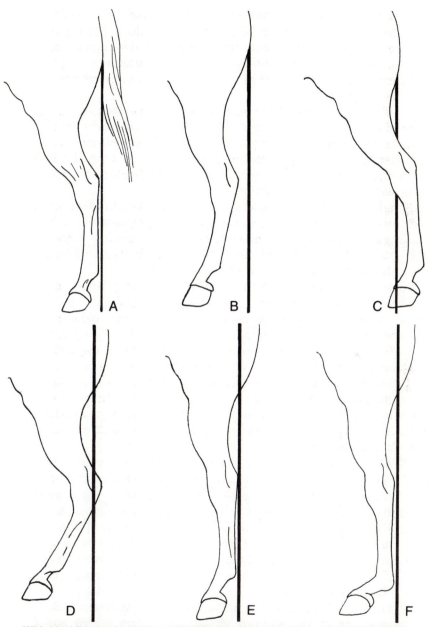

With ideal leg conformation, a straight line drawn from the point of the buttock to the point of the hock continues down the back of the cannons (A). Hind leg conformation faults include camped under (B), camped out (C), sickle hocks (D), straight legs (E), and a too straight leg (post legged) causing stress on the pastern and stifle (F).

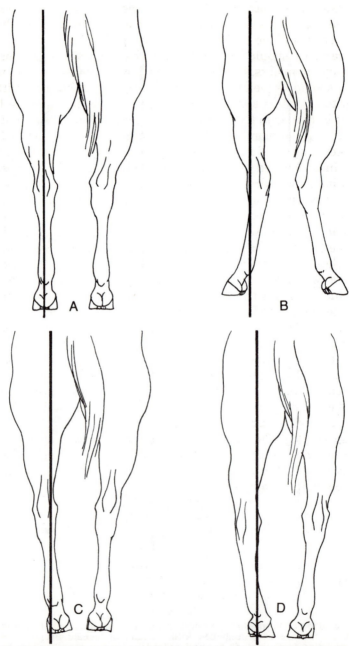

(A) Ideal conformation of the hind legs, when viewed from the rear. Conformation faults include cow-hocked (B), base narrow-toe-in (C), and base narrow-toe-in with bowed legs (D).

cion because they encourage the development of spavins. Turned-out toes often accompany cow hocks. (Many horsemen will accept slightly turned-out toes if the horse's cannons are parallel.)

The cannons should be short, with sharply defined tendons and suspensory ligaments. The fetlock, which connects the cannon to the pastern, should be well set back and sturdy because it joins with the pastern to provide a springy gait. Overall, conformation of the lower hindlegs is similar to that of the lower forelegs. The hindleg pasterns, however, may be more upright and shorter than the front pasterns.

The feet bear all of a horse's weight and must be sound or the horse will be incapacitated. Some horsemen believe that the best way to select a horse is to drape an imaginary curtain over his body so that only the feet are visible. If the feet appear sound, they proceed to examine the lower legs, then the upper legs, etc. But if the feet are unsatisfactory, they waste no time examining the rest of the horse.

The feet should be round at the toes and have broad heels, with no noticeable cracks or blemishes. The hind feet are usually narrower, and have toes that are more pointed, than the front feet. The hoof wall, which includes the horny outer covering that extends from the coronet to the ground, should be thickest at the toe and thinnest at the heel. The hoof should show even wear (it grows about one-quarter to one-half inch per month). Lateral expansion of the hoof wall helps the heel absorb shock by allowing the frog to expand. The frog secures a foothold, and provides a cushioning effect. It should be large, well-developed and resilient. This cushioning is caused hydraulically when the frog expands and traps blood in the hoof's vascular bed. This expansion applies pressure to other structures and helps pump venous blood up the leg. The frog must expand and remain elastic on the ground to function effectively. If it is kept off the ground, it can become atrophied, the heels will contract, and both cushioning and blood flow will be impaired.

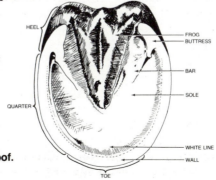

Bottom of hoof.

The sole of the hoof should be arched slightly upward and not touch the ground since its function is to bear internal weight. The sole of the hind foot is more concave than that of the front foot. The bars (continuations of the hoof wall that turn inward to parallel the frog) should be well-developed to allow space for the frog to expand. Because they provide this expansion room, the bars help prevent contraction of the hoof.

The hoof should be sized proportionately to the horse. A "too large" hoof, except in draft breeds, often indicates clumsiness, a heavy step, and poor-quality breeding. A hoof that is too small reduces the area of support and predisposes a horse to bruises, contracted heels, navicular disease, etc. Well-muscled or fat horses with small feet are especially prone to navicular disease since most of their weight is forced onto the navicular bones. Flat-footed horses (the heel is low, the sole is flat and the wall angles out excessively) often have bruised frogs and heels, corns and strained tendons.

Temperament

Temperament may be described as an animal's consistent display of specific reactions to a given situation. It is one of the most important aspects of a horse's character, and yet it is not always considered by the prospective owner. For a horse to work with man, it is vital that he have an agreeable temperament and some tractability.

Temperament is partially selectable as a genetic trait, but it is also molded by physiology, training and past experiences. A horse that has been severely mistreated when young may be distrustful of people. An example of fear induced by physiology may come about if a horse has his vision impaired by a Roman nose and a small eye. This will make him more nervous and excitable because he isn't able to see many things around him. (For more information on the effects of visual problems on behavior, refer to **"Evolution — The Dawn of Today's Horse."**)

Intelligence

Horses are often considered less intelligent than monkeys, pigs, and some other animals, but it is their *reasoning* ability that is deficient and not their learning ability. Horses have no ability to form judgements or conclusions, or to arrive at logical explanations for unexpected events. They have relatively poor concentration (that is why long training sessions are counterproductive). The horse is quite trainable, however, because of his excellent memory. Although he lacks the ability to analyze past experiences, he can recall them rather vividly. A horse in training learns to respond to stimuli by trial and error — he remembers which of his responses elicited punishment and which drew praise. Eventually, most horses (some more quickly than others) develop a set of habits in response to a set of cues. They learn more quickly when the cues are aimed at their strongest senses: touch and hearing. Both a horse's sides and mouth are sensitive, which is why horses respond to leg cues and bits.

As in most animals, a horse's life is governed by the "pain-pleasure" principle. He learns to avoid behavior that results in pain (i.e., spurs in the sides, a quick jerk on the reins) and to develop behavior patterns that result in pleasure (a quiet, relaxed rider and a loose rein). He responds to praise and punishment from man, but cannot logically determine what responses unfamiliar situations demand. The horseman controls behavior by presenting a cue, eliciting a response from the horse, then praising or punishing to encourage or inhibit the response. In this way, the horse learns, through trial and error, what is expected of him. Because of this it is important for the horseman to exercise patience consistently for the horse to learn the desired responses.

Trainability

Trainability is believed to be somewhat heritable, although it depends to a great extent on the horse's intelligence, early experiences and temperament. Early experience may affect a horse's responses: a horse rarely forgets frightening or threatening situations. A bad experience as a young horse may make training more difficult. But his excellent memory can be put to good use since he will also remember pleasant experiences and positive reinforcement. A trainer should school a horse slowly and carefully and be consistent in his rewards and punishment. Horses are creatures of habit and a good trainer will capitalize on that trait. A horse has little need for reasoning ability if he is guided by a consistent, intelligent trainer. ♞

4

PERFORMANCE SELECTION CHARACTERISTICS

Selection on the basis of performance characteristics is important, because a horse's intended use strongly affects the criteria used in its selection. This chapter examines some of the most populous classes of performance horses (race horses, show horses, etc.) as well as some of the most traditionally useful (draft horses, stock horses, etc.). Space limitations restrict the discussions to only a few of the many types of performance horses. The discussions are by no means exhaustive: a thorough study of the selection of the Thoroughbred race horse alone could easily fill a volume of this size. The following chapters, though necessarily brief, discuss some important principles of selection. The horseman will find many of these rules of thumb helpful in the selection of any type of horse.

The Race Horse

Selecting a top-quality race horse is a complex process that has intrigued horsemen for centuries. A horse's pedigree, attitude and conformation are important guides to his future performance, but they are not guarantees. Purchasing, raising and training a race horse is expensive and, for this reason, selection should be considered carefully and the decision made as unemotionally as possible. No matter how the horseman acquires his horse (whether at an auction,

courtesy of The Blood-Horse

All other factors being equal, a strong competitive attitude may enable a race horse to defeat competitors of similar ability. This is often a characteristic of the superior race horse.

from a breeding farm, etc.) he should consider all aspects of selection thoroughly to make the best purchase.

All types of horse racing — from Quarter Horse races on the straightaway to cross-country steeplechases — place strong demands on the equine athlete. In no other area of equine competitive events is good conformation, strong constitution and a competitive attitude more important. Two horses may have similar pedigrees and physical characteristics, but the will to run may make one a champion while the other turns in average performances. A horse may have tremendous speed but, if he has no desire to be in front at the "wire," his chances of becoming a good race horse are diminished. Some horses who do not possess this competitive spirit may show reluctance to pass a leader during a race even though they are capable of doing so. Although attitude is a vital ingredient for a winning horse, it alone will not guarantee that a horse will run. A racing prospect should also possess a strong pedigree, excellent conformation, a high degree of soundness, and, if it is an older horse, a proven record of performance.

PEDIGREE

Pedigree study is an important part of selection. Generally, a good runner's pedigree will contain winners in the first two generations; i.e., parents and grandparents, or close relatives (siblings and half sibs) that have proven themselves on the track. If no outstanding individuals appear within those generations, the horse probably is a bad risk. A horse's potential (and his parents' breeding value) can often be estimated by examining performance records of siblings and half-siblings.

Some Thoroughbred horsemen believe that a good sprinter (speed horse) should appear in the first few generations of a stayer's pedigree. A degree of sprinting blood is necessary to give a stayer the power he needs to capitalize on openings and improve his position in a race. Some stallions are regarded as sires of speed horses, but their offspring will be more likely to be good runners if their pedigrees show speed on the dam's side as well. Since only about 5% of racing colts are retired to stud, they are usually the individuals with outstanding performance records. Some stallions with undistinguished records become successful sires of winners, but they usually are horses that failed to compile good records because they were mismanaged or campaigned lightly (for some reason such as injury). Mares with outstanding pedigrees may be valuable as breeding stock even without outstanding performance records. Even if a mare is a good runner, her value as a broodmare may easily exceed her earning potential as a race horse. (As many as 65% of all mares raced in a given year eventually become broodmares.) For this reason, racing performance in the dam is stressed less than conformation and pedigree. (For a more detailed discussion of pedigree, refer to **"Important Selection Characteristics"** and **"Breeding Aspects."**)

Pedigree balance is an important aspect to consider. Two highly charged, quality horses might be incompatible and could create imbalances in their offspring when mated. Examples of incompatibility are horses at opposite extremes: stayers vs. sprinters, good-tempered vs. intractable horses, etc. A horseman might breed one extreme to another hoping to obtain an intermediate phenotype, but quite often the resulting offspring resembles one of the parents. A stayer bred to a sprinter might produce a stayer with speed, but the cross is just as likely to produce a stayer with speed, a sprinter with no staying power or a horse with neither speed nor stamina. By the same token, matings between horses of very similar breeding will occasionally produce offspring with problems. A stallion highly inbred to a particularly temperamental horse, when mated to a mare with similar breeding, could sire a foal that inherited the worst in temperament from both sides of the pedigree. A highly bred stallion might cross well with a "rough" mare (one with a less admirable pedigree) and produce foals superior to those he might get when bred to a mare in his class, but this is usually not the case. Outstanding parents tend to produce more outstanding offspring than parents of less quality.

PERFORMANCE

Racing ability is thought to be quite heritable, with estimates ranging between 30 and 40%. A proven winner is likely to be a horse with winners in his pedigree and the ability to beget them.

Racing performance is evaluated in several ways, many of which fail to account for such variables as track conditions, weather, trainers, jockeys, etc. In Thoroughbreds, which in this section will be considered representative race horses, the earnings index provides per-

---

Imagine, for example, that heritability of racing ability is in fact 40%. A breeder decides to breed one of his mares who has a PR of 0 (the theoretical average) to a stallion with a PR of +16 (i.e., he could outrun the average horse by 16 lengths). The breeder computes the average predicted performance rate as follows:

average selection differential of parents × heritability =

AVERAGE PREDICTED PERFORMANCE RATE OF PROGENY (APPRP)

The selection differential (SD) is simply the superiority of the parents over the average of the population from which they were selected. In this case:

SD of stallion = 16-0 = 16 lengths
SD of mare = 0-0 = 0 lengths
$$\text{average SD} = \frac{\text{SD stallion} + \text{SD mare}}{2} = \frac{16 + 0}{2} = 8 \text{ lengths}$$

Applying the average performance rate formula given above:

heritability = .40
avg. SD = 8 lengths
APPRP = 8 lengths × .40 = 3.20 lengths

Therefore, on the average, the foals resulting from this mating would be expected to be about three lengths better than average. It should be remembered that performance rate calculations will predict only the *average* potential of the progeny, and will not prescribe upper and lower limits on ability. Many of the horses resulting from the hypothetical mating might fall far below average, while others might exceed the performance of both their sire and dam. Also, these estimates deal only with the foal's *inherited* potential, and do not take environmental influences into consideration.

formance information by comparing a horse's earnings to other horses competing in a given year. (This is derived by dividing the total amount of money won each year by the number of starts. This gives the average earnings of the horses involved. This amount is compared to what each individual won. A horse that earned the same as the average would have an index of 1, while a horse that earned twice as much would have an index of 2.) The index helps compare ability, but is slightly misleading for fillies, who often aren't raced as frequently as colts. Earnings are very often indicative of ability in Quarter Horses and Standardbreds as well. Tests in those breeds rate performance heritability as high as 60%, but that is probably optimistic.

The Performance Rate (PR) developed in 1971 is another method of comparison in Thoroughbreds. The average horse theoretically has a performance rate of 0. A horse with a performance rate of +12 would, in an average race that year, be likely to finish 12 lengths ahead of the average horse. A colt with a −12 rating would be expected to finish about 12 lengths behind the average horse. Performance rates, combined with heritability data, can be useful in predicting the *average* genetic potential of progeny resulting from a given set of matings. Conversely, a horse's genetic potential can be partially estimated by determining this value for his parents.

Races usually are won by the fastest horses, but pure speed can be misleading. A young horse that exhibits speed in early workouts may refuse to be rated over longer courses. Although he might win shorter races early in his career, he may fall behind his competitors over classic distances. Most races with substantial purses for Thoroughbreds demand controlled speed over long distances, and, to excel, a horse must have the physical stamina and courage to be competitive all the way — he must be a stayer. Quarter Horse races are shorter, and therefore pure speed is much more important than staying power.

CONFORMATION

Conformation, to a great extent, determines performance. Certain conformational traits seem to be associated with speed while others are representative of staying power. Generally, a stayer will have a long stride and easy action, and is usually taller than he is long. He tends to have a bold expressive head, a long well-shaped neck, long forearms, a short croup and a long pelvis. The speed horse, on the other hand, has a shorter back, but has more length than height due to extra length in the croup. He has a strong loin, powerful hindquarters and straight legs.

courtesy of The Blood-Horse

Horses running in the Prix de l'Arc de Triomphe, which takes place at Longchamp in Paris, France each October. European owners and breeders consider this race the most important test for three-year-olds and older horses.

Conformation relates closely to soundness as well as performance. While size is usually a secondary consideration in selecting a race horse, excessive height or heaviness may hurt performance. Heavy horses will overstress their legs and feet and may be more predisposed to lameness. Small horses are at somewhat of a performance disadvantage because their strides are normally correspondingly shorter.

Overall, the race horse should possess balance and symmetry. His head should be clean-cut and should reflect intelligence and good breeding. He should have a long, tapering neck that blends into a flat, sloping shoulder. A thin-necked horse may be too delicate, or may lack sufficient room for air intake. A thick-necked horse may not be able to extend himself fully. Foreleg conformation should be faultless, as a young horse will sustain tremendous concussion before his bones are set. Any deviation from the ideal straight column of support will subject the out-of-line area to excessive stress. The horse should have a long, well-muscled forearm, and should have a short cannon bone with definitive creases between the bone and tendons. He should be lean-legged: it takes trim legs to carry a horse at high speeds without tiring, and horses with round, thick legs seldom remain sound. Bone strength and solidarity are vital to sustain the concussion received during a typical race. The average cannon bone can withstand as much as 16,000 lbs. of pressure, but when it is repeatedly overstressed, the bone weakens. Repeated stress could weaken it to the point where, with a misstep, it might break under as little as 9500 lbs. of pressure (close to what it would encounter during a race). Given sufficient rest between periods of stress, bone will rebuild, but if it is stressed again before recovery is complete, the risk

courtesy of The Blood-Horse

Alternating periods of suspension and concussion are illustrated by these runners. Notice how two horses (left and right) are supported momentarily by only one of their limbs, while another horse (center) is suspended in air.

of injury increases with each stride. Ultrasonic techniques are proving helpful in measuring bone strength, and soon may be used to plan training and racing schedules to insure soundness.

The knees of a good runner should be large and flat, and perfectly aligned when viewed from both front and side. Calf-kneed horses (which are back at the knees) are often slow breaking out of the gate, because their weight rests more toward the rear of their feet. Their knees are more subject to fatigue, and these horses often end their careers with bilateral carpal chips. Many of the best Thoroughbred race horses are actually buck-kneed (over at the knees), and their knees generally remain sound. Like the legs, the feet should be set straight, with no outward or inward deviation. One of the Thoroughbred's most common conformational defects is turned-out hind feet. Horses that toe-out in front are especially subject to fracture of the medial sesamoid bone on the inside of the fetlock. Toeing-in subjects them to injury of the lateral sesamoid or suspensory ligament. They may also develop splints.

The hindlegs should be straight from behind, with the hocks set low and straight. They should be wide and strong, and should taper to slender cannons. The pasterns on the forelegs should slope at about a 47 to 50 degree angle, with the angle of the hind pasterns being slightly steeper. Pasterns may be shorter and more upright in sprinters than in distance horses. The feet should be sufficiently large, with healthy frogs and well-built bars. Well-conformed feet and legs will help insure continued soundness.

The withers should have good definition, and the back should be short and strong. A deep heartgirth and well-sprung ribs provide space for the larger heart and lungs essential to the race horse. A strong loin should accompany broad, strong hindquarters that retain their width down toward the hock.

Running Quarter Horses compete over shorter distances, so they need somewhat heavier muscling for quick power and short bursts of speed. Their pasterns might be slightly shorter or more upright and their hindquarters more powerful than in the Thoroughbred, but otherwise their basic conformation is similar.

Any racing prospect should ideally be examined for soundness and condition by a veterinarian, and any predisposition to unsoundness noted. Among the most common unsoundnesses in race horses are bowed tendons, carpal chips (knee injury), osselets, bucked shin, splints and fractured sesamoids, all of which may occur when stress is placed on an improperly built limb. A race horse's legs are subjected to considerable stress in a few years on the track. A horse with poorly built legs may not be able to pass the test. If the condition is heritable, he should not be used for breeding.

In addition to good conformation that permits free, sound action, a race horse needs strong respiratory and circulatory systems. A horse with poor respiration or inadequate lung capacity will have to reduce his speed to reduce his oxygen demand, and one with poor circulation won't be efficient in removing waste products from muscle cells, and therefore will tire easily. Certain circulatory irregularities may be acceptable (many healthy hearts have slight murmers) but most heart problems will eliminate a horse from competition. Some bleeders may race successfully in spite of their diseased lungs, particularly on medication such as furosemide. (These horses rupture blood vessels lining the lungs during a race, and afterward, blood may escape freely at the nose.)

Recently a group of veterinarians have correlated a figure called heart score with heart weight, a factor claimed to influence racing performance. The researchers claim that the larger the heart score, the bigger the heart and the greater the winnings. Heart score is determined through electrocardiogram (ECG) tests performed on horses galloping at racing speeds. Test results on Standardbreds and Thoroughbreds show that heritability of heart scores is highest from dam to son (65%) and lowest from sire to daughter (5%) for an adjusted average of 40%. Heart score could be sex-linked, since dams have greater overall influence than sires. If this work is valid, heart scores might represent a possible basis for the selection of breeding stock.

A blood component, transferrin, also might prove useful in predicting racing ability. Transferrin is one of the principal proteins found in the blood serum, and its main function is to control iron levels in the blood. Transferrin types can be determined by a simple lab test known as gel electrophoresis. Each individual's transferrin type is represented by two letters, such as DD or DF. In the Thoroughbred, five alleles (D, F, H, O and R) have been identified, making 15 combinations or genotypes possible. Transferrin types have proven useful in resolving cases of disputed parentage in horses, and recent evidence indicates that they may help predict racing performance. Tests conducted on sample groups in South Africa have shown that DF and FF horses comprise a much larger proportion of race winners than individuals of other types. The FF individuals particularly seemed to have a performance edge.

ELEMENTS OF SPEED

A horse achieves speed through an efficient stride. One of the keys to this stride is an element known as overlap. Overlap time occurs

courtesy of The Blood-Horse

Secretariat, running in the 1973 Belmont Stakes, demonstrates overlap: his left foreleg and right hindleg are both on the ground. Secretariat achieved his great speed by wasting very little time in overlap. His tremendous reach (note the extended foreleg) coupled with strong push decreases his overlap time and gives him a performance edge.

when two or more feet are on the ground simultaneously. The shorter the overlap time, the faster the horse. Overlap time can be reduced when the horse pushes harder and reaches farther with his legs.

Pushing harder and reaching farther require considerable expenditures of energy, and a horse required to do either will tire faster and revert to a shorter stride. Because reach and push are innate, a horse can only improve them slightly. Push, a function related to leg conformation, depends on the timing of muscular contractions, particularly those in the hind legs. It can be increased through training that strengthens muscles. Although reach is partially influenced by training, it is mainly predetermined by skeletal and muscular conformation. An increased reach lengthens the stride, while an increased push quickens the stride. The right combination of reach and push will reduce overlap time. But if these are not evenly balanced, the deficient factor will limit the stride.

Overlap time seems to be the major distinguishing factor between race horses. The faster horse devotes less stride time to overlap and more time to moving forward. High-speed film comparisons between Triple Crown winner Secretariat and stablemate Riva Ridge in the 1973 running of the Marlboro Cup reveal that Secretariat was able to pass his rival because of greater reach, which led to longer strides and less overlap time (.018 seconds for Secretariat and .115 for Riva Ridge). Fractional differences may seem insignificant, but consider

Secretariat and Riva Ridge come in 1st and 2nd in the 1973 Marlboro Cup. Secretariat was able to pass his rival because of greater reach, which gave him longer strides and reduced his overlap time. Reach, a function of skeletal and muscular conformation, is only partially influenced through training.

this: if two horses with identical stride lengths have .01 seconds difference in stride time, at the end of a ¾ mile race they will be 11 lengths apart. These measurements of overlap time, stride time, stride duration and other elements of stride can, with computer gait analysis, allow prediction of future racing speed.

Similarly, studies showing differences in stride between stance time (time a foot spends on the ground) and swing time (time a foot spends in the air) are helping researchers determine why some horses break down and others remain sound. An instrument called a force plate which measures the force of a horse's footfall has been useful for several years in diagnosing actual or incipient lameness. Now, specially designed horseshoes with similar instrumentation are proving useful in recording the mechanism of lameness and breakdown at racing speeds. Horsemen have long known that excessive stress on legs eventually leads to lameness or breakdown. It now appears that this stress occurs at a particular point in a stride — when an extended leg strikes the ground before it has reached a vertical position and its velocity has dropped to zero.

To understand this cause of stress, visualize a typical stride in which each leg goes through two phases. The first is the stance, or weight-bearing phase, and is a function of the aforementioned push. The second is the swing, or non-weight-bearing phase, which incorporates reach. Swing time, between Thoroughbreds, appears to be rel-

courtesy of The Blood-Horse

Both of the horses pictured above are demonstrating overlap. The horse in the lead has both hind feet on the ground, and the second horse (number 8) has both front feet on the ground. Although some overlap time is necessary, the less time a horse spends in overlap, the greater his speed.

atively constant: about .33 seconds. No matter what the speed, the time required to prepare each leg for the weight-bearing portion of the stride remains about the same. If for some reason, such as fatigue, a horse cannot provide himself sufficient swing time, he will cut short on the retraction phase of his swing. His leg will strike the ground prematurely, causing stress. If the retraction phase were complete, the leg's forward speed would be reduced to zero and far less stress would have resulted upon impact. (Photographs of horses running on a muddy track show mud splashing vertically around the feet of horses that have planted their feet vertically. Mud splashes forward around the feet of horses that have planted their feet prematurely, or while the leg was still extended. This is what a tired horse usually does.)

Swing time is constant between Thoroughbreds, but varies between groups of race horses. Standardbred trotters, for example, average about .35 seconds swing time, while Quarter Horses average between .26 and .29 seconds. Quarter Horses normally have quicker, shorter strides than Thoroughbreds, and complete their strides about 20% faster than the longer distance horses. In a quarter mile race a horse will take only about 61 strides, so he needs tremendous early acceleration to reach full speed quickly. And, because his swing time is shorter, he is able to accelerate by spending more time moving forward through reach and push and less time in retraction. Therefore, his shorter swing time protects him somewhat from the excessive stress affecting Thoroughbreds.

Still, any type of racing places extreme demands on the equine athlete, and the selection process should help the horseman find a horse built to withstand stress. Proper training, handling, feeding, etc. will greatly improve performance, but a horse must have genetic potential before management can make him a winner.

The Harness Race Horse

Harness racing dates back more than two thousand years, to the early Greek Olympiads, when gladiators wheeled their chariots around amphitheaters as proof of their athletic prowess. Today, the horse's athletic ability has taken precedence. Instead of heavy chariots, harness race horses pull light sulkies and compete each year for millions of dollars in purses. Although many breeds can race successfully, the Standardbred has come to be synonymous with harness racing — he has been selectively bred for more than 100 years to perform on the track.

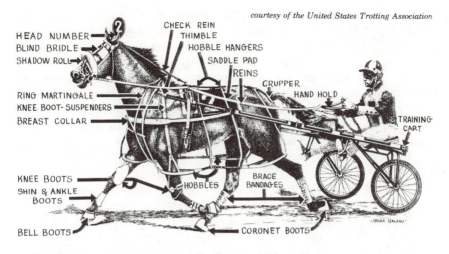

courtesy of the United States Trotting Association

HEAD NUMBER
BLIND BRIDLE
SHADOW ROLL
CHECK REIN THIMBLE
HOBBLE HANGERS
SADDLE PAD
REINS
CRUPPER
HAND HOLD
RING MARTINGALE
KNEE BOOT-SUSPENDERS
BREAST COLLAR
TRAINING CART
KNEE BOOTS
SHIN & ANKLE BOOTS
HOBBLES
BRACE BANDAGES
BELL BOOTS
CORONET BOOTS

Equipment commonly worn by the pacer includes hobbles, or hopples, which help prevent the horse from breaking gait.

Harness race horses fall into two categories, trotters and pacers. Some Standardbreds perform well at both gaits, but generally a young horse will show a tendency toward one, and his subsequent training is aimed toward its development. Pacing allows a horse to travel with minimal concussion, and is slightly faster than the trot. While a trotter's legs move diagonally, a pacer's move laterally: his two nearside legs move forward in unison, followed by his two offside legs. This creates the pacer's characteristic rolling motion. The pace is an efficient gait, but it requires smooth, hard footing for execution. In fact, for a set distance, the speed records of pacers are faster than for trotters. Also, a pacer must wear special hopples to accentuate his action and help prevent him from breaking gait. (Hopples are straps with loops at each end, each loop encircling an upper leg on the same side of the body. They steady the horse as he moves his legs on each side forward and backward.) Many high-strung, but otherwise talented horses have been effectively eliminated from competition because they could not tolerate hopples.

Harness and flat-racing horses have in common a need for speed, but they are separated by one important factor: harness horses do not carry weight, but they must pull a sulky. The emphasis in a good harness horse shifts from weight-carrying ability to speed, so he may have certain characteristics that would be unacceptable in most riding horses. The trotter or pacer has a distinctive higher head and neck set, a more sloping rump and a longer back. Harness horses often have comparatively upright shoulders: unlike riding horses, they do not need either a comfortable ride or a springy step. Despite the fact that harness horses race on hard tracks at speeds up to 30 miles per hour, they have comparatively few concussion-related injuries. Lameness occurs with about equal frequency in forelegs and hindlimbs, probably because harness horses do not carry the weight that makes riding horses more prone to foreleg injuries. This 50-50 division means that harness horses have a higher incidence of *hindlimb* lameness than horses used for other purposes. Pulling weight around improperly banked turns places considerable stress on the hindlimbs.

Free and easy action at the trot or pace is probably the most important selection factor, and any conformational trait that impairs it is a potential basis for rejection. A harness horse has a clean gait if he can move without interfering. Such free movement requires width between the front legs and, above all, straight limbs. The forelegs, when viewed from front and side, should be perfectly straight, with the toes pointing directly forward. (If a horse toes out, especially at the pace, his knees will hit. If he toes in he will strain his tendons and ligaments, and may also paddle and hit his cannons or fetlocks.)

The knees should be broad across the front but not excessively thick: swelling or other evidence of injury indicates reduced flexibility. Knee action is important, and to test it the horseman should pick up the front foot, bend the leg at the knee and make sure the hoof can touch the elbow. Splints on the front legs might be acceptable if they don't extend into the knee, but their presence there could indicate poor flexion. Splints on the back of the splint bone can also be harmful if their location causes them to contact and irritate the suspensory ligaments. Most trainers find slightly buck-kneed horses acceptable, but they discriminate strongly against the weaker calf-kneed condition. A horse's legs tend to bend back at the knee when he is fatigued. If the horse is calf-kneed he is very likely to develop bilateral carpal chips when raced.

Properly conformed hindlegs should have a moderate amount of angulation when viewed from the side. Sickle hocks in a harness horse could be acceptable if his action is unimpaired and if he has not developed curbs. The pasterns may be straighter than those of a flat-racing horse because the harness horse does not need as smooth a stride. Pasterns that are too straight, however, will absorb little concussion and lead to development of osselets and sesamoid problems. Too much slope will strain the suspensory ligaments and lead to the development of bowed tendons, so a moderately sloped pastern is best.

Nevele Pride (Star's Pride x Thankful) in 1969 broke a trotting record set 31 years earlier by the Standardbred Greyhound. Nevele Pride, then four years old, trotted the mile in 1:54 4/5, and went on to win that year's Hambletonian Stake. He retired with 57 victories in 67 starts, Harness Horse of the Year honors three consecutive years and 16 world records.

courtesy of the United States Trotting Association photo by George Smallsreed Jr.

The feet, one of the race horse's major structural components, should be moderately large and rounded at the toes with square, wide heels. Generous heels help protect a horse from navicular disease and coffin bone problems; concussion, combined with contracted heels, can make every step painful. Feet that are badly dished from the front (concave) can also lead to lameness by exerting pressure against inside bones. The hoof wall should be free from evidence of founder rings, and should be of sufficient thickness and strength to withstand frequent reshoeings.

The harness racer needs adequate air intake, so he should have wide, open nostrils, considerable width between his jaws and a long, graceful neck. He needs a wide chest and heartgirth to provide adequate room for heart and lung expansion. The barrel should be wide and long, especially in the trotter, as short-barreled trotters tend to hit their shins when they travel. A harness horse needs a powerful shoulder for a long, bold stride, but the slope can be less than that required of a riding horse. When viewed from behind, the hindquarters should appear muscular and the hips level: a horse with one hip down will be bad-gaited and will run crooked to compensate for the low hip.

The Steeplechaser

Steeplechasers, like harness and flat-racing horses, need considerable strength and staying ability to cover courses as long as four miles at racing speed. A steeplechaser needs good jumping ability, as well as size and substance to meet the challenge of rigorous courses. Even a two-mile steeplechase (the minimum distance) contains at least 12 fences which measure 4 ft. 6 in. or higher, and at least 8 hurdles that are 3 ft. 6 in. from bottom bar to top. A steeplechaser should have a deep chest, prominent muscles and solid bones to help him withstand the frequent concussion that his legs will be subjected to on landing.

In England and France, where steeplechasing is a major sport, horsemen generally contend that the best steeplechasers are sired by winners of flat races. This is probably because jumping ability is thought to be developed through training rather than being an inherited trait. Although a good steeplechaser could pass the required conformational characteristics to his offspring, he could probably not transmit his inclination to jump.

Good flat racers generally are effective sires of steeplechasers because they can pass on the size, substance, and muscle that a chaser needs. Usually, the owner of a promising young colt finds it more practical and profitable to train him for flat racing. The purses are higher and the rewards more immediate, since a flat racer can begin racing at two years of age. Steeplechasers, on the other hand, take years to train and develop. They may not show their full potential until they reach seven or eight. Quite often, the best horses are channeled into flat racing, while those that are less desirable are trained over jumps. Those good flat racers, which have more athletic potential than the steeplechasers, would probably sire the most athletic offspring.

The same usually holds true with mares, although occasionally a steeplechasing mare will give birth to an outstanding winner of classic races.

The Hunter and Jumper

Hunting and jumping, two distinct types of competition, require the horse to possess considerable athletic skill, plus high levels of motivation and conditioning. Hunters and jumpers both need jumping ability, but power and boldness are more important in the jumper (who, in international puissance competition, might clear jumps as high as 8 feet) than in the hunter, who is scored on conformation, style and manners as well as performance over jumps rarely exceeding 4½ feet.

Hunters, originally bred for sturdiness and strength, needed both these qualities to carry their often over-weight English lords across fields and streams during a hunt. Early hunts were essential for survival, as mounted noblemen brought home deer, wild boar and small game to the castle. Later, as the English domesticated cattle, sheep and poultry for primary food sources, the hunt centered on foxes, the pests that plagued the chicken coop and sheep herd. To follow a pack of hounds across the rolling countryside, a horse needed staying power, endurance, and the ability to leap fences and ditches while keeping up with the hounds.

Today, open field hunting remains popular in parts of England, Australia, the United States and Canada. But increasing numbers of horsemen are also selecting hunters for their performance in the show ring. Generally, these horses are shown in three divisions. Hunter broodmares and stallions, shown in hand, receive points for conformation and not performance. They should exhibit outstanding conformation, substance, quiet confidence and the apparent ability to become or beget excellent hunters. Show hunters, a second group, earn between 25 and 60% of their scores on conformation, presence, ride and action, and the remainder on jumping ability. Working hunters, the third group, are judged on performance alone. They are, however, required to remain sound throughout the course, and are disqualified for any unsoundness they exhibit immediately after jumping. (All the horses in a class are usually jogged past the judge before he makes his placings.) Each of the three classes of hunters is divided into two categories: green hunters, for horses in their first or second year of showing over jumps, and open hunters, for more experienced

This jumper is folding his forelegs nicely as he clears an obstacle. He shows both good form and concentration.

horses. Green hunters usually navigate a course of six jumps between 3 ft. 6 in. and 4 ft. in height, while open hunters jump between 4 ft. and 4 ft. 6 in. The course contains obstacles similar to those the horse would encounter on an actual hunt (stone walls, hedges, ditches, banks and rails) and gives him an opportunity to demonstrate his abilities and training. Consistency, and an even pace over the course, are more important in hunters than in jumpers: hunters must provide a quiet ride, so they are graded on smoothness and good manners. Because they need to cover ground quickly and efficiently, hunters should excel at the gallop: it should be extended and effortless.

Most hunters are of Thoroughbred breeding because Thoroughbreds have the size, speed and athletic ability to carry riders of various weights across the country during a season's hunting. Many horsemen favor some cold (draft horse) blood in their hunters, because they believe it contributes to size and stamina and exerts a calming influence on the more volatile Thoroughbred temperament. Some of the world's best hunters and jumpers are predominantly Thoroughbred, but have some Shire or Irish draft horses in their pedigrees. These crosses are called heavy hunters. In recent years, Thoroughbred matings with the German Holstein and Trakehner horses have become popular, because the horses produced by such crosses seem to have increased endurance and strength. (Holsteins and Trakehners are European warm-bloods noted for their jumping ability.)

Breeders who want pony hunters (less than 14-2 hands tall) often cross small Thoroughbreds or Arabians with Welsh Cob, Connemara or Highland ponies to obtain hardy, tractable mounts. These animals might lack the Thoroughbred's grace and speed, but compensate with strength, stamina and proportionately greater weight-carrying ability. Huntsmen in many rugged, mountainous areas consider their tenacity and sure-footedness invaluable.

Show jumpers, like hunters, must be powerful enough to carry a rider over a prescribed series of jumps. Both types must complete the course smoothly and cleanly. They receive faults for knocking down or touching fences, for falls or disobediences (i.e., refusals, run-outs, loss of gait, circling, etc.). Speed and boldness are important for jumpers because they must compete against time and also leap obstacles of varying heights and widths. Because of the need for speed and agility, Thoroughbred blood dominates in jumpers as well as in hunters. Many powerfully built, leggy Quarter Horses have also become excellent stadium jumpers.

Whether selecting a hunter or a jumper, a horseman first looks for overall balance. The front half of the body should be proportional to the back half, because a horse uses his head and neck for jumping balance and propels himself with his hindquarters. A strong, medium-length back and well-built forelegs indicate that a horse has weight-carrying ability and can withstand the impact of landing. (A long back makes collection for jumping easier, but tires more easily.) The jumper should be a substantial-looking athlete with a long, sloping shoulder, high withers and muscular hindquarters. He also should have a thick stifle, short cannons, clean joints and a high hip. Most trainers prefer a long loin in a jumper, and point out that animals geared for jumping, such as cats, have that trait. A goose rump and a low tail set would be considered faults in many show horses, but could help a jumper by enabling him to draw his legs under his body and clear jumps more efficiently.

Size is more of a practical consideration than a requirement, but small horses (or those with choppy strides) have difficulty pacing themselves on courses designed for lanky Thoroughbred types. Most successful jumpers measure 16 to 16-2 hands at the withers. Even though a smaller horse may be willing and able to jump amazingly high, he usually jumps with less grace and style than a larger horse. Taller horses have efficient, ground-covering strides to carry them across the course; obstacles are less imposing to them. A horse with an effortless stride, rather than a short and choppy one, subjects his legs to less impact-related stress. Such stress can lead to unsoundnesses (navicular disease, shoulder lameness, stringhalt, sole bruises, etc.), all of which can hinder performance. A prospective jumper should be examined before exercise for "cold" lamenesses, then exercised for half an hour and inspected for "warm" lamenesses. He should stand straight and not toe-out or toe-in. (Pigeon-toed horses place excessive strain on their fetlocks, while splay-footed horses interfere.) The forelegs are especially important because they bear much of a horse's weight on landing. In cases where ringbone, navicular disease or other undesirable bony changes are suspected, radiographs of the front feet might be indicated.

Because jumpers are prone to stress-related injuries, most trainers wait until their horses are five years old to begin intensive training. Before five years, all work may be flat work and cavalletti (a series of low jumps). By five, a horse has developed sufficient muscle, and his joints and ligaments are set. Pushing young horses through training is sometimes done for economic reasons, but it can prove damaging and is usually unwise. Besides, hunters and jumpers don't reach their performance peaks until between 8 and 12 years of age. They often remain serviceable into their mid-teens.

A jumper's forelegs must be well-conformed and strong-boned to withstand the tremendous concussion of landing. Note the stress on the over-extended fetlock joint.

Maneuverability is especially important in the open field hunter, who might be called upon to twist in mid-air and change direction before landing. Elasticity and freedom of movement characterize the superior jumper: alternate collection and extension is required between and over jumps for a horse to excel at this event.

The horseman can assess athletic ability by watching a horse work at liberty, both on a longe line and in a jumping chute or lane. When a horse is free from rider influence he can reveal his natural ability through his balance, form and timing over jumps. Ideally, he should gauge his take-off point so that he clears each jump in a smooth arc. He should collect himself at the point of take-off and remain collected, with his neck low and extended and his back arched, throughout

The hunter and jumper should jump in a smooth arc — his take off and landing points should be approximately equidistant from the jump. He should jump just high enough to clear the obstacle — a horse that jumps too high wastes effort and tires more quickly.

the jump. He achieves a symmetrical arc by adjusting his take-off and landing points to account for the height and width of the jump, thereby eliminating wasted effort.

A horse should be examined over jumps that are comparable to those he will be expected to negotiate. A pony hunter, for example, should be jumped over 3 and 4-foot jumps, and should not be expected to tackle 35-foot spreads. Larger, more experienced open jumpers can be tried over more challenging fences and walls. An excellent jumper might grow bored leaping small fences, and he might exhibit poor form, reveal a bad attitude or even knock over a jump.

Faults in jumpers include hanging at the knees (not folding the forelegs adequately) or dropping a leg (lack of symmetry). A horse with these faults might compensate for his dangling leg by jumping higher, but this requires more effort and subjects his forelegs to greater concussion on landing. Athletic ability alone will not make a great hunter or jumper: a horse must have courage and heart. Horses, unlike cats or deer, are not natural jumpers; in fact, performance usually is so dependent on man's urgings that even a world class jumper might consider himself trapped inside a 4-foot paddock fence. The horse must respond to his rider's cues and be eager to please. Under the saddle, the jumper should exhibit confidence in his rider and a willingness to attempt all jumps without hesitation.

Combined Training

Combined training, which offers competition from novice levels to the esteemed international three-day event, is a demanding sport originally designed to test the speed and fitness of cavalry mounts. (The three-day event consists of three days of competition in dressage, cross-country work and show jumping.) Today, combined-training horses need suppleness and discipline for dressage, endurance and speed for cross-country work, and jumping ability. They need an inherent natural balance and impulsion (forward, driving motion) that training cannot provide.

A good prospect is one with strong hindquarters, a maneuverable forehand (forelegs, shoulder, head and neck) and the general conformation of a good jumper. He should have strong, clean-cut hocks and long muscling in the gaskin. He should be long from the base of the tail to the point of the hock, and short from the hock to the ground. His shoulder should be long and sloping; his legs, clean and straight. A horse participating in combined training will need excellent heart and wind, so he should have room across the chest and depth through the girth. A long back is acceptable, but a short back usually is not: a back that is too short gives a horse little room to collect his hindlegs for a jump.

Size will affect both speed over a course and timing between jumps. A horse that is too tall may have trouble handling tricky distances between obstacles. In evaluating performance over jumps, the horseman should not discriminate against the cautious and awkward beginning jumper. With proper training, he probably will make a better jumper than the horse that flings himself uncontrollably toward every obstacle he encounters. Under the saddle, however, the prospect should transmit a sense of power and strength, and his movement should be free and effortless. A good choice might be a four or five-year-old Thoroughbred type that has been ridden lightly (possibly as a green hunter). He should stand about 16 hands tall, and should have good bone and substance.

A horse used for dressage (the refined execution of a set pattern of movements) should show evenness of pace, willingness to perform and the ability to extend and collect. Because most of the stress he encounters will fall on the hock, stifle and hip, he needs sound limbs, free movement of his joints, and powerful hindquarters. A long, sloping shoulder will allow him full extension of his forelegs.

A rider beginning at the lower levels of dressage might benefit from experience on a settled, well-trained horse. But highly trained dressage horses are very much in demand, so a neophyte rider might choose a veteran horse with some minor, acquired unsoundness (a horse unfit for serious competition but still able to perform). A talented beginner might prefer purchasing a green horse and developing some proficiency in schooling the animal. Promising young horses also are in demand, but a green horse with certain physical limitations that might prevent him from achieving perfection would suffice most riders for several years. Obedience and manners are more important than brilliance in the early stages of training. A successful dressage horse should move well at all three gaits, and at the walk, should be fluid enough to overstep his front hoofprints by 18 to 20 inches. His trot should be free and suspended, and when cued to go faster he should lengthen his stride rather than quicken his pace.

The Stock Horse

Stock horses have been helping man tend his flocks and herds for centuries. These hardy workers remain popular today. Their economic importance contributes to their popularity, because even today mechanical devices cannot consistently equal a horse's ability to work cattle.

A good cutting horse is agressive and focuses his attention on his calf without cues from the rider.

The multi-talented stock horse performs many jobs on a modern day ranch. He helps his rider herd cattle, cut individual animals from the herd, and rope them for branding or doctoring. A good stock horse is versatile and possesses both innate ability and trainability.

Various breeds (including the Morgan, Appaloosa, Arabian and Paint) have proven successful as stock horses, but it is the Quarter Horse that has emerged as the ideal, owing to its short coupling, agility, speed, and reputation for having "cow sense." Cattlemen believe that a horse must have the ability to sense a cow's movements to be a good stock horse. They value it especially in cutting horses, who must be able to anticipate the efforts cut cattle make to rejoin the herd.

Some trainers believe that certain horses are born with cow sense, and although it can be strengthened through training, no amount of schooling can develop cow savy in a disinterested mount. Geneticists have conducted little research on the trait's heritability: some rank it from medium to high, while most rank it much lower. Regardless of scientific estimates, some horsemen consider cow sense highly heritable, and have been attempting to breed it into their horses since the Spanish first drove cattle into North America.

Cow sense, like other qualitative traits, is difficult to evaluate, especially in the young, untrained horse. Generally, a good stock horse is intelligent, and takes an interest in his surroundings and in cattle. He also is aggressive: he will work cattle with minimal cueing from his rider. (Working without cues is important for the exhibition cutting horse, since he is faulted for receiving aids from his rider.)

Versatility and stamina are trademarks of the working stock horse. A horseman selecting for this type should look for a horse with an excellent heart and respiratory system. Cutting, roping and carrying a rider for long periods of time place heavy demands on a horse; he needs considerable athletic ability to endure the strain; speed, to overtake fast calves; weight, to brace himself against the pull of a heavy steer at the end of a rope; and endurance, to withstand long hours under the saddle.

Short-coupling gives a cow horse greater weight-carrying ability, and short, strong cannons combined with clean joints and straight legs reduce his chances for lameness. His forearms should be well-muscled, but not excessively heavy or too wide through the chest since the forehand should be light for maneuverability. Perhaps most importantly, the stock horse needs good muscling in his stifle, gaskin and loin. He "works off" his hindquarters and is provided propulsion by them. To balance these powerful hindquarters, the horse needs a good head and a fairly long neck.

The stock horse should be free from any structural defects or lamenesses that would detract from his stamina. Because he is judged only on performance, certain blemishes (scars, superficial conditions, etc.) may be acceptable. Strong hindquarters and physical ability are important, but the cow horse should also be manageable and should approach his work eagerly. Historically, cow horses have been companions as well as workers. Although today's ranches are more businesslike than before, the cowboy still values a horse with a calm disposition, tremendous heart, and courage. Many cowmen ride geldings because of their reputation for more consistent day-to-day performance.

Competitions for cutting and roping horses differ markedly from actual ranch work. While the working stock horse is versatile, the competition cutting or roping horse is a specialist: his early handling and training are tailored toward developing skill in his event. The exhibition cutting horse usually learns early that he can work with a free head, while the young roping horse is taught to "follow" calves or steers.

Competitive cutting horses are scored on how well they handle (or out-maneuver) cattle within a specific time period (often 2½ minutes). The horse should enter the herd quietly, selecting a calf that his rider indicates and easing it out of the herd. Once he has separated the calf, the cutting horse positions himself between it and the herd and "works" the calf. High scores go to the horse that can get the most action out of his calf and still maintain a defensive position. To goad the calf to action, the cutting horse might taunt it by "dancing" on his forelegs. He usually keeps his body low to the ground, with his head low and outstretched and his attention focused on the calf.

Well-developed, powerful hindquarters give the cutting horse a point of balance over which he can pivot and spin. They also provide power for short bursts of speed the horse needs to keep from losing the calf. The cutting horse also needs fairly prominent withers to anchor the saddle. Size preferences vary: some horsemen prefer large, strong horses, but others argue that smaller horses (less than 15 hands) are more maneuverable.

A good horse usually demonstrates cow sense early in his training. One that has been worked around cattle for several weeks and fails to take any interest in his work probably does not have cow sense.

A good roping horse needs muscular hindquarters and balance to execute sliding stops and hold the calf.

Roping horses are also highly specialized. Calf roping, a popular rodeo event, dates back more than 100 years to the early frontier days, when a cowboy had to rope and throw calves for branding, doctoring, castrating, etc. The working cowboy and the professional calf roper still need a rugged horse that can keep them within throwing distance of speedy calves. Like cutting horses, roping horses need strong hindquarters for bursts of speed, but they also need to execute sliding stops. The roping horse chasing a calf at full speed must slide to a stop on cue, then hold the rope taut while a cowboy ties his calf. Competitors in timed roping events should concentrate on finding a horse that has power for quick starts and speed over short distances, and that can get his hindlegs well under his body to stop.

A good roping horse needs the intelligence to rate the calf (gauge its speed) and to keep the rider in throwing position behind it. He needs agility to follow the turning, ducking calf, and steadiness to hold the rope taut after his rider dismounts. Steer roping horses need considerable strength and weight to brace themselves against the pull of heavier steers. Small, quick horses (less than 15 hands) may be maneuverable and fast, but generally taller, stouter horses make better roping horses because they have a height (and usually weight) advantage over cattle.

Competitors in team roping need two kinds of horses: heading horses, for the riders who rope the steers' horns; and heeling horses, for the riders who rope the hind feet. A good heading horse is usually tall (about 15-2 hands) and normally weighs between 1200 and 1350 lbs. He is a big-boned, solid horse with considerable substance.

A heeling horse can be somewhat lighter, because he needs agility to maneuver into position for a follow-up throw. Most heelers stand about 15 hands tall and weigh in the 1100 lbs. range. Professional steer ropers usually prefer working with mature (at least six or seven years old), experienced horses.

Overall, the stock horse needs to be an alert, energetic worker, with sheer physical strength, speed, agility, and a desire to get the job done.

The Endurance and Competitive Trail Horse

Endurance and competitive trail riding test the horse's ability to withstand stress by giving horse and rider an opportunity to demonstrate soundness, hardiness, good wind and staying ability. Endurance horses must carry weight (saddles and riders) over rough terrain through temperature extremes that may vary from below freezing to 120 degrees — all on one ride.

Two types of trail rides have emerged in recent years: endurance rides, which place a premium on speed, and competitive trail rides, which reward good condition. Typical ride lengths are 50 to 100 miles. Endurance rides are actually races, and the first horse across the finish line wins provided that he remains sound. Only horses in excellent condition can successfully complete endurance rides. Horses in competitive trail rides are scored on condition and not speed, although they must complete each day's riding (between 25 and 40 miles per day) within a specific time limit (usually 6½ to 7 hours). Horses in competitive rides are scored on soundness (40%), condition (40%), manners (15%) and way of going (5%).

Both types of rides require periodic veterinary checks. Endurance riders stop their horses for three compulsory one-hour stops each race. Competitive riders stop more frequently, since their placings are based on veterinary and layman evaluations at the stops. These checks are designed to protect the horse, and the vets will pull any horse that could injure himself by continuing. Veterinarians at the checks will disqualify a horse for lameness, excessive dehydration, signs of illness, poor attitude, or poor recovery from stress as indicated by temperature, pulse and respiration rate (TPR) checks.

The prospective trail horse should be bright, alert and eager: dull, lackluster eyes and a tired stance indicate overuse and overtraining. At a walk and trot, the horse should move freely, with spring, elasticity and sufficient impulsion. A ground-covering gait is important because the trail horse must cover many miles effortlessly and quickly. A competitive trail horse needs a good trot — at an extended trot he will cover about 6 to 7 miles per hour with a minimum of exertion. (This pace may be too slow for a winning endurance horse, which must average between 10 and 15 miles per hour. For endurance horses, a good gallop is more important.)

A lean, lightly muscled horse will fare better than a heavily muscled animal. Heavy muscles can "tie up" (become swollen and tender from lactic acid build-up) during a ride, rendering the horse unfit to continue. Also, heavy muscle or excessive fat requires the horse to carry additional weight, and he will tire more readily. Endurance comes from long, slim muscles.

The size of a trail horse is a definite consideration, although it is less critical when the horse will travel only short distances. Smaller horses sometimes have the advantage because they can shed heat more effectively than large horses. A small horse has proportionately more body area for evaporation than a horse that is bulky. Since evaporation of sweat cools the body, a smaller horse can shed excess heat more effectively than a larger horse with more bulk but a pro-

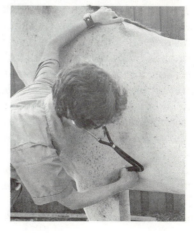

T.P.R's (temperature, pulse and respiration) are taken prior to and at several intervals during endurance and competition trail rides. These measurements monitor each horse's conditioning; well-conditioned horses return to a normal T.P.R more quickly than do poorly conditioned horses.

portionately smaller evaporation area. A smaller horse may also be a better feed converter and have greater weight-carrying capacity in proportion to its size.

The sex of a trail horse has little bearing on its performance, but many experienced competitors prefer geldings. Most endurance rides prohibit mares in foal from competing, and some smaller rides may also restrict the entry of stallions.

There are persuasive arguments for waiting until a horse turns five before piling weight on his back and forcing him to carry it over rugged terrain. Few rides allow horses less than five to compete. Their bones are not yet hardened, neither are their joints, tendons and ligaments sufficiently developed. The best endurance horses are often between 7 and 12 years old. Youthful vigor alone will not make a good trail horse; it must be tempered with experience. Even horses well into their teens can serve as excellent mounts for serious competition if the animals are sound and of the proper type.

Assessment of a trail horse should begin at the feet and work upward. The hooves should be large but not out of proportion, and should be wider at the base than around the coronary band. A relatively thick hoofwall can tolerate the frequent reshoeing necessary in the trail horse. The heels should be open, and should be examined for strike marks which could indicate that the horse is over-reaching and subjecting himself to injury. Inside, the hoof horn should be dark and dense, the bars firm. The frog, the hoof's spongy cushion of support, should be elastic and untrimmed. (Trimming the frog improves its appearance but reduces the hoof's ability to absorb shock.)

The pasterns should be sloping, the cannon bones short, and the forearms long to allow a swinging gait and effortless stride. All the bones should be dense and flat, and the joints should be large (but not filled with fluid, which can result from trauma, overwork, bad conformation, etc.) Well-defined tendons, with deep grooves separating them, indicate the absence of previous tendon strain.

The legs should be straight from both front and rear views. Any deviation or minor unsoundness will lead to serious problems when pushed over a 100-mile course. The horseman should probably reject a horse that shows striking or forging marks on the back legs. On a horse with a narrow base he should look for marks on the fetlocks and pasterns of the forelegs, and should reject the horse if its gait is not clean. The legs should be free from all other blemishes; old splints, thoroughpins or bowed tendons could become active again. Because weight-carrying ability is so important, a trail horse needs a short back, a strong loin and a long underline. High, muscular withers will hold the saddle in place on taxing rides. A trim throatlatch and long neck indicate maneuverability, and large nostrils are a sign of good oxygen-carrying capacity.

A deep, wide chest and well-sprung ribs provide sufficient room for the expansion of heart and lungs. This is necessary for the horse to cover great distances at speed and to recover for the obligatory vet checks. (For a horse to remain in a ride, his pulse/respiration ratio — normally about 3:1 at rest — must stay above 1:1. Also, his pulse must drop below 70 and his p/r ratio must return to 2:1 following a 45-minute rest. A horse with a resting pulse of 36 and a respiration rate of 12 might elevate his pulse to 100 during part of a ride. But once he has had sufficient time to rest, his pulse must drop below 70. His respiration rate must drop correspondingly within 45 minutes or he will be removed from competition.

One of the most important characteristics a trail horse can have is a willingness to push himself — he should be energetic and should move out with little prodding. It is difficult for a rider to urge a recalcitrant mount over thousands of miles. Some riders like peppy horses, but one that is too high-strung will easily tire both himself and his rider, and will have a poor recovery rate.

A good trail horse should retain his appetite when worked hard and should drink water from many sources. A horse that refuses to drink strange water is likely to dehydrate even on a short ride.

Certain breeds have proven especially suited to trail riding: the Thoroughbred for his courage and lengthy stride; and the Arabian for his smaller size, denser bone, low feed consumption and reputation for remaining clean-limbed into old age. The Arab also has smaller, more numerous erythrocytes with increased surface area and the ability to maintain oxygen levels in the tissues. Still, the wise horseman will consider all breeds, and various crosses, to select a horse that can best withstand the rigors of the trail: temperature extremes, precipitous footing, risk of dehydration, exhaustion, etc.

The Draft Horse

Since his domestication the horse has borne many of man's burdens. During Medieval times, the Great War Horse carried armor-laden European knights into battle, often packing as much as 400 lbs. From this sturdy stock sprang the work horses that for centuries helped farmers plant and harvest their crops and transport them to market. These horses served for years as one of the primary means of commercial transportation across Europe, Asia and the Americas.

courtesy of the Percheron Horse Association of America

Laet, a Percheron stallion foaled in 1916, became a premier sire of his breed in the 1920s. He sired more grand champion stallions at the Chicago International than any other sire in the history of his breed.

Five major draft breeds developed from these horses: the Shire and Suffolk Punch in England, the Percheron in France, the Belgian in Belgium and the Clydesdale in Scotland. The breeds today differ somewhat in function as well as in form. Although some European countries still use working draft horses, their popularity as such began to decline with the advent of the gasoline-powered engine. Today, draft horses are shown at fairs, pulling contests and other exhibitions where (much as their cart and delivery horse ancestors) they serve as walking advertisements for their owners.

Although experts on each of the breeds differ markedly in their opinions of the "ideal," certain conformational and behavioral qualities are common to all good draft horses. Overall, working animals should be deep, broad and muscular: they should have the rugged construction necessary to pull heavy loads at a walk. A good drafter should possess size and substance. Most draft horses weigh more than 1600 lbs., and stand between 16 and 19 hands tall. Weight in the forehand is especially important, since the draft horse pulls loads greater than his weight by pushing his heavy neck, shoulder and breast against his collar.

Draft horses are deep, long-bodied individuals with legs that are short in relation to body height.

At the walk and trot, the draft horse's most important gaits, his action should be swift and elastic — power is not necessarily accompanied by sluggishness. The Clydesdale, for example, has considerable animation and high action. Its springy gaits stem from longer than average pasterns, and its collected motion from being slightly cow-hocked.

Animation and action are desirable, especially now that draft horses have become so popular in public relations work. Any breed should appear alert and energetic, but should remain manageable. While a dull, phlegmatic horse would not be a crowd pleaser, he probably has a gentler disposition and would be more tractable than one that is more spirited. Because geldings are often more stable and well-mannered than mares or stallions, horsemen with driving hitches generally prefer them for parades or exhibitions. The flashy specimens that make the best parade horses also might be the best breeding prospects. A breeder might stand a young horse at stud for several years, then geld him for use at exhibitions when he is five or six.

The ideal prospect will have a deep, wide chest, a massive neck and a heavy forehand. (A chest that is too wide will increase concussion and predispose the horse to ringbone.) A well-muscled croup and gaskin, a thick stifle and a heavy forearm — accompanied by a strong, flat cannon bone — will give the horse pulling power. The loin and topline should be strong, the back short and the middle and flanks deep (long from top to bottom): his build should be powerful and compact.

Certain conformational traits that would normally be undesirable in a riding horse may be tolerated in a draft horse since he would not be ridden. His shoulder, for example, may be straighter than normal, his croup more vertical and his withers less prominent than is desirable on most riding horses. He should not be faulted for standing close behind, as this conformation will give him a firmer column of support and enable him to keep power under his body. He also will have more power if his legs are short in proportion to body height. Ideally, the body length should be several inches more than its height at the withers. The body depth from withers to underline should be several inches more than the distance from the brisket to the ground. The draft horse needs good underpinning for support, and his short, straight legs should taper down to lean ankles and adequately angled pasterns. (Again, horsemen will tolerate a more vertical angle than they would in a riding horse.) On Shires, the leg feathers, which extend from the hock to the ground, should be long, flowing and silky but not excessively abundant. Clydesdales should have less feathering than Shires. Short, very upright pasterns should be discriminated against. Such conformation, combined with the stress encountered in pulling heavy loads, accounts for a high incidence of sidebone in draft animals.

The feet, the single most important part of a drafter's body, should be large and rounded, full at the toe and quarters, and wide and deep at the heel. (Wide heels give when the foot hits the ground, decreasing the concussion.) Flat soles, contracted heel and other problems, especially of the front feet, should be discriminated against. A horse with flat soles loses full benefit of the frog as a "cushion" for delicate foot bones, and thus is more likely to contract navicular disease.

Strength in the body and legs is vital, but the horseman should not sacrifice quality. A draft horse should be balanced and possess style and refinement: his head should be well-shaped (not coarse), his eyes large and bright, his ears alert and his face intelligent. Small heads in foals and yearlings often indicate early maturity, and an ultimate lack of size, so they should be discriminated against. Dependability and an amiable temperament are characteristic of the giant drafters, and the horseman should promptly reject an animal that fails to exhibit either.

The Show Horse

With the large number of breeds and the many show classes and events within each breed, this text is obviously not the place to attempt a detailed discussion of each. There are a few practical aspects relative to the show horse, however, that are of general importance to the fanciers of all breeds.

The ideal show horse possesses beauty, impeccable manners, and a smooth, stylish way of going, along with animation and spirit. He should display a high degree of poise and execute his assigned maneuvers willingly and in a manner that appears virtually effortless.

The show horse should be well-suited and highly trained for the task he will be expected to perform. If a horse is not an above average specimen of his breed, and/or an above average performer in the area in which he is to compete, he is simply not a show horse.

Often, a show horse is taken on extensive trips and put through an exhaustive show schedule. Because of this, he should be hardy and adaptable to different environments. Specifically, a good show horse should be able to eat and drink under varying circumstances without going off his feed or otherwise upsetting his system. He should be calm and unruffled by crowd noises, activity of other horses, bustling, confusion, etc.

A show horse should be able to tolerate changes in his routine since grooming and exercise schedules can vary widely between shows. He should be especially amenable to handling, and should stand quietly while being saddled, hitched or groomed.

It is important for a show horse to have a quiet, compliant attitude and a willingness to learn. Good manners are important, both inside

Quarter Horse trail class. Note the concentration of both horse and rider.

For superior barrel racing performance the horse should drop low in front and lean into the turn, allowing the fastest possible run of the barrel pattern.

and outside the show ring. A show horse will probably spend a great deal of time surrounded by spectators and competitors, and he should remain amiable with no tendency to kick, strike or bite.

Horses that are judged on conformation must excel in those traits, but they must also adhere closely to breed type. A horse shown in hand should approach the ideal description of his breed. He should have a "typey" head, and those specific conformational characteristics that distinguish his breed or type. Color may also be a factor: it should be representative of the breed, both in shade and intensity. In some breeds, "flashy" colors and markings are desirable, while in others, subdued colors win more classes. (For specific information, see ideal breed descriptions published by individual associations, and American Horse Show Association [AHSA] rules.)

Two horses may have essentially the same conformational traits, but if one has been better conditioned than the other, this will often influence the judges. A horse's coat should be well-groomed, his hooves should be well cared for, and he should appear sleek and solid. Even though conditioning may influence the judges, excellent conformation and overall balance should be a major factor in breed classes since halter championships have a significant impact on breeder goals. An attractive head and a long, graceful neck will create a favorable impression, but should be mounted on a body that has depth, good bone structure, and muscling.

In selecting a finished show horse, training is obviously important. The horseman should consider his prospect's performance record, noting the size, competition and location of listed shows. He should use that information to evaluate placings.

section III.

applying genetics to selection

5

BASIC GENETIC
DEFINITIONS

The remaining chapters of **EQUINE GENETICS & SELECTION PROCEDURES** contain a number of technical terms that apply to the science of genetics. These terms are defined throughout the book and again in the glossary. In addition, some of the most frequently used terms are defined below so that the reader may recognize and understand their use in later discussions.

Chromosome

The *chromosomes* are the rod-like structures that contain the genetic information carried by an individual. They are located within each cell of the body, inside the nucleus, which is an area of highly concentrated material. The number of chromosomes in each cell is the same for all members of a species, but it differs between species. For example, man has 46 chromosomes in each cell, while the horse has 64. The chromosomes are found in the *somatic cells* (i.e., all body cells; note that the sperm and egg are *not* body cells). Humans have 23 chromosome pairs, while horses have 32 pairs. The sperm and the egg, known as the sex cells or *gametes,* contain only one chromosome from each pair. Human gametes have 23 chromosomes, while equine gametes have 32 chromosomes. The equine sperm and egg unite during conception to form a new individual with 32 pairs of chromosomes (32 from the sperm plus 32 from the egg = 32 pairs = 64 chromosomes).

The sex of the offspring (i.e., whether it will be male or female) is determined by the *sex chromosomes,* which are known as the X and Y chromosomes. An individual with both an X and a Y is a male, while a female has two X chromosomes. All of the other chromosomes are collectively called the *autosomes*, or autosomal chromosomes.

Gene

The genetic inheritance carried by the chromosomes is divided into basic units known as *genes*. Although the exact number is not yet known, there are believed to be thousands of genes located on each chromosome. Each gene is responsible for determining one or more characteristics, or traits.

An individual's total genetic makeup is known as his *genotype*. This is what he carries in each cell, and what he can pass on to his offspring. The genotype cannot be completely judged from the horse's outward appearance. The visible and measurable characteristics of an individual are known as his *phenotype*. The difference between a genotype (genetic inheritance) and a phenotype (outward appearance) can be illustrated with coat color. For example, two grey horses which phenotypically appear the same may have different genotypes, i.e., *Gg* and *GG*, where G stands for grey and g represents nongrey (refer to **"Coat Color and Texture."**) Although the environment cannot alter the horse's genetic makeup (genotype), it may influence his outward appearance (phenotype), and determine how closely he approximates his genetic potential. For example, a horse may have the genotype to reach a height of 16 hands but, due to poor early nutrition, only attains 15-2 hands at maturity. In this case, his phenotype (15-2 hands) is not the same as his genotype (16 hands) due to environmental conditions (poor nutrition).

Each gene occupies a fixed position, or *locus,* on a chromosome. Since two similar chromosomes form a pair, there is also a pair of genes for each trait carried by the horse, with one gene located on each chromosome at the same position, or locus. There can be more than one form of the gene for a specific trait. For example, in the horse, there are four forms of the gene at the *A* locus. Each of these forms is known as an *allele*. These alleles are A^+ (for a coat pattern like that of the Przewalski's horse), A (for a bay coat pattern), a^t (for seal brown), and a (for black or liver). Although there are four alleles for this locus, the horse can only carry two. For instance, the horse could carry A and a, or a^t and a^t. If the two alleles are identical (such as $a^t\ a^t$), they are in the *homozygous* condition (*homo*-same, *zygous*-egg). If they are different (such as $A\ a$), the condition is *heterozygous* (*hetero*-different, *zygous*-egg).

Sometimes, one allele at a certain locus will hide the presence of another allele at the same locus. For example, if a horse carries the

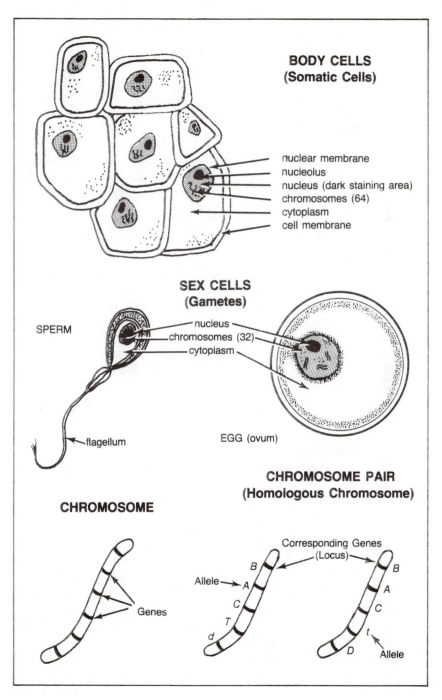

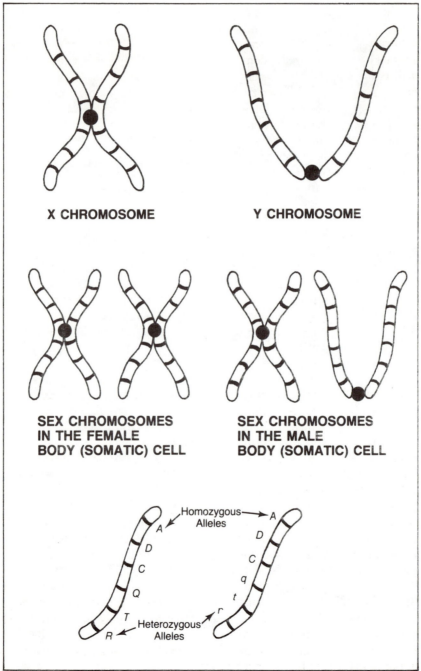

X CHROMOSOME

Y CHROMOSOME

SEX CHROMOSOMES
IN THE FEMALE
BODY (SOMATIC) CELL

SEX CHROMOSOMES
IN THE MALE
BODY (SOMATIC) CELL

Homozygous Alleles

Heterozygous Alleles

alleles *G* and *g,* where *G* is grey and *g* is nongrey, the grey allele will hide the nongrey allele, and the horse will turn grey as he ages. In other words, grey (*G*) is *dominant* over nongrey (*g*), and nongrey is *recessive* to grey. A simple dominant allele on one chromosome of a pair will cover up a simple recessive allele on the corresponding chromosome. Therefore, the homozygous dominant horse (i.e., the horse that carries two dominant alleles) appears the same as the heterozygous horse (i.e., the horse that carries one dominant and one recessive). A homozygous dominant horse (GG) would be grey, as would a heterozygous horse (Gg). A nongrey horse would be a homozygous recessive (gg). A recessive allele reveals its presence only when there is no dominant allele to mask it. For this reason, some grey horses produce only grey offspring.

---

homozygous grey (GG) × nongrey (gg) = all grey offspring (Gg)

A gene may also be *epistatic,* meaning that it is dominant over a gene at another locus. The epistatic gene hides the effects of genes at other loci. For example, the dominant white gene (*W*) is epistatic to all other color genes. No matter what color gene the horse is carrying (e.g., genes for bay, black, palomino, etc.), the horse will be white if he has the dominant white allele (*W*). This is true even though the dominant white allele is only at the *W* locus, and the other genes are at their respective loci (e.g., bay is at the *A* locus, black at the *B* locus, etc.). The genes that are hidden by an epistatic gene at another locus are known as *hypostatic*. So bay, black, palomino, etc., are hypostatic to dominant white, and dominant white is epistatic to bay, black, palomino, etc.

When one trait is controlled by many different genes, it is referred to as a *quantitative* trait. Due to complex genetic control (i.e., the action of many genes), a *quantitative* trait often varies in degree between individuals. The phenotypes for a quantitative trait can usually be measured. Height, for example, is controlled by many different genes, varies in degree between individuals, and is measured in hands. (Examples also include fertility, speed and growth rate.) Unlike quantitative traits, *qualitative* traits are controlled by one, or just a few, genes. These traits do not vary continuously and cannot be measured. They are either expressed, or they are not seen at all. The somewhat rare, nonfading black coat color is a qualitative trait; a horse is either jet black or he is not.

Heredity

Heredity is the transmission of genetic information from parents to offspring through the sex cells (i.e., sperm and egg). Half of the offspring's chromosomes are from his sire, and half are from his dam. Therefore, he has a genotype that differs from either parent. This genotype remains unchanged throughout life, even if the phenotype (external appearance) changes somewhat, due to environmental conditions. On rare occasions, a change may occur in the chemical composition of a gene or in the structure of a chromosome. Such a change is known as a *mutation*. If the change occurs in a somatic cell, a change in phenotype may result. If the mutation occurs in a gamete (sperm or egg), it can be passed on to the offspring. The mutation then becomes a part of the offspring's genotype and can be passed to his progeny. Although mutations are usually harmful, they may (in rare instances) cause a desirable change and subsequent improvement within a species.

Environment

The surrounding conditions which influence the growth and development of an individual are known as the *environment*. Both external factors (such as living conditions) and internal factors (such as hormones) are a part of the environment. The environment is capable of drastically changing the phenotype of an individual. For example, a foal with the genotype for straight legs may receive a nutritionally imbalanced diet and developed crooked legs. In this case, early correction of the foal's diet (i.e., environment) may cure the problem. Alternately, a foal with a genotype for crooked legs will not develop straight legs even with a sound diet. (This condition can sometimes be corrected by surgery and other veterinary remedies.)

6

BREEDING ASPECTS

Within the past few decades, the horse industry has become a multimillion dollar business. In many instances, operational income depends on carefully managed breeding programs and the selection of at least average, but hopefully better than average, breeding stock. The importance of careful breeding stock selection within all facets of the horse industry (e.g., racing, jumping, dressage, etc.) is recognized by all serious horse breeders. Within each breed or performance category, there are specific selection guidelines (refer to **"Performance Selection Characteristics"**). Of equal importance, however, is the breeding potential (physical capacity) of each horse in a breeding role.

Emphasis on the different steps involved in selecting a stallion and selecting mares will vary according to the individual breeder's goals. For example, the strictly non-market breeder (whose goal is to breed for private use) might accidentally establish an unknown sire as a good producer, since he can place less emphasis on the popularity of a horse's pedigree. The market breeder, on the other hand, hopes to obtain a profit from his yearling sales by emphasizing pedigrees and early maturing lines. Breeders will not always fit neatly into one category, but an ability to identify breeding goals and practical limitations is the first, and perhaps most important, step in selecting breeding stock.

Selecting a Stallion

Within each generation of foals, it is estimated that only 5% of the males are selected for breeding purposes. Because the stallion has greater influence over future populations through his ability to sire many foals per reproductive year (as compared to the mare's ability to produce only one foal per reproductive year), selection pressure is high. Stallion evaluation should include a thorough reproductive examination, a detailed analysis of conformation, and a study of both pedigree and available performance records.

REPRODUCTIVE ANATOMY

The testes, normally positioned between the thighs and suspended by a strand of elastic tissue called the scrotal ligament, contain the sperm-producing tissue. Sperm production is highly sensitive to both heat and cold. Nature has accounted for this by surrounding the testes with a thin protective sac, or scrotum, which is designed for climatic response. In the winter, muscular contractions within the scrotal walls (dartos muscle) and along the suspension tissue (cremaster muscle) move the testes next to the body, where they receive sufficient heat to continue sperm production. On hot days, the scrotum relaxes to help lower the testicular temperature.

After the sperm are formed within the testes, they move into the epididymis, a tube with complex coils, where they mature. Later, muscle contractions within the epididymis move the sperm into the next structure, the vas deferens. This tubular structure carries the sperm to the seminal vesicles, where they mix with fluid from the accessory sex glands. From there, the sperm move to the urethra, which leads them to the penis.

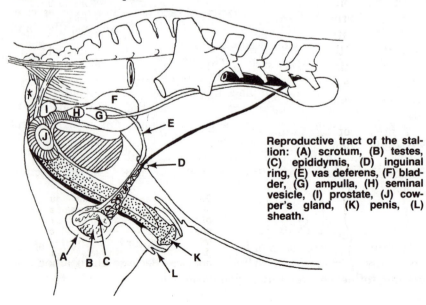

Reproductive tract of the stallion: (A) scrotum, (B) testes, (C) epididymis, (D) inguinal ring, (E) vas deferens, (F) bladder, (G) ampulla, (H) seminal vesicle, (I) prostate, (J) cowper's gland, (K) penis, (L) sheath.

The stallion's self-propelling sperm pass on his genetic contribution to each of his foals. A study on how this transmission is made possible will be discussed in **"Mitosis"** and **"Meiosis"** under **CYTOGENETICS AND PROBABILITY**.

REPRODUCTIVE EXAMINATION

An examination of the stallion's genitalia includes palpation of the testes. Small flabby testes indicate hypoplasia (or underdevelopment) and probably sterility. Frequently, the right testis may be slightly smaller and higher than the left; this has no effect on fertility. Hard testes (caused by a tumor or injury) indicate possible sterility.

Bilateral cryptorchidism (failure of both testes to descend to their normal position) results in complete sterility; unilateral cryptorchidism (failure of one testis to descend) does not affect fertility significantly. Because cryptorchidism is an inherited defect (refer to **"Reproductive System"** under **"Inherited Abnormalities"**), the merits of a unilateral cryptorchid stallion should be weighed against the possibility that he might propagate bilateral cryptorchidism. In rare instances, what appears to be a unilateral cryptorchid is actually a unilaterally castrated stallion (e.g., due to trauma or infection). These stallions often have up to 95% of their normal reproductive capacity. Complications caused by the presence of a scrotal hernia (refer to **"Reproductive System"**) may also affect the stallion's reproductive performance.

The stallion's penis should be washed and checked for signs of injury or infection. An injury to this structure (e.g., due to the kick of an uncooperative mare) may result in swelling which hinders normal retraction of the penis into its protective covering or sheath. Stallions injured in this manner often lose their sex drive and their ability to cover a mare naturally. The penis should also be checked for any indication of venereal disease, squamous cell carcinoma, summer sore, etc.

A microscopic examination of the stallion's semen (sperm and fluid ejaculate) should be made to determine his actual reproductive capa-

Equine spermatozoa quality is judged on the basis of both shape and ability of the tail-like flagellum to move the sperm head. (A) normal, (B) coiled tail, (C) double tail, (D) pyriform head, (E) protoplasmic droplet, (F) tail abnormality, (G) bent tail, (H) loose tail.

bilities. For example, the sperm quantity (number/unit volume) and quality (motility and form) reflect the stallion's fertility and, therefore, his ability to settle a mare. Semen quantity and quality are highest during the summer months and lowest during cold weather. In some instances, buyers may request that a stallion be test bred and settle several mares to insure his productivity.

Other important characteristics to consider are libido, manners, and physical ability to mount a mare. The stallion's libido, or sex drive, should be active and vigorous when a mare in estrus is present. Arching the neck, raising the tail, prancing, neighing, and erection are characteristic of the stallion's normal sex drive. Negative associations with a breeding accident or mishandling, and subsequent loss of libido, may permanently damage a stallion's breeding career.

The quiet, well-mannered stallion usually mounts from the mare's left rear, enters quickly, and waits until the penis is retracted before dismounting. The inexperienced stallion may mount from any direction and become excessively excited. Some overly aggressive stallions may savage their mares as they mount.

The breeder should also check for any lameness of the hind legs which might limit the stallion's ability to mount a mare. A stallion's refusal to mount, although sexually excited, could be caused by his physical inability to stand on his hind legs (e.g., he may have painful arthritis).

REPRODUCTIVE RECORDS

All available records indicating the stallion's past performance should be considered during the selection process. The stallion's past reproductive performance (breeding record) indicates his siring capacity. By providing information on the stallion's reproductive successes and failures, with respect to a large number of mares, these records provide a somewhat objective view of his overall fertility. Records should include:

1) the number of mares that he bred per year,

2) the number of mares that he settled per year,

3) the number of times that he served a mare during one heat period (average),

4) the number of times that he served a mare for one pregnancy (average) and

5) the number of estrus cycles during which he served a mare for one pregnancy (average).

PERFORMANCE RECORDS

A study of the stallion's athletic achievements is another important step in selecting a sire. Performance categories and level of competition should be considered when analyzing the horse's achievements. Advertisements, registry records, and published statistics are sometimes helpful in determining the stallion's past performance. Although a successful performance career will not always indicate the stallion's ability to transmit superior genes to his offspring, physical achievement should be one in a series of selection guidelines.

An evaluation of the stallion's conformation, his good points as well as his weaknesses, is another important step in the selection process. The stallion should be judged with respect to his breed type and the goals of the breeder. An outstanding cutting horse, for example, might be accepted with slight cow hocks, while on a halter horse the same trait is discriminated against. The relative importance of a conformational fault in a prospective sire also depends on whether the characteristic is inherited. (Refer to the discussion on conformation under **"Inherited Abnormalities."**)

SIRING RECORDS

When a stallion's records indicate that he is fertile, that he is an outstanding athlete, and that his conformation fits specific needs, the siring record should be examined. Siring records indicate how well the stallion's offspring produced or performed. The stallion's prepotency, or ability to pass his desirable traits to his offspring, is measured by these records. A Thoroughbred racing sire's production records, for example, show his Average Earnings Index (AEI). This is a comparison between the average earnings of his progeny and the average earnings of all race horses for one season. A high AEI indicates that the stallion is successfully passing on the desired characteristic.

I. Average Earnings of Sire A's progeny for 1978 $=$ $\dfrac{\text{Total earnings of Sire A's progeny for 1978}}{\text{Total number of runners in 1978 produced by Sire A}}$

II.   Average earnings
       for all runners   = $$\frac{\text{Combined earnings of all runners in 1978}}{\text{Total number of runners in 1978}}$$
       in 1978

III.   Sire A's 1978
       Average Earnings   = $$\frac{\text{Average earnings of Sire A's progeny in 1978 (see I.)}}{\text{Average earnings of all 1978 runners (see II.)}}$$
       Index

Occasionally, a breeder or a geneticist will progeny test a stallion. The test involves mating the stallion back to a group of his own daughters to determine whether he carries any hidden (recessive) genes for undesirable traits. The principles behind this test are:

1) If a stallion carries one recessive allele, it will not be expressed, but will be passed on to some (approximately 50%) of his offspring.

2) If this carrier stallion is mated to a group of his own daughters (some of which may carry the recessive allele), some of the resulting offspring may receive the recessive allele from both parents and will therefore be homozygous recessive at that locus (meaning that they will carry the recessive allele on both corresponding points of a chromosome pair).

3) The homozygous recessive offspring will express the trait controlled by the recessive allele (since there is no dominant allele to hide its expression), and the stallion will be identified as a carrier of the recessive trait.

In other words, if a stallion is bred to a group of his own daughters, as the number of resulting offspring free from undesirable recessive traits increases (e.g., umbilical hernia, combined immunodeficiency disease, etc.), the stallion's breeding value increases. This might be especially helpful if any of the stallion's close relatives are known to carry an undesirable recessive gene. For example, an Arabian stallion is a proven CID (combined immunodeficiency disease) carrier if he has produced any CID foals. The owners of a full brother to this carrier stallion wish to progeny test their stallion, and hopefully show that he is probably not a CID carrier. Although progeny testing is never 100% accurate, its accuracy increases with the numbr of father-daughter matings. Refer to the discussion on **"Selection for a Trait"** under **"Breed Improvement Through Applied Genetics."**

Progeny tests are not often performed in horses (expense and time are too great) but, when such records can be derived from pedigree analysis, the actual strength or weakness of a stallion's genotype can be revealed through father-daughter matings.

PEDIGREE

If a colt, whose athletic performance and production ability have not yet been determined, is selected as a prospective addition to a breeding farm, the breeder should look closely at the performance and production ability of his close relatives (usually within three generations). A complete pedigree should show how each relative performs as well as that relative's ability to produce winners. (The importance of an ancestor's influence decreases markedly with each generation removed.) The degree of inbreeding within an individual can also be calculated through the pedigree study. (Refer to **"Inbreeding and Outbreeding."**) Incomplete pedigrees, which tend to over-emphasize popular names, should be viewed cautiously.

Because the Y chromosome carries the genes that cause the developing fetus to become a male, the stallion always gives a copy of his Y chromosome to his sons who, in turn, give their sons a copy of the same. Any other traits controlled by genes located on the Y chromosome will only be passed from sire to son to grandson, etc. This passage of certain traits through the family males is referred to as "one-sided inheritance" and is believed, by some breeders, to be the basis of tail-male influence. Many geneticists argue this concept, however, since Y-linked traits (other than sex) have not been identified in the horse. (Refer to **"Sex Chromosomes"** under **CYTOGENETICS AND PROBABILITY**.)

STUD FEE

The stallion's stud fee may influence the breeder's final decision, but should not be used as a guideline for breeding potential. Based on the principle that the mare also limits and contributes to the value of a foal, many breeders set a limit on stud fees depending on the value of the prospective dam. For example, a breeder may require that a stallion's foals average three times his stud fee at the yearling market. Other breeders may require that the stud fee costs no more than

20% or, at most, 50% of the broodmare's market value. Although the stud fee might influence the breeder's selection decision for financial reasons, it does not necessarily indicate the stallion's performance ability, fertility, prepotency, etc. The previously mentioned selection aids are more important considerations.

THEORIES ON SELECTING A SIRE

Nicks

A nick might be defined as the ability of certain crosses to consistently produce high-quality offspring. Although it would be very difficult to predict these exceptional matings, nicks are usually crosses between two unrelated bloodlines or families. For example, a stallion might sire outstanding foals primarily when mated to the daughters of a certain unrelated stallion. Some breeders feel that a nick is simply a cross between individuals with complementary strengths and weaknesses. In other words, if each parent is homozygous (i.e., prepotent) for desirable characteristics that the other parent lacks, the offspring would tend to be superior to either parent. Nicks might also result from an increase in heterozygosity. When two unrelated lines are crossed, the offspring tend to receive different alleles from each parent, causing an increase in the number of heterozygous gene pairs. This increase in heterozygosity causes many undesirable traits to be masked, overdominant (superior) traits to be expressed, and overall vigor and fertility to improve. (Refer to **"Inbreeding and Outbreeding"** under **APPLYING GENETICS TO SELECTION**.) For this reason, it is not surprising that nicks might occur, when members from two unrelated families (i.e., horses with very different genotypes) are crossed. Offspring from such a cross will tend to have a high level of heterozygosity and, therefore, the up-graded characteristics associated with heterozygosity.

Broodmare Sires

Some breeders place importance on broodmare sires. They feel that certain stallions, although unsuccessful in producing numerous winners, sire daughters which consistently produce superior offspring. From a genetic viewpoint, one explanation is that genes for many desirable traits are located on the X chromosome.

(Note: The male normally carries, within each body cell, an X and a Y chromosome; the female carries two X chromosomes. During fertilization, the embryo's sex is determined by whether a sperm with an X or a sperm with a Y chromosome combines with the X chromosome of the dam's ovum. These chromosomes, which are collectively called the sex chromosomes, will be examined in more detail in the discussion **"Sex Chromosomes"** under **CYTOGENETICS AND PROBABILITY**.)

Since the presence of a Y chromosome results in the development of a male, the stallion can only transmit his Y chromosome to his sons and his X chromosome to his daughters. If genes for desirable traits are located on the stallion's X chromosome (i.e., X-linked traits) his sons cannot inherit those traits. His daughters, on the other hand, will inherit the X-linked genes and, even if they cannot express the traits, are capable of passing them to both male and female offspring.

Families

Many breeders rely on pedigrees and the reputation of a stallion's family when selecting a sire. Although the stallion's athletic capacity and siring ability are part of his heritage, distant relatives have little influence over his actual genetic makeup (except in the case of linebreeding). A stallion advertised as the "great-grandson of a world famous sire" may have inherited very few of his great-grandfather's genes. Therefore, the breeder should use other selection aids before selecting or rejecting the stallion as a prospective sire.

Selection of the best stallions from the best families offers a far greater chance of producing superior offspring than if the breeder uses an outstanding athlete from a relatively obscure family. Horses from successful families usually produce above-average offspring more consistently than do outstanding individuals with heterosis (i.e., individuals which are significantly superior to their mediocre parents). (Refer to **"Inbreeding and Outbreeding."**)

Selecting a Broodmare

Both the dam's maternal influence (i.e., effects of her uterus, milk, attitude, etc.) and her genetic contribution outline her importance on the breeding farm. Compared to stallion numbers, a larger broodmare population is required to maintain production needs within the horse industry. For this reason, the selection pressure upon mares is not nearly as high as that for stallions. An estimated 40-60% from each generation of fillies are selected for breeding purposes, while only about 5% of the colts are selected. Due to rising costs of equine management and a gradual increase in the overall horse population, selection of broodmares is becoming more discriminatory. Breeders are beginning to place more emphasis on quality rather than quantity in the broodmare band.

REPRODUCTIVE ANATOMY

The vulva is the external entry to the mare's reproductive tract. It is located beneath the anus and consists of two lips which should form a tight, protective seal. The vagina is the reproductive canal which lies directly in front of the vulva. This area receives the stallion's penis and serves as the birth canal. It is separated from the uterus by the cervix, a tight constriction which protects both the mare and the developing fetus from infection. In response to changing hormone (body chemical) levels, the cervix relaxes and opens, allowing sperm to enter or a foal to be born.

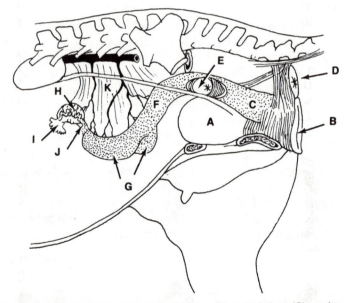

Reproductive tract of the mare: (A) bladder, (B) vulval lips, (C) vagina, (D) anus, (E) cervix, (F) body of the uterus, (G) uterine horns, (H) ovary, (I) infundibulum, (J) oviduct, (K) broad ligaments.

Directly in front of the cervix lies the uterine body which divides into the left and right uterine horns. The normal mare has two kidney-shaped ovaries, each about the size of a large walnut. Each ovary is connected to a uterine horn by connective tissue and a tubular structure called an oviduct. The ovaries, prompted by a delicate and complex balance of hormones, respond by producing an ovum (egg)

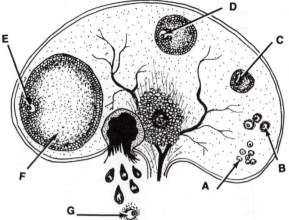

Stages of germ cell maturation in the ovary: A through E show that, as the primary egg cell develops, it is surrounded and nourished by a follicle cell (F). With proper hormone levels, the follicle bursts, expelling the ovum (G) into the oviduct.

which carries the mare's genetic contribution to her offspring. The ovum falls into the oviduct where it can be fertilized by the stallion's sperm.

When an ovum is penetrated by one sperm, a protective membrane is formed so that no other sperm can enter. At this point, the genetic makeup of the future foal is determined. For example, the foal's sex depends on whether the sperm contributes an X or a Y chromosome to the ovum. Chromosomes from both the sire and dam carry genes which direct complex cell division and detailed development of tissue.

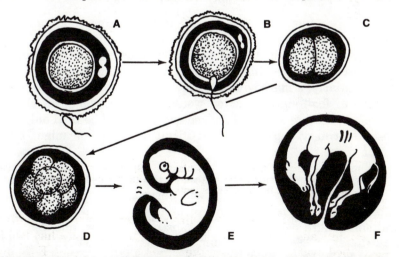

Development of the fetus: (A) Sperm approaches the ovum. (B) Sperm head enters and fertilizes the ovum, forming the zygote. (C) Cell division begins. (D) The zygote divides into a multi-cellular body. (E) Cellular differentiation results in the formation of the embryo. (F) The embryo develops into the fetus.

REPRODUCTIVE EXAMINATION

During the reproductive examination, the veterinarian identifies abnormalities of the reproductive tract which can affect the mare's breeding career. The mare's external genitalia should be checked for deformity or immaturity. The vulva should be positioned vertically, to avoid contamination by fecal matter. Thin mares, very old mares, and mares with high, level croups tend to have slanted vulvas. This conformation trait causes a poor vulval seal and permits air and fecal matter to enter the vagina resulting in infection and lowered fertility. Also, a damaged vulva predisposes the mare's vagina to infection and, subsequently, lowers her fertility. (Refer to the discussion on pneumovagina within **"Reproductive System."**) Underdeveloped mammary glands, a small vulval slit and an exaggerated clitoris sometimes indicate chromosomal abnormalities. (Refer to **"Sex Chromosomes."**)

With the aid of a speculum, the veterinarian checks for the presence of urine pooling within a deep, slanted vagina. A rectovaginal laceration (opening between the vagina and the rectum) may also be detected. This condition is sometimes caused during delivery when a foal's hoof penetrates the roof of the vagina and forces an opening into the mare's rectum. The laceration allows fecal contamination and subsequent vaginal infection. Although this problem can be corrected surgically, the chances are high that it will recur and, at best, a reproductive year is lost during the recuperative process.

If the speculum exam reveals cervical discharge or inflammation, the mare's breeding capacity may be limited by infection. When the cervix is relaxed to facilitate fertilization or delivery, strict hygiene is of utmost importance; the mare is highly susceptible to uterine infection. Similarly, a mare with a torn cervix would be almost impossible to keep in foal, since infection would probably trigger an abortion. Therefore, a mare with a damaged cervix should be bred only under close veterinary supervision. (Artificial insemination might benefit such a mare.)

Using rectal palpation, the veterinarian can check the condition of the uterus and ovaries. A large soft uterus after foaling, for example, indicates scarring or infection. An infantile (underdeveloped) uterus and small hard ovaries could be caused by chromosome abnormalities. (Refer to **"Sex Chromosomes"** under **CYTOGENETICS AND PROBABILITY**.) The veterinarian can also detect an improper position of the uterus, caused by stretched broad ligaments (these ligaments are connective tissue bands which control the position of the uterus during pregnancy). The presence of ovarian cysts or tumors should be considered potential hazards to reproductive performance.

DISPOSITION

Because the mare's attitude can influence her reproductive success, behavior should be taken into account during the selection process. Good manners and a quiet disposition enhance the breeding value of any mare.

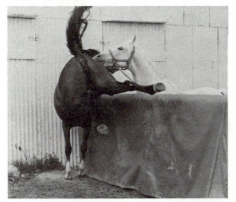

A negative response to teasing. This mare shows resentment to the stallion and is not in heat.

The mare's acceptance of the teasing stallion shows that she is in estrus. Note the mare's tail carriage, relaxed vulva and wide-set leg stance. The stallion is showing the Flehmen posture (upward curling of the upper lip).

The breeder should also analyze the mare's estrus behavior (e.g., through breeding records). Silent heat complicates the breeding process. Inconsistent behavior might endanger the stallion, and extreme sex drive or refusal during estrus could be caused by nymphomania or a form of intersex.

Mothering ability, another important broodmare characteristic, affects the health and behavior of the young foal; some dams are more protective of their offspring than others. Also, foals may imitate their dam's vices, such as wood chewing, kicking, etc.

BREEDING RECORDS

The mare's past reproductive performance can be examined by reviewing her foaling records. (The accuracy of these records should always be considered.) The age of the mare should be noted. The risk involved in acquiring a young maiden mare is much higher than that of an older mare with an established reproductive career. Obviously, a history of early fetal loss or retained placentas should be regarded cautiously. Ease or difficulty in settling should be considered with respect to the reputation of the stallion and the efficiency of the management.

The mare's gestation period (period between conception and birth) may indicate the longevity of her breeding career. Mares with longer gestation periods seem to have longer reproductive lives than mares with short gestation periods. (The average gestation period is 336 days, but varies from 310 to 370 days.) Any disease or lameness which might affect breeding performance should of course also be considered.

The breeder should check the mare's records for details which might indicate past reproductive success or failure:

1) Does she produce every year?

2) Will she settle with a foal at side?

3) How often does she return to cycle after covering?

4) Does the length of time between breeding and the next estrus indicate failure to conceive or early fetal loss?

UTERINE BIOPSY

Information on the presence of uterine infection, the ability of a mare to get in foal, and the possibility of fetal loss can be gained through uterine biopsy. This technical procedure eliminates the risk of keeping a mare who gets "in foal" but cannot maintain the pregnancy; a careful check of her history might reveal that the mare has never carried to full term.

Small snips of uterine lining (endometrium) are removed by means of surgical forceps introduced through the cervix. Laboratory analysis of these tissue samples can reveal the presence of a uterine infection (both mild and severe) and its causative organisms. Tests on the biopsy sample may also indicate the presence of tissue atrophy, immaturity, or excessive scarring (which causes poor attachment of the protective fetal membrane to the uterus and subsequent fetal death).

PEDIGREE

If a mare's athletic, show, or production performance have not been established, the reputation of close relatives should be considered. The pedigree should include performance records of the mare's sire, dam, grandsires, and granddams. As discussed previously, emphasis on distant relatives is misleading since, as individuals, they have little influence upon the mare's genetic makeup. Available information on the success or failure of half and full siblings would also aid the selection process. The pedigree will also indicate if the mare is inbred. Again, incomplete pedigrees, which place emphasis on distant ancestors, should be viewed cautiously.

Because the mare's tail-female line is located below the other pedigree lines, it is often called the bottom line:

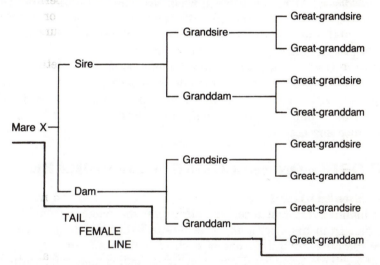

This line will often trace the ancestry of a mare to a superior broodmare (tap-root mare or "blue hen") who produced during the early history of the breed registry. Unless the mare was closely linebred (refer to **"Inbreeding and Outbreeding"**) to the tap-root mare, she will have inherited few genes from the famous ancestor.

PERFORMANCE RECORDS

The mare's overall appearance (breed type, femininity, conformation, and condition) is an important consideration during the selection process. Each of these characteristics is important to the mare's overall value and influences her ability to produce sound foals. Physical abnormalities that might be transmitted from generation to generation will be examined in **"Conformation"** under **"Inherited Abnormalities."** Lamenesses or blemishes which are caused by injury, malnutrition, etc. should be considered only with respect to the mare's physical ability to produce a foal. Because it is a common practice to place unsound fillies into a broodmare band, the breeder should closely scrutinize "injury tales" to determine whether the so-called "injury" resulted from an inherited weakness. A lameness due to carpal (knee bone) chips, for example, might have been caused by an inherited abnormality in leg set (e.g., calf knees).

Because the dam contributes to the genetic makeup of the foal, her athletic ability is just as important as the sire's achievements. If, however, she was placed in a reproductive role early in her career, production records (indicating achievements of her offspring) might be considered. Although a mare produces only a limited number of foals in her lifetime, her reputation can be enhanced by superior male offspring. If these stallions consistently produce quality foals, the mare's production record is strengthened. Should a significant percentage of the mare's granddaughters and grandsons be outstanding performers or producers, her own value — and the value of her offspring — will increase. Furthermore, the production records of a mare's full or half siblings can help the breeder to predict her ability to produce successfully.

THEORIES ON SELECTING A BROODMARE

Maternal Influence

Although every foal receives half his genetic makeup from his dam and half from his sire, many breeders feel that the dam contributes more (55-60%) to the nature of each of her foals than does the sire (40-45%). This additional contribution is called maternal influence and is attributed to three main characteristics.

First, the mare provides protection and nourishment to the fetus during gestation. The mare's health and the condition of her uterus influence the viability of the developing fetus (e.g., excessive scarring on the endometrium limits the passage of nutrients from dam to fetus and, therefore, results in prenatal malnutrition). Also, the size of the dam's uterus limits the mature size of her offspring, as indicated by Shire-Shetland crosses. (Refer to **"Environment and Heredity"** under **APPLYING GENETICS TO SELECTION.**)

Another characteristic which might contribute to maternal influence is mothering ability. Milking ability (an inherited trait) and maternal instinct should provide the foal with adequate nutrition and

San Diego Zoo Photo, Hartman's Mountain Zebra

Maternal Influence: During the suckling period, the foal may mimic its dam and later develop similar habits and attitudes.

protection. The tendency for foals to mimic their dam's behavior and attitude toward other horses is also of importance.

The relatively new science of cytoplasmic genetics also supports the concept of maternal influence. The cytoplasm is the cellular material which surrounds the nucleus and contains important protein production sites. Because the female's ovum contains far more cytoplasm than the male's sperm, the dam could provide more constructive material and, thereby, have more influence over the success or failure of her foals.

The mare's ovum (left) contains more cytoplasm than the stallion's sperm (right) — a possible contributing factor to maternal influence.

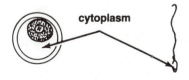

cytoplasm

Performance vs. Pedigree

When selecting a dam, some breeders place more emphasis on the mare's pedigree than on her own athletic and reproductive performance. Because the mare's reproductive career is limited to the production of a few foals, extensive athletic competition would limit production even further. These breeders believe that the performance of close relatives, illustrated by the mare's pedigree, should be the primary consideration during the selection process.

On the other hand, some breeders feel that the mare's athletic ability should be tested prior to placement within a band. These breeders may require that a mare have attained some measure of performance success to insure the commercial value of subsequent foals. An early breeding career is warranted by these breeders only when injury or illness limits athletic performance without affecting reproductive ability.

Fertility

As the horse evolved over the years, he persistently adapted to changes in his environment. Throughout his evolution, the horse continued to propagate his species without serious fertility problems. If a mare was a problem breeder, she simply did not breed; the hereditary problems which may have caused infertility stopped there. In the past, only the strongest stallions with the greatest sex drive sired the future generation. Nature was objectively practical, selecting only the most capable reproducers. Now that man has become provider and selector for the equine species, the reproductive role of today's horse has been drastically altered.

The national conception average (USA) for horses is presently only about 60% and dropping. Why? Inadequate breeding management, poor nutrition, infection, and limited selection against problem genes are only a few of the reasons for today's fertility problem.

The brevity of the mare's natural breeding season (late spring to mid-summer) is an important cause of the low conception average. Most horse registries designate January 1 as the birthday for all horses born within that entire year. This, coupled with early performance competition, encourages early conception for an earlier foaling date.

Certain steroids, used to build muscle or to suppress the mare's reproductive cycle during showing or racing, can also cause infertility problems in both the mare and stallion. Because these steroids are of the male type, they cause masculinizing effects in the mare. For example, she may cease to produce eggs (ovulate) for an extended period after training. This may significantly increase the time necessary for the mare's "let down." Although the administration of male steroids may cause increased libido in the stallion, sperm production is frequently inhibited.

In many respects, fertility is an inherited trait (influenced by many traits) which could (theoretically) improve through a careful long-term selection program. For most practical purposes, however, the breeder cannot afford to cull less fertile horses from a breeding herd.

Because fertility is also influenced by the environment, proper management can be an important tool for improving the conception rate within a breeding program. Occasionally, well-managed horse farms report conception rates of up to 85%. Good teasing programs seem to be especially helpful. Some breeders have found that fertility is enhanced by pasture breeding, but this is usually impractical and must be weighed against the chance that a valuable stallion might be injured by an uncooperative mare, and vice versa. There are many management practices to enhance breeding performance in the horse, but a discussion of these practices is beyond the scope of this book. (Refer to the complete text on this subject, **BREEDING MANAGE-MENT AND FOAL DEVELOPMENT.**)

INHERITANCE OF FERTILITY

Fertility is a trait which encompasses the coordinated action of many genetically controlled systems. In this respect, fertility is classified as an inherited quantitative trait.

Inherited defects of the reproductive tract are important causes of lowered fertility in both the mare and stallion. Tipped vulva, for example, is an inherited conformation trait which increases the mare's susceptibility to infections and, subsequently, lowers her fertility. Other reproductive problems, such as scrotal hernias and cryptorchidism, are also inherited. For a complete description of these, and other defects which might influence conception, refer to **"Reproductive System"** under **"Inherited Abnormalities."**

Inherited hormonal problems may also cause infertility in the mare and stallion. Hormones are the body chemicals necessary for certain reactions within the body. Everyone has felt the effects of adrenalin, a hormone which causes an exaggeration of muscular strength and heart rate, when they are frightened or angry. Other hormones cause the ovary to produce eggs, the testes to produce sperm, and the cervix to open. An attempt to describe all the functions of the reproductive hormones will not be made here, but it should be understood that, without this delicate and complicated balance of hormones, the reproductive systems will not function properly. A hormonal malfunction in the stallion may appear as lowered sex drive or lead to the production of weak, deformed sperm. The mare with a hormone imbalance is inconsistent; she may fail to produce an egg even though she readily accepts the stallion, or she may never accept the stallion.

The presence of lethal genes in the genetic makeup of the mare or stallion may also account for lowered fertility. A lethal gene is one which, by its presence on a chromosome, causes the death of the fetus (true lethal) or foal (delayed lethal). To understand why these lethal genes kill offspring without affecting their parents, a review of conception is necessary.

Most lethal traits are caused by recessive alleles. Therefore, in the heterozygous state, most lethals are not expressed. (The corresponding dominant allele masks the effects of the recessive lethal allele.) If a sire and dam both carry the same unexpressed lethal (i.e., if they are both lethal carriers), and if they both transmit this lethal allele to their offspring, the resulting fetus will die. (In the homozygous state, the recessive lethal allele is expressed.) True lethal genes usually cause early embryonic death or abortion.

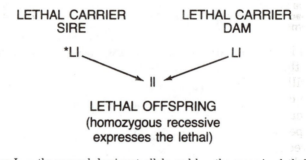

**LETHAL CARRIER
SIRE**   **LETHAL CARRIER
DAM**

*Ll   Ll

ll

LETHAL OFFSPRING
(homozygous recessive
expresses the lethal)

*Where L = the normal dominant allele and l = the recessive lethal allele.

Another inherited characteristic which contributes to lowered fertility is twinning. When the mare's ovaries ovulate two eggs simultaneously, there is a good chance that both ova will be fertilized within their respective oviducts. Since the mare's reproductive tract is not constructed to properly care for more than one fetus, twin conceptions are considered undesirable. If a mare conceives twins, competition for nutrients and space in the uterus often results in the death of one twin. This triggers uterine contractions to abort the dead fetus and can cause abortion of the live fetus at the same time. Sometimes, one fetus will mummify and remain in the uterus, preventing future conception. Even if the mare successfully supports the two fetuses, a chance remains that the foals will be delivered simultaneously causing dystocia (difficult or impossible foaling). Weakness and higher susceptibility to disease also result in a decreased survival rate for twins that do survive birth. Often one twin is remarkably smaller in

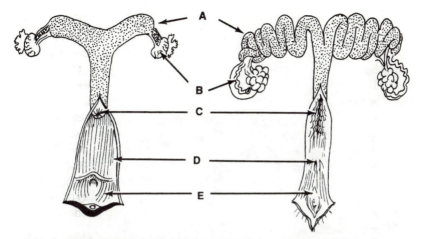

Animals which commonly give birth to more than one young, such as the sow (right), have proportionately longer uterine horns than the mare (left). Limited uterine space is a detrimental factor during twin gestation in the mare. (A) uterine horns, (B) ovary, (C) cervix, (D) vagina, (E) vulva.

size. Twins in horses are fraternal twins and are no more related to each other than any other full siblings. If the twins are male and female, the filly should mature to a normal fertile mare, unlike male and female cattle twins where the female twin will be sterile.

Approximately 2% of all conceptions are twins, but only half of these escape prenatal death. The draft breeds, and perhaps the Arabian breed, seem to show a higher frequency of live twins. Pony mares usually produce only one egg per cycle, suggesting that twinning has been naturally selected against in the smaller breeds.

An unusual case of successful twinning. These healthy twin Arabian foals were produced by the mare Ro Habibi on February 3, 1978.

courtesy of Lairmore Arabians, Wichita Falls, Texas

courtesy of Lairmore Arabians, Wichita Falls, Texas

The two-day-old twin foals, Rallye (bay filly) and Rockye (smaller black colt) keep pace with their dam.

courtesy of Lairmore Arabians, Wichita Falls, Texas

Twins at two weeks of age. Although limited uterine space usually causes a significant size difference between newborn twins, these twins soon reached comparable sizes due to similar genetic potential and adequate supplemental nutrition.

7

INBREEDING AND OUTBREEDING

For years, horsemen have questioned the advantages and disadvantages of inbreeding. Some breeders believe a limited degree of inbreeding is desirable (and even unavoidable), but that it is detrimental to productivity if carried to the extreme. Others claim that intensive inbreeding is important to breed improvement. Still others try to avoid inbreeding and remain within the boundaries of an outbreeding system. This section will examine the advantages and the disadvantages of the various inbreeding and outbreeding systems, relating each aspect to basic genetic principles.

The value of any breeding system depends upon careful analysis of a breeder's goals. It is not the purpose of this study to imply that one breeding system is preferable to another. Rather, it will explain why available finances, experience of the breeding manager, size of the breeding operation, and quality of the foundation stock should be carefully considered prior to the adoption of any breeding program.

Culling

Culling refers to the "elimination" of undesirable animals from a breeding herd. "Elimination" simply means that those individuals which do not meet the breeder's standards are removed from the breeding herd and sold, or used in another capacity. A colt that shows little promise as a sire may be gelded. In a large breeding operation, broodmares which fail to produce a significant percentage of average or above average foals (with respect to some desirable trait) might be culled from the breeding herd.

The extent to which the breeder can cull his stock depends on the size of the operation. A small operation might involve limited culling and, possibly, the use of a well-known stallion from another farm. On the other hand, a large scale breeder might cull 40-60% of his fillies and 90-95% of his colts, using only the best foals as future breeding stock.

Culling is a necessary tool for breed improvement. Without natural or artificial selection, each generation would be a direct reflection of their immediate ancestors. The average size, speed, and endurance within a herd would remain unchanged from generation to generation. Culling and selecting for above average performers and producers results in improvement within the breed.

SELECTION

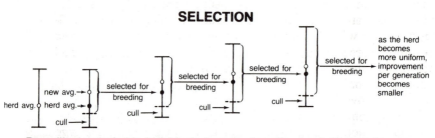

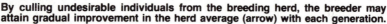

By culling undesirable individuals from the breeding herd, the breeder may attain gradual improvement in the herd average (arrow) with each generation.

Both genetics and the environment influence the overall character of a horse. A horse might be smaller than the average for his breed because of inherited genes (genetic variation) or because of a nutritional imbalance during his formative years (environmental variation). Variation is often caused by a combination of genetic and environmental aspects.

The success of the culling process depends on 1) the ability to distinguish between genetic and environmental variation, and 2) selecting only those individuals with superior gene types for breeding. Suppose a breeder checks the conformation and soundness of his recent foal crop and finds that most of the foals are acceptable, but that it also includes a few undesirable types. Since these foals live in a similar environment (similar nutrition, exercise, exposure to disease, etc.), their physical differences are probably caused by different gene types. If the breeder keeps only the most outstanding foals as prospective breeders, and if he evaluates the dams and sires of this foal crop to determine whether they have made a significant contribution to the breeding program, he will gradually improve his breeding herd.

This improvement is possible only if the quality of the parents is caused by genes rather than by better treatment, nutrition, etc. Should the variation (improvement or decline from generation to generation) be caused by environmental factors, obviously culling will have very little effect on herd improvement. (Refer to **"Environment and Heredity."**)

Inbreeding

Inbreeding is the mating of relatives. Its purpose is to fix certain traits or the influence of certain ancestors upon the progeny. This procedure varies in degree from intense closebreeding (i.e., the mating of close relatives) to mild linebreeding. Although inbreeding can be detrimental to fertility, vigor, and athletic ability within the offspring, it can also result in true-breeding strains of horses (i.e., horses that consistently pass important traits to their offspring). Because most breeds were formed by a process of inbreeding, the breeding of purebred horses is, by definition, a form of inbreeding. Some breeds are more inbred than others. (Degree of inbreeding depends on the number of common ancestors, how far back in the pedigree they appear, and how often each common ancestor occurs.)

From a genetic viewpoint, inbreeding results in an increase in the number of homozygous gene pairs in the offspring. Homozygous refers to a condition where two paired chromosomes have the same allele (gene type) at a corresponding point. Because two close relatives tend to have more of the same alleles (by virtue of inheritance) than two unrelated individuals, their mating provides a greater chance for identical alleles to be paired within their offspring. This increase in homozygosity is directly related to the appearance of both desirable and detrimental characteristics that were not necessarily apparent in the sire and dam. The following section will explain the principles involved in an inbreeding system and examine both its advantages and disadvantages.

DISADVANTAGES OF INBREEDING

When horses are inbred haphazardly, without culling of inferior stock, many undesirable traits may become predominant in their offspring. For example, the inbred horse's ability to resist disease (vigor) and his overall performance capacity are often depressed. The growth rate of the inbred foal, and the average mature size within the inbred herd, frequently decrease. Many breeders have reported that nonselective inbreeding is directly related to a depressed fertility level and to an increase in abortion and stillbirth. Physical deformities may also become more predominant in the inbred offspring. Some basic principles of genetics show why these traits are directly related to inbreeding.

When two unrelated horses are mated, the chances of unidentical alleles combining within the resulting embryo are high. On the other hand, mating close relatives increases the pairing of identical alleles (increases homozygosity). The effect of increased homozygosity is a decrease in the number of heterozygous gene pairs and, subsequently, a decline in heterosis (i.e., loss of vigor and fertility).

When inbreeding decreases heterozygosity, it eliminates many desirable heterozygous gene combinations and contributes to an apparent decline in the quality of inbred offspring. These desirable heterozygous gene combinations are the result of a genetic interaction called overdominance. Overdominance is simply the ability of heterozygous alleles to produce superior results (as compared to what each allele produces in its homozygous state). Although the reason for this allelic interaction is not clear, geneticists believe that its presence contributes to the overall quality of an individual. Therefore, as homozygosity increases within the inbred herd, physical quality controlled by overdominant alleles declines.

CHROMOSOME SECTIONS
ILLUSTRATING THE A LOCUS

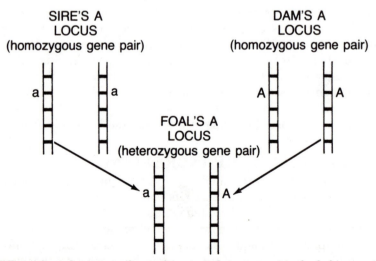

Effects of overdominance: the resulting trait (heterozygous) in the foal is superior to the trait (homozygous) in either parent.

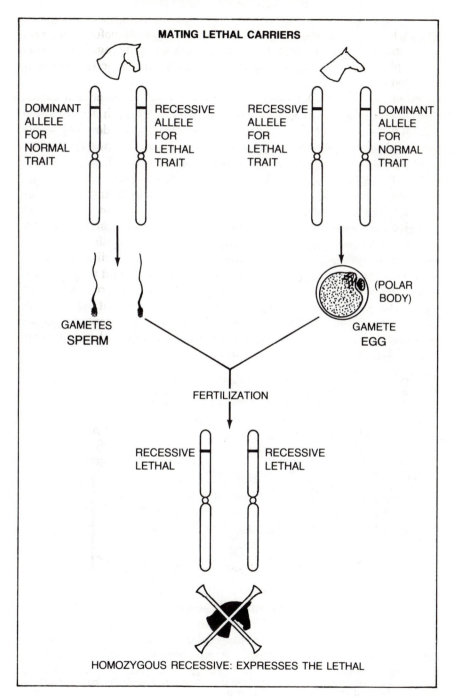

MATING LETHAL CARRIERS

DOMINANT ALLELE FOR NORMAL TRAIT

RECESSIVE ALLELE FOR LETHAL TRAIT

RECESSIVE ALLELE FOR LETHAL TRAIT

DOMINANT ALLELE FOR NORMAL TRAIT

(POLAR BODY)

GAMETES SPERM

GAMETE EGG

FERTILIZATION

RECESSIVE LETHAL

RECESSIVE LETHAL

HOMOZYGOUS RECESSIVE: EXPRESSES THE LETHAL

Many undesirable genes affecting the horse's overall vigor and fertility are recessive. Fortunately, they have no influence in the heterozygous state, since the effect of the recessive allele is completely hidden by the effect of the corresponding dominant allele. Because the overall effect of inbreeding is an increase in homozygosity, it increases the number of homozygous recessives. Hence, the effects of undesirable recessive genes begin to surface. The increase in abortion and stillbirth, associated with inbreeding, may be the result of certain lethal alleles in the homozygous state. If a mare and a stallion both carry the same recessive lethal allele, there is a one in four chance that they will produce an embryo with the homozygous lethal condition. Early fetal death due to lethal alleles might be interpreted as infertility or as an abortion.

Similarly, recessive genes which depress performance, size, or growth rate, or which cause deformities and undesirable conformation characteristics, may be expressed more frequently through an increase in homozygosity. In other words, inbreeding does not create undesirable traits, it exposes recessive alleles for hidden weaknesses which are present within the sire and dam.

Because successful inbreeding demands the culling of inferior breeding stock over many generations (to help eliminate some of the undesirable recessive genes from the herd), it may not be feasible for some breeders. Not only is the time factor impractical for most breeders, the intense culling often necessary may be an economic problem. Additionally, the traits which tend to surface within the inbred herd (such as depressed growth rate and decreased size) contrast sharply with what many breeders select for. Therefore, the breeder must be objective when the need to cull arises.

ADVANTAGES OF INBREEDING

Perhaps the greatest advantage of inbreeding is that it increases the prepotency of individuals within a herd and consequently helps to create distinct true-breeding strains or families. This prepotency (the ability of a stallion or a broodmare to stamp desirable characteristics upon their offspring with a high degree of predictability) is the result of the parent being homozygous for important desirable traits. When such a parent carries two identical alleles on corresponding points of a chromosome pair, he transmits that allele to the same chromosome point within his offspring. If two such parents are mated, the offspring will always possess the same desirable trait. Therefore, as inbreeding increases homozygosity, it also enhances prepotency. (This is advantageous only if the parents are homozygous for *desirable* traits.)

As mentioned previously, inbreeding exposes certain weaknesses within the inbred herd. Uncovering these undesirable traits can be an important tool for the overall improvement within a large breeding program. By setting certain selection guidelines, and by carefully eliminating inbred individuals which show inherent weaknesses, the breeder can slowly remove many undesirable recessive genes from his herd. He will find that vigor and fertility are actually improved when inbreeding is accompanied by careful selection.

A successful inbreeding program requires good foundation stock and severe culling over many years. For this reason, inbreeding is usually practiced by experienced breeders who operate large farms for the production of superior prepotent breeding stock. It can also be used to establish breeds, or true-breeding types, with respect to certain characteristics, such as color or size.

CLOSEBREEDING

A breeding system which uses extreme inbreeding, such as mating between siblings or between parents and offspring, is referred to as closebreeding. The detrimental effects of inbreeding (such as a decrease in vigor, fertility, athletic ability, and size) are usually exaggerated in a closebreeding system. This is especially true when average breeding stock are used and little culling has been implemented. The economic importance of possible lowered conception rates, barren mares, and infertile stallions should be considered by those who practice closebreeding. On the other hand, closebreeding members of a superior family, and culling individuals with apparent weaknesses from each generation, can produce exceptional prepotent foals.

Closebreeding can produce extremely good, or extremely poor, results. Success and failure depend on factors such as planning, foundation stock, emphasis on culling, and completeness of records (pedigrees, performance records, etc.). Haphazard closebreeding could be very detrimental to the overall quality of the resulting offspring. To avoid disaster, a careful study of the merits and weaknesses of the breeding stock should precede a closebreeding program. Only the most outstanding mares and stallions can be used with any degree of safety in a long term closebreeding program.

Closebreeding is a valuable tool in genetic research, since it quickly exposes hidden gene types that an individual carries. Because of its extreme nature and the chance that it may suddenly cause undesirable effects in the offspring, closebreeding is not often used by horse breeders. Some breeders, who operate large and well-organized breeding programs, might utilize closebreeding if they progeny test their

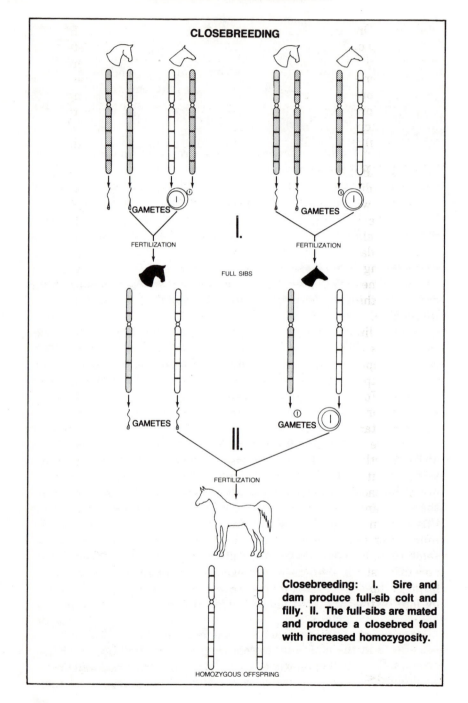

CLOSEBREEDING

I.

GAMETES

FERTILIZATION

FULL SIBS

GAMETES

FERTILIZATION

II.

GAMETES

GAMETES

FERTILIZATION

Closebreeding: I. Sire and dam produce full-sib colt and filly. II. The full-sibs are mated and produce a closebred foal with increased homozygosity.

HOMOZYGOUS OFFSPRING

stallions. (One method of progeny testing a sire is to mate him to a large group of his own daughters. A study of the offspring determines whether he carries undesirable genes hidden in the heterozygous state.) After a stallion proves that he is of superior gene type, the experienced breeder may choose to continue the closebreeding to increase the prepotency of future breeding stock. He should do this only after balancing each of the economic factors (possible losses and possible assets) and deciding that the possible gains are worth the risks.

LINEBREEDING

Linebreeding, the most conservative form of inbreeding, is usually associated with slower improvement and limited risk of producing undesirable individuals. It can involve matings between closely or distantly related horses, but it does not emphasize continuous sire-daughter, dam-son, or brother-sister matings. The main purpose of linebreeding is to transmit a large percentage of one outstanding ancestor's genes from generation to generation without causing an increase in the frequency of undesirable traits often associated with inbreeding.

Because linebreeding is not based strictly on mating closely related individuals (with very similar gene types), it does not necessarily cause a rapid increase in homozygous gene pairs. Consequently, it will not expose undesirable recessive genes as extensively as closebreeding. For this reason, linebreeding is generally a safer inbreeding program for most breeders (especially those who must rely on producing marketable foals).

Intensive inbreeding (and resulting increased homozygosity) is often directly related to an increase in the expression of many undesirable traits. Therefore, the linebreeder should carefully study pedigrees for each prospective mating and determine if, and how closely, the mare and stallion are related. (Refer to **"Relationships"** within this section.) By following certain guidelines, the breeder can limit inbreeding (and, therefore, homozygosity) within his herd. At the same time, he may increase the influence of a common ancestor upon an entire strain or family. For example, a quality sire is bred to a large group of unrelated mares. Four superior offspring are produced by four of the unrelated mares:

COLT A ⟨ OUTSTANDING SIRE I / MARE #1

COLT C ⟨ OUTSTANDING SIRE I / MARE #3

FILLY B ⟨ OUTSTANDING SIRE I / MARE #2

FILLY D ⟨ OUTSTANDING SIRE I / MARE #4

These half-siblings (A, B, C & D) are selected by the breeder as future breeding stock. When half-sibs are mated, their offspring are inbred:

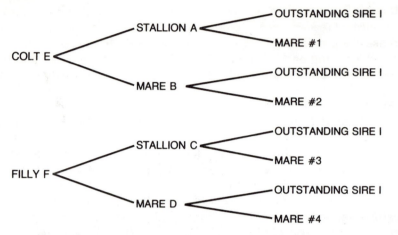

Colt E is inbred since his sire and dam (A & B) may have contributed some of the same gene types. In the same way, filly F is inbred. If the breeder decides that E and F do not show any inherent weakness, he may choose to mate them together. He may find that some of their offspring closely resemble the original outstanding sire, even though the stallion is, at this point, only a great-grandsire:

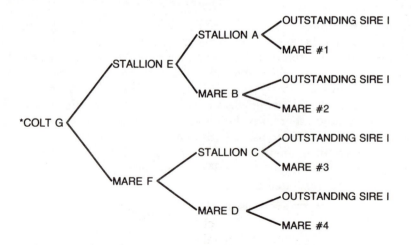

*This is only one example of linebreeding. There are actually many types of linebreeding programs.

Because colt G is related to the original sire four different ways, he may have inherited many genes from his famous great-grandsire. The colt is not heavily inbred, however, since the common ancestor is removed by three generations. If colt G eventually proves to be an outstanding performer and prepotent sire, the breeder may choose to repeat the entire linebreeding process, using G as the common ancestor for future linebred offspring (especially if the original sire is no longer available). On the other hand, the breeder may continue linebreeding to the original sire, until a large population of closely related, but only slightly inbred, progeny are produced. These linebred offspring might provide the breeder with foundation stock for a uniform strain of horses, based on the quality of one superior sire.

Careful selection of breeding stock and sound breeding management are essential in any breeding operation, but the key to successful linebreeding is an ability to monitor the degree of homozygosity within each of the linebred offspring. The following section will explain methods of determining the degree of relationship between two horses and the degree of inbreeding (increased homozygosity) within linebred or closebred offspring.

Relationships

When an individual inherits genes from another, or when two individuals inherit some of the same genes from a common ancestor, they are said to be related. The degree of their "relationship" and the percentage of genes they share can be measured by a fairly simple process. For example, each horse receives half of his genes from his sire and half from his dam. He is, therefore, 50% related to each of his parents:

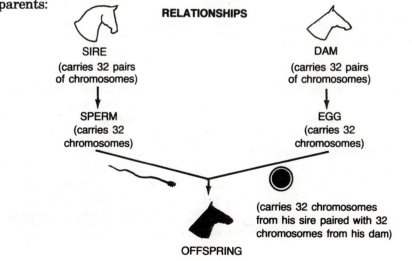

RELATIONSHIPS

SIRE
(carries 32 pairs
of chromosomes)

DAM
(carries 32 pairs
of chromosomes)

SPERM
(carries 32
chromosomes)

EGG
(carries 32
chromosomes)

(carries 32 chromosomes
from his sire paired with 32
chromosomes from his dam)

OFFSPRING

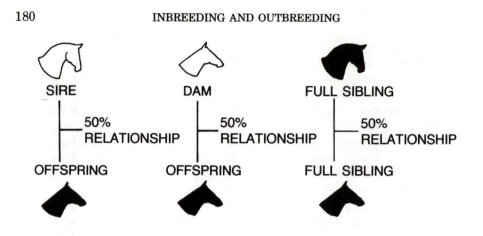

The relationships between collateral relatives (brothers, sisters, cousins, and so on) can also be determined mathematically. This relationship table shows the results of several such calculations.

COLLATERAL RELATIVE	RELATIONSHIP
FULL BROTHER	50%
FULL SISTER	50%
HALF BROTHER	25%
HALF SISTER	25%
FRATERNAL TWIN (NOT IDENTICAL)*	50%
FIRST COUSIN	12.5%
SECOND COUSIN	3.13%
AUNT	25%
UNCLE	25%

*Identical twins have not been reported in horses. If they did occur, their relationship would be 100%.

Because the relationship shows how closely two horses are related, it also shows the percentage of gene types which they might have in common. Relationships are, therefore, a good indication of how well one horse's records (production, athletic, etc.) can be used to predict the value of another related horse.

Relationships are also important guidelines for an inbreeding system, since they indicate the possibility of similar gene types from two related horses combining within their inbred offspring. For the benefit of the serious breeder, the following discussion will examine these calculations and their practical applications.

Inbreeding Coefficient

A horse's inbreeding coefficient measures the number of homozygous gene pairs which he carries because his sire and dam are related. In other words, the sire and dam received identical alleles from a common parent or ancestor and then passed some of these on to their offspring, who became a little more homozygous than either of his parents. The breeder who wishes to tackle a relatively complicated mathematical formula can monitor the amount of homozygosity within his herd, and thereby limit the appearance of undesirable traits caused by homozygous recessive alleles.

For the purposes of this discussion, a five generation pedigree will be considered. Although common ancestors within the sixth and seventh generations, and the degree of inbreeding within the foundation stock of specific breeds, also affect a horse's inbreeding coefficient, computations involving these factors are complicated and tedious. For this reason, an analysis of five generations is presented to provide the breeder with a practical means of determining fairly accurate inbreeding coefficients.

To explain the basic principles which define the mathematic formula, a simple problem will be examined step by step. Two important aspects to remember throughout this process are:

a) The purpose is to measure the number of identical alleles two parents could have possibly contributed to their offspring.

b) Each parent contributes just one allele from each allelic pair. Therefore, the computation is based on the principle that related parents probably contribute only half of their common alleles to their offspring (i.e., inbreeding coefficient = half the relationship between the two parents).

Example

Step 1) To determine the inbreeding coefficient for one horse, the breeder must first check its five generation pedigree. The number of common relatives (e.g., ancestors that are repeated more than once) should be noted. Common relatives are the primary concern, since they cause increased homozygosity by introducing their alleles into the pedigree more than once.

Step 2) In the example given on the following page, the sire and dam are half sibs with only one common ancestor (D) in five generations. The relationship between half sibs is 25% (refer to the table under **"Relationships"**).

Five Generation Pedigree for horse X:

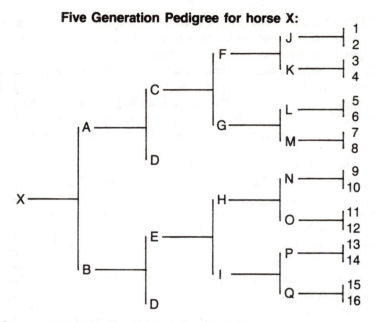

One common relative: **D**   (Note: relatives beyond a common ancestor are not counted.)

Step 3) The 25% relationship between Sire A and Dam B indicates that 25% of A's alleles are identical to 25% of B's alleles (by virtue of inheritance). Therefore, when A contributes a copy of one chromosome from each chromosome pair (and, hence, one allele from each allelic pair) to his offspring, chances are good that he will contribute half of the alleles that are also carried by B (i.e., half of the inbred alleles). Similarly, B contributes half of her inbred alleles to her offspring. Consequently, the foal receives approximately half of his parents' inbred alleles (i.e., half of their relationship percentage). (Note: This is only an approximation; precise computation requires the use of correction factors which will not be presented at this time.)

Step 4) A simplification of the inbreeding coefficient formula is:

> Inbreeding Coefficient = The Sire-Dam relationship
> divided by 2

For the purposes of this example (one common ancestor appearing only twice in the five generation pedigree), this formula can be applied without significant error:

Inbreeding Coefficient for X = The relationship between A & B
divided by 2
I.C. for X = Half sib relationship divided by 2
I.C. for X = 25% divided by 2
I.C. for X = 12.5%

Step 5) A comparison of inbreeding coefficients should consider that certain breeds are more inbred than others. An inbreeding coefficient of 15% might be moderate in one breed, and high in another. Thoroughbreds have been estimated to be 8.4% inbred (by virtue of foundation stock and history). An inbreeding coefficient of less than 8% would be low, and 8-15% moderate, when compared to the breed as a whole. Because the Quarter Horse has been designated as approximately 1.2% inbred, individuals with inbreeding coefficients over 10% would be considered highly inbred compared to the breed average.

BREED	INBREEDING COEFFICIENT*
Thoroughbred	8.4% (1941)
Quarter Horse	1.2% (1958)
Standardbred	4.4% (1940)
American Saddle Horse	3.2% (1935)
Clydesdale	6.2% (1925)

*values determined 1925-1958. Because inbreeding coefficients for a specific breed will fluctuate from year to year, these values are only approximations.

Another important consideration is the presence of undesirable recessive alleles within the breeding herd. Designations of low, moderate, and high for each inbreeding coefficient depend on the quality of the breeding stock, and should be carefully determined by the individual breeder. The Arabian breeder, for example, might be especially concerned about the appearance of CID within his inbred herd. (CID is a homozygous recessive condition which causes a deficiency of important antibodies and early death of the foal.)

The previous example was presented to explain the basic principle of computing inbreeding coefficients. Most mating systems are much more complicated, however. An inbred horse frequently has more than one common ancestor, and the relationships within his pedigree are often complex. When the possibility of error increases due to the presence of more than one common ancestor or because an ancestor appears more than twice in the five generation pedigree, another formula must be used. This computation requires an understanding of probability and simple algebra. (Refer to **"Population Genetics"** under **CYTOGENETICS AND PROBABILITY.**)

WRIGHT'S FORMULA: $\sum [\, (\tfrac{1}{2})^{n_1 + n_2 + 1} (1 + F_A) \,]$

Description:

1) $[\, n_1 + n_2 + 1 \,]$ designates the number of generations between a common ancestor and the horse whose inbreeding coefficient is questioned: n_1 = the number of generations between the horse's sire and the common ancestor

n_2 = the number of generations between the horse's dam and the common ancestor

1 = the generation between the horse and his parents

2) $(\tfrac{1}{2})$ is the probability that an allele is transmitted from parent to offspring. (Remember that there are two alleles at each locus and that the parent contributes one allele from each allelic pair — a one out of two chance for each allele.)

3) $(\tfrac{1}{2})^{n_1 + n_2 + 1}$ indicates the total probability that an allele is passed through the pedigree to the horse in question. For example, in the following pedigree, X is related to common ancestor C by three generations at the second generation level, and four generations at the third generation level.

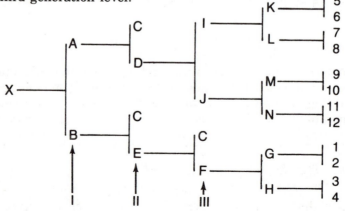

For the first path (2nd generation level), $(\frac{1}{2})^{n_1 + n_2 + 1} = (\frac{1}{2})^3$; $(\frac{1}{2})^3$ means that $(\frac{1}{2})$ is multiplied three times:

$$
\begin{aligned}
(\tfrac{1}{2})^3 &= (\tfrac{1}{2})\,(\tfrac{1}{2})\,(\tfrac{1}{2}) \\
&= 1/8 \\
&= .125 \\
&= 12.5\%
\end{aligned}
$$

For the second path (3rd generation level), $(\frac{1}{2})^{n_1 + n_2 + 1} = (\frac{1}{2})^4$; $(\frac{1}{2})^4$ means that $(\frac{1}{2})$ is multiplied four times:

$$
\begin{aligned}
(\tfrac{1}{2})^4 &= (\tfrac{1}{2})\,(\tfrac{1}{2})\,(\tfrac{1}{2})\,(\tfrac{1}{2}) \\
&= 1/16 \\
&= .063 \\
&= 6.3\%
\end{aligned}
$$

The sum of the results from the two paths is:

$$
\begin{aligned}
.125 + .063 &= .188 \\
&= 18.8\%
\end{aligned}
$$

The figure 18.8% is the probability that X received homozygous alleles from his common ancestor C.

4) Multiplying $(\frac{1}{2})^{n_1 + n_2 + 1}$ by a correction factor, $1 + F_A$, compensates for any previous inbreeding in the common ancestor. F_A equals the common ancestor's inbreeding coefficient. This is sometimes taken as the average inbreeding coefficient for the breed as a whole. Supposing that F_A equals 4%, the corrected inbreeding coefficient would be:

$$
\begin{aligned}
.188\,(1 + .04) &= .188\,(1.04) \\
&= .1955 \\
&= 19.6\%
\end{aligned}
$$

5) If more than one common ancestor is involved, this entire process must be repeated for each ancestor. The symbol $\sum$ indicates that the resulting probabilities for each common ancestor should be added to determine the final answer.

Determining the inbreeding coefficient for every member of an inbred herd is obviously a long, tedious procedure. Nevertheless, it is important that the breeder keep a close watch over the degree of inbreeding and, subsequently, the degree of increased homozygosity within his herd. Calculators and computers can be helpful aids to this process, especially when using a seven generation pedigree for greater accuracy. A rough "common sense" estimate could also be made by carefully studying a horse's seven generation pedigree and eliminating the correction factor from the formula.

INBREEDING COEFFICIENT

RELATIONSHIP (PARENTS)	INBREEDING COEFFICIENT
SIRE TO DAUGHTER	25%
DAM TO SON	25%
FULL BROTHER TO FULL SISTER	25%
HALF BROTHER TO HALF SISTER	12.5%
GRANDSIRE TO GRANDDAUGHTER	12.5%
GRANDDAM TO GRANDSON	12.5%
UNCLE TO NIECE	12.5%
AUNT TO NEPHEW	12.5%
COMMON GRANDPARENT (FIRST COUSIN ONCE REMOVED)	3.1%
FIRST COUSINS	6.3%
ONE COMMON GREAT-GRANDPARENT	1.6%
TWO COMMON GREAT-GRANDPARENTS	3.1%
THREE COMMON GREAT-GRANDPARENTS	4.7%

Outbreeding

Outbreeding involves several systems of mating either unrelated or at least distantly related individuals (those which have no common ancestors within seven generations, for example). Outbreeding is perhaps the most common breeding system used within the horse industry today. A breeder might search for a stallion that is complementary with, but unrelated to, his broodmare. He hopes to bring in "new blood," so to speak. Actually he is introducing new gene types into his herd and causing important genetic changes. In the following discussion, these changes will be studied and the five major types of outbreeding will be examined: outcrossing, linecrossing, grading, crossbreeding, and species hybridization.

From a genetic viewpoint, outbreeding involves the mating of individuals which have significantly different genotypes. Because these unrelated horses contribute different gene types to many of the corresponding chromosome positions, the number of heterozygous gene pairs increases within their offspring. This genetic condition causes the exact opposite effect of inbreeding. For example, the undesirable recessive genes (remember, most genes for abnormal or lethal traits are recessive) are hidden by the heterozygous state. Hence, the overall fertility, vigor, and athletic ability within the outbred herd often improve as the heterozygosity increases. This improvement, which is an important characteristic of outbreeding, is referred to by geneticists as heterosis, or hybrid vigor.

Another important cause of heterosis in the outbred herd is overdominance. As discussed previously under **"Inbreeding Disadvantages,"** overdominance simply means that certain traits are especially outstanding when their controlling alleles are in the heterozygous state. Because outbreeding causes an increase in the number of heterozygous gene pairs, it allows for more expression of the overdominant traits.

Not to be overlooked is the fact that outbreeding increases genetic variability, as opposed to inbreeding (which increases homozygosity and decreases genetic variability). Horses within an inbred herd may have only one kind of allele for many of their corresponding genetic points (loci). These horses contribute similar genotypes to each of their offspring. Consequently, each generation shows little difference from the herd average, and the breeder can no longer improve the genetic characteristics within his homozygous breeding stock.

When genetic variability is exhausted, outbreeding will help to revive it by increasing the heterozygosity within the herd. By mating an unrelated stallion with a highly inbred broodmare, the possibility of different gene types combining within the offspring is very high. Hence, the heterozygosity and consequent heterosis of the outbred offspring increase. These offspring will someday produce another generation that exhibits far more genetic variation than did the original breeding herd.

OUTCROSSING

Outcrossing is defined as the mating of unrelated animals within the same breed. It is often used by horse breeders, since it is a relatively safe and effective breeding system. Outcrossing is safe because it is improbable that two unrelated horses will carry the same undesirable gene types (alleles). It is, therefore, unlikely that undesirable recessive alleles will become homozygous to any great extent in the outbred offspring. (Members of a breed are usually inbred to some degree, but inbreeding is assumed to be insignificant unless it occurred in the first seven previous generations.) Because outcrossing does not usually expose these undesirable recessives, it is an especially safe system to use when breeding average stock. (These types are more likely to carry undesirable recessive alleles.) Although outcrossing cannot guarantee improvement, it keeps the herd safe from the loss of fertility and vigor which is common in an inbreeding system.

Outcrossing is especially helpful in improving quantitative traits, those which are controlled by many genes (e.g., fertility, vigor, growth rate, size, speed, etc.). (Inbreeding is especially detrimental to quantitative traits.) Outcrossing may also be used to introduce new traits into a herd or, perhaps, to complement one horse's weaknesses with desirable traits (and, hence, desirable genes) from an unrelated horse. Selecting breeding stock which are complementary but unrelated is the key to successful outcrossing.

Outcrossing is also a helpful tool within an inbreeding program, since an occasional outcross helps mask undesirable recessive alleles in the inbred herd. A breeder who operates an inbreeding program may discover that he has created a homozygous recessive condition for some undesirable trait (when a large percentage of foal crop inherits that characteristic). He decides to search for an unrelated stallion exceptionally strong in that area to outcross with his mares. By introducing different genes into the herd, he increases the heterozygosity of the next generation. Consequently, the undesirable trait may become less apparent (the recessive allele might be hidden by a corresponding dominant allele).

LINECROSSING

Linecrossing is a type of outcrossing in which members from two distinct family lines are mated. A non-market breeder may work for years to develop an inbred line that is especially strong in conformation and soundness. To improve several other below average traits, he may cross his line with another distinct family that is notably strong in the areas he desires. (As long as the parents are not inbred to the same ancestors, two inbred horses can be mated and still produce linecross offspring.)

Linecrosses that consistently produce quality foals are sometimes referred to as "nicks." Market breeders may take advantage of an established "nick" to insure yearling market values. Although linecross offspring may excel in athletic ability and/or appearance, their prepotency is often low (due to the increase in heterozygosity). Therefore, "nicked" foals do not always become consistent producers. Their offspring, however, may benefit from the merits of overdominance and hidden recessive alleles.

GRADING

Grading is the outbreeding of a superior stallion to below average mares, in the hope of raising the quality of each mare's offspring. Although extreme grading is common in the cattle industry, it is not often used with horses.

Grading might be helpful to some horse breeders, depending on their situations and goals. The breeder should keep in mind that the mare is capable of limiting the quality of her foal, regardless of the quality of its sire. Therefore, the foal sired by a famous stallion, and out of a below average mare, might not be valuable enough to compensate for the stud fee.

Because of this, many market and non-market breeders place certain requirements on the quality of a stallion as compared to the limitations of the mare he is to be bred to, and vice versa. There are three important guidelines used by many breeders which might be helpful in establishing your own "rule of thumb:"

1) Stud fee should be limited to half the market value of the mare.

2) Stud fee should not exceed 50%, or preferably 20%, of the average market value of the sire's offspring.

3) The stallion's siring record should be protected by allowing him to receive only mares of at least comparable quality.

CROSSBREEDING

Crossbreeding, the mating of horses from different breeds, is useful in several respects. Special performance horses, such as hunters and park horses, are produced by crossing distinct breed types. The heavy hunter, for example, might be the product of a Thoroughbred-Cleveland Bay or a Thoroughbred-draft horse cross. The Half-Arab park horse, which results from an Arabian-Saddlebred cross, is another example of crossbreeding for performance purposes.

Crossbreeding may also be used to produce heterosis, the sudden increase in vigor and fertility caused by a sudden increase in heterozygosity. Because horses from separate breeds usually carry very different genotypes, crossbreeding causes a more extreme form of heterosis than that achieved by linecrossing or outcrossing. (The possibility of each parent contributing identical alleles to their offspring is remote.) Heterosis from crossbreeding often appears as a sudden improvement in physical characteristics, such as size, endurance, disease resistance, etc.

New breeds are sometimes established by crossing members of two or more breeds and carefully inbreeding the original crossbred offspring. Crossbreeding initiates the desired change, while inbreeding increases the ability of each generation to breed "true to type."

SPECIES HYBRIDIZATION

The most extreme form of outbreeding is species hybridization, or the crossing of members of two separate species. For example, the horse, the ass, and the zebra are equine species which can be crossed to produce foals with extreme heterosis, or hybrid vigor. Crosses between the jack and the mare produce the mule, a remarkably vigorous individual often capable of withstanding harsher conditions than either parent. It should be noted that crosses between different species do not always result in outstanding hybrids. The stallion-jenny (female donkey) cross, for example, produces the less popular hinny. The zebra-horse, zebra-pony, and zebra-ass hybrids are also examples of species hybridization. Although breeders originally sought to hybridize the disease resistance of the zebra with the manageability of the domestic species, the unusually marked hybrids have not been established in a working capacity.

Species hybridization resulting from a zebra-donkey cross. The resulting hybrid (zeonkey) shows characteristics of both species.

Crosses between the jack (Equus asinus) and the mare (Equus caballus) result in the hybrid mule. The mule's size is usually a reflection of the size of its dam (e.g., crosses between a jack and a Shetland mare produce smaller offspring than do crosses between a jack and a draft mare).

Although it is possible for the chromosomes from these equine species to mix and eventually form healthy offspring, the hybrid individuals are usually unable to reproduce. The reason for this sterility is that each species has its own characteristic number of chromosomes. The horse has 32 pairs of chromosomes, while the ass has 31 pairs. When the jack is mated to the mare to produce a mule, he contributes 31 chromosomes (via the sperm), while she contributes 32 (via the ovum). Successful gametogenesis (formation of egg and sperm which carry half the original chromosome number) within these hybrids might be inhibited by chromosome incompatibility. (During normal gametogenesis, similar chromosomes must pair, or synapse. An odd number of chromosomes, or the presence of incompatible pairs of chromosomes, may alter normal gamete formation. Refer to **"Mitosis"** and **"Meiosis"** under **CYTOGENETICS AND PROBABILITY.**)

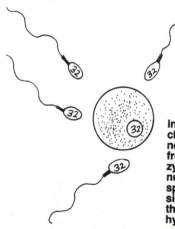

In the modern horse (Equus caballus), 32 chromosomes from the sire's sperm normally combine with 32 chromosomes from the dam's ovum, forming (within the zygote) the species specific chromosome number, 64. When two different equine species are bred, chromosomes from the sire and dam do not usually pair evenly in the hybrid offspring. For this reason, the hybrid offspring are usually sterile.

Male mules are always sterile, and mule-mule crosses have never been successful. There have, however, been reports of horse-like foals being produced by female mules. One source estimates that the occurrence of fertile female mules is approximately one in 200,000. The reasons for sterility and occasional fertility are not clear. In each reported case of fertility, the female mules conceived only when bred to horses. This suggests that their ovaries somehow produced ovum containing only their horse-related chromosomes. When bred to stallions, these mules produced horse-like foals. Some scientists question the validity of these reports, arguing that so-called "fertile female mules" were never identified as mules by chromosome analysis. (Refer to **"Karyotyping"** under **CYTOGENETICS AND PROBABILITY.**)

8

ENVIRONMENT
AND HEREDITY

Without proper management, a genetically superior horse will not reach his full genetic potential. Therefore, his inherent ability could be overlooked, and his genes never introduced into future generations. For this reason, the effects of the environment upon genetic potential must be considered by anyone who selects a horse, whether for breeding or performance purposes. The ability to distinguish between genetic and environmental variation is an important step in selecting genetically superior horses.

An example of how environmental factors can mask genetic potential. Note the muscle wasting of this emaciated mare.

Genetic Variation

Throughout evolution, the prehistoric horse adapted to changes in his environment and developed specialized features which helped him to survive. These changes were possible because nature culled the undesirable types (weak, diseased, sterile, etc.) from the population. Only those individuals that were fertile and able to survive their harsh environment were selected as breeding stock. Because horses that varied in gene type also varied in physical fitness, some were better equipped for survival than others. Hence, nature's selection-culling process depended on genetic variation which, in turn, depended on mutations and the normal reproductive process.

A mutation is defined as a sudden change in the structure of a gene or chromosome. (The reasons for these changes will be discussed in **CYTOGENETICS AND PROBABILITY**.) These structural changes can affect the expression of certain traits. For example, a mutation may cause a severe physical deformity or a desirable change in appearance. On the other hand, a mutation may have no external effect, but may cause a fatal dysfunction of an internal organ, such as the heart or kidney. Although mutations may have been an important source of genetic variation throughout evolution, they are not significant for the improvement of horses today. Why? Mutations are very rare. When they do occur, their effects are often undesirable.

Genetic variation can also be caused by the normal reproductive process, when chromosomes multiply and cells divide to form the egg or the sperm (refer to **"Mitosis"** and **"Meiosis"** under **CYTOGENETICS AND PROBABILITY**.) During this process, both the egg and sperm receive 32 chromosomes, which is half the horse's characteristic chromosome number. There are literally thousands of genes located on each chromosome, and there are usually many possible mutations (alleles) of the same gene for each corresponding chromosome point. Therefore, the combination of egg and sperm (fertilization) usually produces an embryo with a unique set of genes, allowing the resulting offspring to vary, to some extent, from both of his parents and any full siblings. (Exception: If the parents are extremely inbred along the same lines, they may be capable of producing offspring with genotypes similar to their own. Refer to **"Inbreeding and Outbreeding."**) Any physical difference between parent and offspring that is caused by new allelic combinations is referred to as normal genetic variation.

Environmental Variation

The effects of "man-made" environmental aspects, such as nutrition, management, conditioning, training, etc., can cause horses of comparable genetic capacity (genotype) to vary in appearance (phenotype). Qualitative traits, such as coat and eye color, are controlled by only a few genes. These traits are often highly heritable and are not significantly affected by environmental pressures. For example, the spotted pattern on the tobiano Paint is caused by the horse's genotype, not by diet or treatment. On the other hand, quantitative traits are controlled by many different genes and are influenced very heavily by the environment. During any selection process, quantitative traits (growth rate, size, behavior, fertility, soundness, etc.) should be carefully studied with respect to possible environmental pressure. As a case in point, a man may try to sell an unsound mare to a breeder by excusing her fault on the basis of an "old injury." It is the breeder's responsibility, to himself and to the breed as a whole, to determine whether or not the "injury" is the result of an inherent weakness. The interaction between genetics and environment should be carefully considered:

—Is the horse's poor appearance the result of poor management or of a below average genotype?

—If an unsoundness was caused by an injury, did an inherited weakness in conformation increase the horse's susceptibility to breakdown?

—If the weakness was inherited, will the merits of the breeding prospect outweigh the risk that he may pass the undesirable trait on to his offspring?

—On the other hand, is the horse outstanding in his herd because he carries superior genes or because he received better treatment than the horses around him?

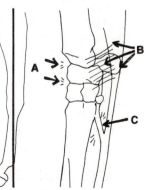

Interaction of genetics and the environment. Calf knees, an inherited forelimb deviation, increases the affected horse's susceptibility to injury. Heavy work causes stress upon the carpal joint (A), the carpal ligaments (B) and the inferior check ligaments (C). When coupled with physical exertion, the inherent weakness may result in fractures or chips of the anterior carpal bones.

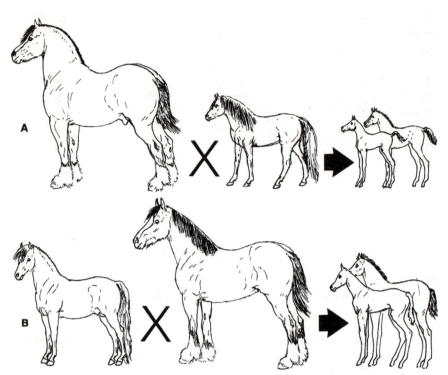

Effects of maternal environment upon genetic potential. Shire-Shetland crosses have been made to show how important the size of the dam's uterus is to the size of her offspring. The Shire stallion-Shetland mare crosses (A) produced offspring that reached a much smaller mature size than did the Shetland stallion-Shire mare crosses (B).

Judging Genetic Potential

During selection, it can be misleading to compare horses without considering possible environmental influence. Every horse is the result of both his genes and his environment:

GENOTYPE + ENVIRONMENT = PHENOTYPE

Perhaps the best guide for anyone in search of genetically superior breeding or performance stock is the following:

1) Use the selection guidelines presented within this and the preceding section.

2) Consider the horse's appearance, performance, and production with respect to his environmental background (nutrition, training, management, etc.).

3) Analyze the outstanding or defective traits under consideration by checking the heritability estimates for each. Remember that the variation of a quantitative trait (i.e., its strength or weakness) is probably, to a great extent, environmental variation. (Refer to the heritability table in **"Breed Improvement Through Applied Genetics."**)

INFERTILITY

Environmental Factors	Inherited Factors
1. Difficulty in detecting estrus	1. Twinning
2. Breeding season (length of daylight)	2. Inherited lethals (lethal roan, lethal white, etc.)
3. Breeding time (foal heat, day of estrus)	3. Chromosomal abnormalities (i.e., pseudohermaphrodite)
4. Breeding technique (hand, pasture, AI, etc.)	4. Pneumovagina
5. Condition (health, injuries, etc.)	5. Inherited abnormalities of the reproductive tract (i.e., scrotal hernia)
6. Handling	6. Inherited hormonal imbalance (e.g., hypothyroid mare)
7. Effects of anabolic steroids (training aids)	7. Inherited sperm quality and quantity
8. Infections	8. Bilateral cryptorchid
9. Overuse of the stallion	
10. Fever (effects upon sperm production)	
11. Nutrition	
12. Certain drugs during gestation period	

DEPRESSED GROWTH RATE

Environmental Factors	Inherited Factors
1. Nutritional imbalance (growing foal may need up to 50% more protein and minerals than the mature horse) 2. Nutritional deficiency (mare's milk may not supply suckling foal with adequate nutrients) 3. Parasites (may result in digestive disturbances and poor appetite) 4. Lack of exercise (stall confinement)	1. Quantitative trait (controlled by many genes)

DEPRESSED SIZE AT MATURITY

Environmental Factors	Inherited Factors
1. Nutritional deficiency or imbalance through maturity 2. Size of the dam's uterus 3. Testosterone (the male hormone allows greater muscle development but inhibits growth of long bones; therefore, the stallion is slightly shorter and more muscular than the gelding with similar genes for size)	1. Quantitative trait (controlled by many genes)

BEHAVIOR

Environmental Factors	Inherited Factors
1. Poor nutrition 2. Castration or spaying 3. Influence of other horses (especially the dam) 4. Early experiences 5. Training 6. Daily handling 7. Illness, etc.	1. Genes for intelligence 2. Inherent learning ability 3. Hormonal imbalance (may cause abnormal behavior such as that seen in nymphomania)

9

BREED IMPROVEMENT THROUGH APPLIED GENETICS

Many factors influence the degree and rate of progress achieved through a particular selection and breeding program. Such factors include the breeder's goals, the selection methods used, and the types of traits involved. While the selection methods discussed in this chapter are of value to all breeders, the information is perhaps most useful when applied to larger breeding herds.

The breeder should have definite goals in mind when selecting breeding stock. These goals should be directed toward creating a type of horse whose characteristics will be useful and desirable in both the immediate and distant future. Since it takes several horse generations for a selection and breeding program to have any lasting effect on an equine population, changing goal directions (modifying the breed standard, for example) will greatly slow progress.

In addition, breeders should select only for important traits which have substantial influence on the horse's conformation, performance, productivity, etc. The more traits considered in selection, the less improvement can be made for each individual trait. For example, it may not be too difficult to find a stallion that meets a certain minimum speed standard. Yet if certain standards for height, weight, color, disposition, etc. are also required, it becomes more difficult to find a horse that fulfills all of the requirements.

Selection Methods

Another important factor which affects the amount of progress that a breeder can attain with a breeding program is the selection practice used. There are different methods that allow the breeder to select with emphasis on either one trait at a time, or more than one trait simultaneously. The number of traits chosen, their relative economic importance, and emphasis placed on each will all have an effect on progress. The ability of the breeder to recognize and observe each horse's individual merit for important characteristics and to weigh their relative importance is necessary. The breeder must especially realize that increased culling for one specific trait will reduce the degree of culling that can be done for all other traits (otherwise, all horses might eventually be eliminated from consideration). Strict selection for a certain minimum height, for example, would reduce the degree of selection for speed, since many fast, but short, horses would have been culled.

Many horsemen instinctively use one or more of these selection methods without realizing it. For example, the horseman who selects a horse by 1) establishing minimum requirements and 2) appraising each horse on the basis of its individual strengths and weaknesses is using a combination of minimum culling level and selection index methods. The horseman who selects for the leopard Appaloosa pattern until his entire herd is leopard, and then selects for conformation, is using the tandem selection method. All of these methods will be examined within this chapter.

NUMBER OF TRAITS

As the number of traits considered important increases, the progress made in improving a particular trait decreases. In addition, breeders should keep in mind that some traits are more difficult to perfect than others. For example, it may be fairly easy to introduce a desirable trait into a herd if it is due to the action of one dominant allele. If the trait is the result of a recessive allele, or of many gene pairs, it may take several generations to establish it within the herd. For example, it may take a large number of generations to establish a herd of miniature horses when beginning with normal-sized horses, since height is due to the action of many gene pairs. When selecting for these quantitative traits (traits controlled by many gene pairs), progress is slow, and consequently, a high level of consistency in the breeding and selection program is necessary.

The following formula can be used to compare the rate of progress when selecting for one trait vs. the rate when selecting for two or more traits simultaneously. Let n designate the number of traits under consideration:

$$\text{Rate of progress is affected by a factor of} \quad \frac{1}{\sqrt{n}}$$

Example: If speed, temperament, color and height are selected for at the same time (four traits simultaneously), the rate of progress for a particular trait is only half as fast as it would have been had selection over the same time period been for that trait alone.

$$\text{One trait: } \frac{1}{\sqrt{n}} = \frac{1}{\sqrt{1}} = \frac{1}{1} = 1$$

$$\text{Four traits: } \frac{1}{\sqrt{n}} = \frac{1}{\sqrt{4}} = \frac{1}{2}$$

TANDEM METHOD

In the tandem method of selection, the breeder culls all horses that do not have one specified trait, and then begins to work on perfection of a second trait. Since this method does not consider the fact that the culled horses could have many other desirable characteristics, it is not an ideal selection method. In addition, progress is made with only one trait at a time, so overall improvement of the herd with respect to any other trait is normally negligible. An example of tandem selection would be choosing Appaloosa breeding stock strictly on the basis of speed, without considering conformation or color. Next, when the breeder is satisfied with speed, he may decide to try to fix an attractive blanket pattern in his running stock.

When using this method, the breeder should first select for the most important trait. This characteristic should be selected for during several generations, until it is at an acceptable level within the herd. For instance, if speed were considered the most important trait, the breeder would select for speed first when choosing breeding stock. Less important traits, such as height, amount of bone, etc., should be selected for in later generations, and without as much emphasis.

The following formula can be used to compare the rate of progress when selecting for one trait vs. the rate of progress when selecting for two or more traits separately, each a certain predetermined number of generations. Let n designate the number of traits under consideration:

Rate of progress is
affected by a factor
of $\dfrac{1}{n}$

Example: Suppose a breeder allows six generations to improve his herd with respect to three separate traits. If he selects for the first trait the first two generations, the second trait the second two generations and the third trait the third two generations, he is using the tandem method of selection. Since n=3 traits, the average improvement for the first trait is only $\dfrac{1}{n}$ or $\dfrac{1}{3}$ the average improvement that could have been obtained if only that trait had been selected for during the six-generation period. In the same way, the second and third traits show only $\dfrac{1}{3}$ the optimum rate of progress.

An important problem with tandem selection is that some breeders select for a relatively subjective and unimportant characteristic as the primary trait, or select for a minor trait for too many generations. For example, a breeder may select for a trait such as disposition at first, when characteristics such as conformation and speed are more important to his race horse breeding program.

When using the tandem method, the breeder must also consider whether the traits are positively or negatively associated. With positive association, selecting for one characteristic automatically selects for the other as well. An example might be heart score and racing ability. Some studies indicate that as heart score increases, racing ability improves. (Refer to **"Performance Selection Characteristics."**) With negative association, an increase in one trait results in a decrease in the other. For instance, suppose that speed and endurance are negatively associated. If this were true, it would be impossible to breed a brilliantly fast horse that also had stamina since, as speed increased, stamina would decrease. On the other hand, if the horse were bred for endurance, he would not have speed. Obviously, if two traits are negatively associated, optimum results will never be achieved. (The reasons for positive and negative association are discussed in **CYTOGENETICS AND PROBABILITY**.)

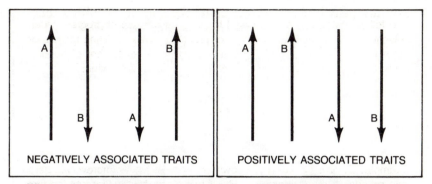

When traits are negatively associated (left), one will increase as the other decreases. When positively associated (right), the traits increase and decrease together.

MINIMUM CULLING LEVEL

Using the minimum culling level method (also known as the independent culling level method), the breeder sets certain standards for traits such as conformation, performance, and productivity. Any horse that fails to meet one or more standards is culled. This method is superior to tandem selection in that more than one trait is considered at one time, resulting in faster *overall* progress. (As mentioned earlier, it is important for the breeder to restrict the number of desired traits to meet a realistic goal. Once the desired traits have been chosen, it may be more practical to select for several traits together than to select for only one characteristic at a time.) The minimum culling level has the disadvantage of culling individuals that could be greatly superior in several traits and only slightly inferior in one.

	Stallion A	Stallion B	Stallion C
Speed Index	92	106	94
height	16 hands	15-3½ hands	16-1 hands
money earned	$116,000	$273,000	$135,000

If the breeder set minimum culling levels of Speed Index - 90, height - 16 hands, and money earned - $100,000, Stallions A and C would both be acceptable, while Stallion B (a superior performer and money earner) would be culled.

Culling heavily for one trait limits the amount of culling that can be practiced for other important traits. For example, the palomino color can be selected for by the minimum culling level method but, since it is due to a heterozygous allelic pair (designated as $C\ c^{cr}$), the palomino-palomino cross can only result in approximately 50% palomino offspring:

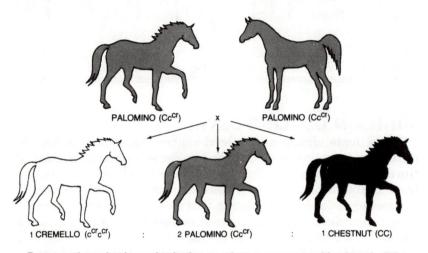

Because the palomino color is due to a heterozygous combination of alleles ($C\ c^{cr}$), the palomino-palomino cross will result in an approximate 1:2:1 ratio of cremello:palomino:chestnut offspring.

With the minimum culling level selection method, approximately half of the offspring are automatically culled since they do not meet the color requirement. This greatly reduces the amount of selection that can be practiced for other traits.

The advantage of the minimum culling level method of selection is that culling is possible at a young age, or as soon as it becomes apparent that the horse will not meet one of the standards. In addition, this type of selection is used for qualitative traits (such as color) in which the horse either possesses the trait or does not. As with the other methods of selection, those traits chosen should be as economically important and as highly heritable as possible. The breeder should also consider the number of animals that must be kept, since the minimum culling level is most effective when only a small percentage of horses are retained (a high selection pressure). In addition, only a small number of traits should be considered (preferably only one or two) when using this method.

SELECTION INDEX

The best way to select breeding stock is with the use of a selection index. With this method, the breeder determines which traits should be considered in selection and the relative importance of each trait. In this way, the breeder weighs the index to select for the characteristics that are most important to his program.

The animals that score highest according to the selection index are retained for breeding purposes. This system allows the breeder to keep horses that are superior in most traits and inferior in only a few. The amount of weight given each characteristic depends on the economic value of that trait, its heritability (there is no point in heavily weighing a trait that is only slightly affected by heredity), and any genetic and environmental correlations between traits. Other factors the breeder should consider when determining the index include the need for each specific trait in the herd, and the amount of deviation from the average that is acceptable for each characteristic.

For example, a breeder could decide on the ten traits that are most important to his breeding program. These could include characteristics such as general conformation, disposition, and height. Then, each trait is assigned a point value according to its relative importance. General conformation may be given 20 points out of 100, disposition 8 points, etc. More vital traits are given a larger number of points, while lesser traits are given fewer. A characteristic such as tail set, for instance, may be worth only 2 points. When the breeder is comparing horses during selection, he can judge them against this ideal of 100 points, giving each horse under consideration a score and retaining only those individuals who score higher than some minimum value (e.g., 80). If a horse has perfect conformation, for instance, he receives 20 points, while a horse with a minor flaw may get 17. Many experienced successful horsemen use a system similar to this when they are evaluating a horse without ever realizing that they are using a system at all.

The breeder must also consider the heritability of each trait under evaluation. A highly heritable trait should be given more points than a less heritable one, since the latter is more easily improved through proper management and a good environment.

Although the selection index is the most efficient method of selection, it does have some disadvantages. For instance, a horse cannot be completely "scored" and compared to other horses until all animals under consideration have matured enough to allow realistic appraisals of their conformation, performance, etc. In addition, qualitative traits (which have distinctly different phenotypes) cannot be included in an index. For example, a certain color may be required for a breed, but this factor cannot be indexed since a horse is either the desired color or he is not. A horse could not be 50% palomino, and still receive half of the maximum points that a palomino would get. Instead, the breeder should use a minimum culling level for qualitative traits.

As with all methods of selection, the index is only as accurate as the information that determines it. Errors may be made by the breeder when deciding the relative importance of characteristics and when estimating heritabilities. The traits that are emphasized should be those that will be needed and in demand both immediately and in the future. A mistake in estimating heritability could result in emphasis on, and selection for, a trait that could more easily be improved through changes in environment. However, if the selection index method is correctly used, it will be much more effective than either the minimum culling level or the tandem method.

The following formula can be used to compare the efficiency of the selection index and minimum culling methods. Let n equal the number of traits under consideration:

$$\text{Rate of progress using the selection index method} = \sqrt{n} \times \text{Rate of progress using the minimum culling method}$$

Example: If speed, height, money earned and stakes won are considered by the breeder, n=4 traits. If he uses the selection index method, his rate of progress will be $\sqrt{4}$, or 2 times the rate of progress he would obtain using the minimum culling method:

$$\text{Rate of progress using the selection index method} = \sqrt{n} \times \text{Rate of progress using the minimum culling method}$$

$$= \sqrt{4} \times \text{Rate of progress using the minimum culling method}$$

$$= 2 \times \text{Rate of progress using the minimum culling method}$$

Selection Pressure

The degree of selection intensity, or selection pressure, also affects the progress the breeder can expect. If the pressure is high, the breeder selects only a small number from the available horses, thereby using only superior animals in the breeding program. If the pressure is low (e.g., due to economic circumstances), the breeder may use at least some breeding stock that is not much better, if at all, than average. For example, most breeders practice much more stringent selection (higher selection pressure) of a stallion than a mare. If the top 5-10% (high selection pressure) of all stallions and the top 60% of all mares are used, the breeder has essentially selected the top third of the available horses. If a high degree of selection pressure is used in choosing breeding stock, the breeder will initially make greater and more rapid progress in improving the herd. With low selection pressure, the rate of improvement is slowed and many generations may be required to meet predetermined goals.

Heritability Estimates

Heritability estimates reveal how closely the phenotype of a horse resembles its genotype. It is the percent of what the breeder selects for (in the stallion and mare) that is actually expressed in the resulting offspring. Therefore, it tells how much improvement can be made by a given amount of selection pressure. The amount of progress that can be obtained in a breeding program is definitely limited by heritability of the important traits. The higher the heritability of a characteristic is, the greater the amount of variation (selection differential) due to heredity and the less due to environment. Selection differential is the amount of difference between selected horses and the average of the horse population from which they came. If the heritability is high, then the offspring will closely resemble the parents regardless of the environment. If heritability is low, most of the selection differential may be due to environmental factors (such as nutrition, handling, etc.). In this case, parents and offspring raised under similar conditions may tend to resemble one another even though the trait under consideration is only lowly heritable.

Heritability estimates consider the degree of heredity (as opposed to environment) that affects a particular trait, usually in terms of a percentage. These estimates are obtained by determining how closely relatives resemble one another for a certain trait, as compared to non-relatives. For example, there should be greater overall similarity within groups of half siblings than between unrelated horses. The greater the degree of similarity for a particular trait, the higher the heritability of that trait. Because heritability estimates show the percentage of phenotypic variation that is due to heredity, they can be used to calculate the amount of progress that can be expected when selecting for a certain trait.

HERITABILITY ESTIMATES*

Running speed	40%
Body weight	25-30%
Cannon bone circumference	28%
Pulling power (at 4 yrs.)	14%
(at 5 yrs.)	24%
Walking speed	41%
Trotting speed	43%
Height at the withers	26%

*According to Dusek, Ericksson, Johansson, Rendel and Varo.

PREDICTING BREEDING VALUES

The breeding value of a horse can be approximated by determining the difference between his performance and the average performance of the overall population, and then multiplying that difference by the trait's heritability estimate. For example, suppose that a breeder wanted to determine the breeding value of a race stallion. He would first determine the difference between the stallion's racing ability and the average race horse's ability:

Stallion's racing ability   =   20 sec. to run
a quarter mile

Overall racing average   =   25 sec. to run
a quarter mile

At the quarter mile, the stallion is 5 seconds faster than the average race horse. With this information, the breeder should then consider the heritability for racing ability. If racing ability is 40% heritable, the breeder should realize that it is probable that only 40% of the stallion's 5-second difference can be transmitted to his offspring. Therefore, his breeding value is calculated as follows:

$$5 \text{ sec.} \times 40\% = 5 \text{ sec.} \times .40 = 2 \text{ sec.}$$

The stallion's breeding value for racing ability at the quarter mile is 2 seconds. This means that his offspring are estimated to be at least 2 seconds faster than the average race horse at the quarter mile.

(This, of course, also depends on the dam's breeding value and assumes that she is an average race mare.)

The higher the heritability of the trait under consideration, the more accurate the breeding value, and the less need there is to use information other than the individual horse's own record to predict his breeding value. If the heritability of a trait is less than 30%, however, the individual's record plus that of relatives must be used to predict the breeding value. Characteristics that are lowly heritable will not improve significantly through selection. (These traits will respond readily to superior environment, however.) Characteristics that are over 30% heritable can be improved quite rapidly by selection. A specific horse's racing ability, for example, can be predicted (approximately) by comparing the performance of his sire and dam with respect to the relatively high heritability of racing ability.

For example:

If a Standardbred stallion trots the mile in 2:00 (2 minutes), and a mare trots the same distance in 2:10 (2 minutes, 10 seconds), their offspring should (theoretically) complete the mile somewhere between 2:00 and 2:10. Although the stallion is 10 seconds faster than the mare (the selection differential is 10 seconds), he can only contribute 40% of this difference to his offspring (assuming that trotting speed is 40% heritable). Therefore, the offspring's racing speed is predicted as follows:

1) 10 seconds × 40% = 10 seconds × .40 = 4 seconds

2) The stallion's 4-second contribution allows his offspring to be 4 seconds (:04) faster than their dam: (2:10 − :04) = 2:06 (2 minutes, 6 seconds).

3) Thus, the offspring should be genetically capable of trotting the mile in 2:06.

If the offspring, a mare, is bred to a second Standardbred stallion of the same caliber as the first (i.e., trots the mile in 2 minutes), the selection differential will now be only 6 seconds since the stallion is 6 seconds faster than the mare. Of that 6-second differential, 40% can be contributed to the second generation offspring.

1) 6 seconds × 40% = 6 × .40 = 2.4 seconds

2) The sire's 2.4-second contribution allows his offspring to be 2.4 seconds (:02.4) faster than the dam: (2:06 − :02.4) = 2:03.6 (2 minutes, 3.6 seconds).

3) Thus, the offspring should be genetically capable of trotting the mile in 2:03.6.

It is important to note that as the selection differential decreases for each succeeding generation, the amount of improvement possible for each generation will unfortunately also decrease.

Quantitative Inheritance

Quantitative inheritance refers to the inheritance of traits that are controlled by many pairs of genes (polygenic inheritance). The exact number of gene pairs involved, however, is seldom known. There is a continuous variation from one phenotypic extreme to the other, with no sharp distinction between phenotypes. The measurement of a quantitative trait in a large population results in a bell-shaped curve, with most individuals falling near the middle. The environment has much more effect on quantitative traits than on qualitative traits, making the heritability of these characteristics more difficult to measure. Examples of quantitative traits include racing speed, milk production, pulling power, growth rate and weight.

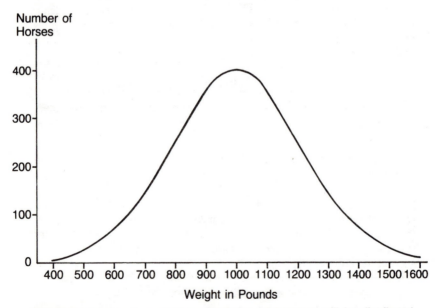

Bell-shaped curve. In a random population, phenotypes will be distributed along an average curve. In this example, weight is distributed along the bell-shaped curve with the largest number of horses corresponding to an average weight (1000 lbs.).

Using the bell-shaped curve, individual horses can be compared to the average value for a specific group of horses. For example, the performance of a race horse could be compared to the average performance of other race horses of his same class. The differences between horses in a group or population can be described in terms of variation from the average. These differences are important to the breeder because variation between horses allows the improvement of the breed by selection of the superior animals. If there were no variation between horses, there would be no opportunity to improve genetically.

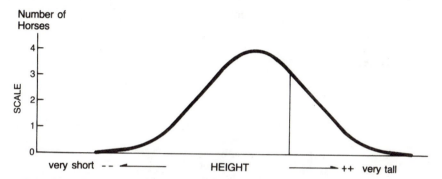

Selection of superior animals illustrated by the bell-shaped curve. If an increase in height is desired, the breeder could compare each individual to the population average (high point on the curve) and select only those to the right of the designated point (e.g., to the right of the vertical line.)

A geneticist's approach to estimating the number of gene pairs affecting a quantitative trait would be to cross two individuals of opposite extremes, then intermate the offspring to produce a second generation. The number of offspring in the second generation that are of an extreme phenotype would be used to determine the number of gene pairs that influence the trait. Suppose that horses of two extreme heights are mated (Shetland ponies and Shire draft horses, for instance), and that the offspring from this cross are intermated. If there are 32 offspring in the second generation, of which 2 are very small and 2 are very large, the number of gene pairs involved is estimated at 2. This is computed by dividing the 32 by 2 (2 is the number of one of the extremes), which equals 16. Looking at the following chart, it can be seen that a 1:16 ratio of extreme to nonextreme offspring implies that there are 2 gene pairs involved. The larger the number of gene pairs involved, the slower the progress, and the longer the period of time over which progress can be made.

DETERMINING THE NUMBER OF GENES AFFECTING A TRAIT

Conversion factors: n = the number of alleles affecting a trait

$\dfrac{n}{2}$ = the number of loci

$(\frac{1}{2})^n$ = the fraction of the population that expresses the extreme phenotype

Loci	Alleles	Conversion	Fraction of the population with the extreme phenotype
1	$n=2$	$(\frac{1}{2})^2$	$\dfrac{1}{4}$
2	$n=4$	$(\frac{1}{2})^4$	$\dfrac{1}{16}$
3	$n=6$	$(\frac{1}{2})^6$	$\dfrac{1}{64}$
4	$n=8$	$(\frac{1}{2})^8$	$\dfrac{1}{256}$
5	$n=10$	$(\frac{1}{2})^{10}$	$\dfrac{1}{1024}$
10	$n=20$	$(\frac{1}{2})^{20}$	$\dfrac{1}{1,048,576}$
20	$n=40$	$(\frac{1}{2})^{40}$	$\dfrac{1}{1,099,511,627,776}$

The many genes that affect a quantitative trait are sometimes referred to as polygenes ("poly" = many). The genetic interaction between polygenes is usually additive, meaning that each gene contributes slightly to the trait. For example, if H designates the gene for tall, and h is the gene for short:

$$
\begin{aligned}
HHHHHHHH &= \text{very tall} \\
HHHHhhhh &= \text{intermediate} \\
hhhhhhhh &= \text{very short}
\end{aligned}
$$

The possible genotypes could range from one extreme to the other, with corresponding changes in the phenotypes (e.g., *HHHHHHHh* = fairly tall, *HHhhhhhh* = fairly short, etc.).

Qualitative Inheritance

Qualitative traits are those that are due to the action of only one or very few gene pairs. There is a definite and sharp distinction between phenotypes. Coat color is an example of a qualitative characteristic, because a horse is either a certain color (such as chestnut) or he is not. The environment has very little, or no, effect on this type of trait. Some qualitative characteristics, such as coat color, are easy to measure with visual appraisal, although shades of color may be affected slightly by sunlight, nutrition or season.

When describing a qualitative trait controlled by one pair of genes, the following terms are used:

1) A homozygous individual has two genes of the same type (two identical alleles). For example, a horse with a white chin spot has two matching recessive alleles for white chin spotting.

2) The term homozygous recessive is sometimes used to describe individuals with recessive qualitative traits.

3) Homozygous dominant refers to individuals possessing two identical dominant alleles (for example, two B alleles in a black horse).

4) A heterozygous individual has two different types of alleles (for example, a black horse with Bb). If one allele is completely dominant over the other, the heterozygous individual will phenotypically appear the same as the homozygous dominant (for example, the heterozygous Bb horse and the homozygous dominant BB horse are both black).

With qualitative traits, there are six possible crosses, each of which produces certain ratios of offspring with and without the trait, depending on whether the parents are dominant or recessive, homozygous or heterozygous. For example, the greying gene (G) is completely dominant over the nongreying gene (g). Therefore, a grey horse may be either GG or Gg, while a nongrey horse is always gg. The six possible crosses of grey and nongrey horses are as follows:

1.	G	G
g	Gg	Gg
g	Gg	Gg

3.	G	G
G	GG	GG
G	GG	GG

5.	g	g
G	Gg	Gg
g	gg	gg

2.	G	G
G	GG	GG
g	Gg	Gg

4.	g	g
g	gg	gg
g	gg	gg

6.	G	g
G	GG	Gg
g	Gg	gg

1.)

HOMOZYGOUS GREY   ×   HOMOZYGOUS NON-GREY
(GG)   (gg)

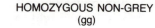

HETEROZYGOUS GREY
(Gg)

2.)

HOMOZYGOUS GREY   ×   HETEROZYGOUS GREY
(GG)   (Gg)

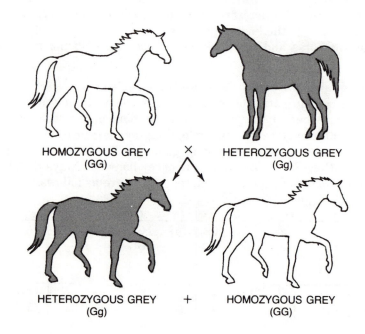

HETEROZYGOUS GREY   +   HOMOZYGOUS GREY
(Gg)   (GG)

3.)

HOMOZYGOUS GREY × HOMOZYGOUS GREY
(GG)   (GG)

HOMOZYGOUS GREY
(GG)

4.)

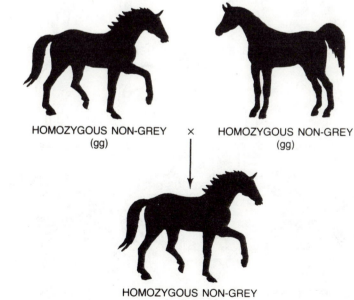

HOMOZYGOUS NON-GREY × HOMOZYGOUS NON-GREY
(gg)   (gg)

HOMOZYGOUS NON-GREY
(gg)

5.)

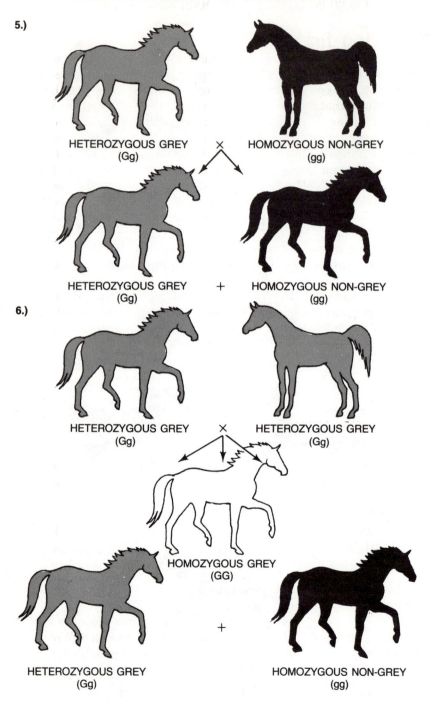

HETEROZYGOUS GREY   ×   HOMOZYGOUS NON-GREY
(Gg)   (gg)

HETEROZYGOUS GREY   +   HOMOZYGOUS NON-GREY
(Gg)   (gg)

6.)

HETEROZYGOUS GREY   ×   HETEROZYGOUS GREY
(Gg)   (Gg)

HOMOZYGOUS GREY
(GG)

HETEROZYGOUS GREY   +   HOMOZYGOUS NON-GREY
(Gg)   (gg)

Selection for a Trait

When introducing a trait into a herd, the amount of progress that can be expected is partially determined by the type of gene action involved. It is easy to select for, or against, a dominant gene because every animal that carries the gene exhibits the trait. For this reason, dominant genes which cause undesirable traits are easily removed from the population. It is difficult to distinguish between homozygous dominant animals (which carry the dominant on both chromosomes) and heterozygous animals (which have a recessive allele that is masked by the effects of a corresponding dominant allele). Heterozygous individuals transmit hidden recessive alleles through their heterozygous offspring; these undesirable recessive alleles are eliminated from a population only by eliminating all heterozygous individuals. Because recessive traits are much harder to remove from a herd, many of the present day abnormalities and lethals are controlled by recessive genes.

In many instances, the only way to differentiate between a homozygous dominant and a heterozygous horse is by their respective offspring. If the horse is homozygous dominant for a certain trait, his offspring will always inherit that trait:

$$\text{Black} = BB \text{ or } Bb$$
$$\text{Liver} = bb$$
$$\text{and } B \text{ is dominant over } b$$

Homozygous dominant				
BB	$\times$	bb	$=$	Bb
Black	$\times$	Liver	$=$	all black offspring
Homozygous dominant				
BB	$\times$	Bb	$=$	$1\ BB : 1\ Bb$
Black	$\times$	Black	$=$	all black offspring

On the other hand, the heterozygous individual will not necessarily transmit the trait to his offspring.

Heterozygous				Homozygous recessive offspring show a different form of the parent's trait.
Bb	$\times$	Bb	$=$	$1\ BB : 2\ Bb : 1\ bb$
Black	$\times$	Black	$=$	3 black : 1 liver

Even if only one recessive is produced, the horse is proven to be a carrier of the recessive gene and therefore heterozygous. However, it takes many offspring to prove that an animal is homozygous dominant. For example, if a sire is mated to five recessive mares and produces no recessive offspring, there is a 95% chance that the stallion is homozygous dominant (according to the laws of probability). On the other hand, there is still a 5% chance that the stallion is heterozygous. If the sire was bred to known heterozygous mares, 11 matings with no resulting recessive offspring would be necessary to achieve this 95% accuracy rate.

NUMBERS AND KINDS OF MATINGS REQUIRED TO TEST A MALE TO DETERMINE THAT HE IS NOT A CARRIER OF A RECESSIVE GENE

Kinds of Females	Probability Test Females are Carriers*	No. of Matings Required Without the Production of Homozygous Recessive Offspring at Odds of:	
		95/100	99/100
1. Homozygous recessive	1.00	5	7
2. Known heterozygotes**	1.00	11	16
3. Phenotype normal and both sire and dam known heterozygous	0.67	17	26
4. Phenotype normal and at least one parent known heterozygous	0.50	24	35
5. Test mated to own daughters	0.50	24	35

*Females in groups 3, 4 and 5 must be a random sample with no single female counted twice.
**Known heterozygotes may be of two types — those that have produced at least one homozygous recessive offspring or those having one parent known to be homozygous recessive.

BACKCROSSING

Introducing a desirable trait is most effectively accomplished by backcrossing. In this practice, a stallion with the desired trait is crossed to a herd of mares without it. All female offspring carrying the trait are bred to stallions of the original breed that lack it. Once again, the resulting female offspring that carry the trait are retained for breeding. This continues for several generations.

Introducing a Dominant Trait

To illustrate selection for a dominant trait, assume that tobiano spotting is completely dominant to solid color, and that the following breeding program was implemented to introduce tobiano spotting.

$$
\begin{array}{lcl}
\text{tobiano spotting} & = & TT \text{ or } Tt \\
\text{solid-colored} & = & tt
\end{array}
$$

TOBIANO STALLION   ×    SOLID-COLORED MARE
(TT)   *(tt)*

TOBIANO OFFSPRING
(Tt)

The female tobiano offspring are then mated to solid-colored stallions:

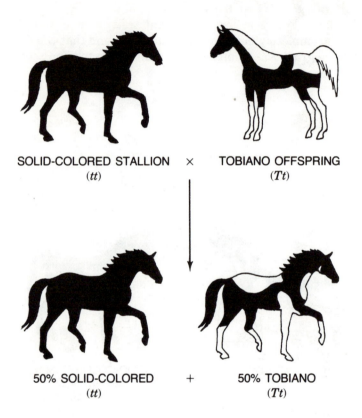

SOLID-COLORED STALLION   ×   TOBIANO OFFSPRING
(*tt*)   (*Tt*)

50% SOLID-COLORED   +   50% TOBIANO
(*tt*)   (*Tt*)

The tobiano offspring from that cross are then crossed back to solid colored horses, and the pattern is repeated.

Backcrossing is sometimes used to introduce a characteristic into a herd while retaining the original breed type. For example, an Arabian stallion is bred to a group of tobiano mares. By continuous backcrossing, the amount of Arabian blood in the offspring could be increased in each generation while still retaining the tobiano color. After six generations, the result would be essentially a tobiano-colored horse of Arabian type:

1.)

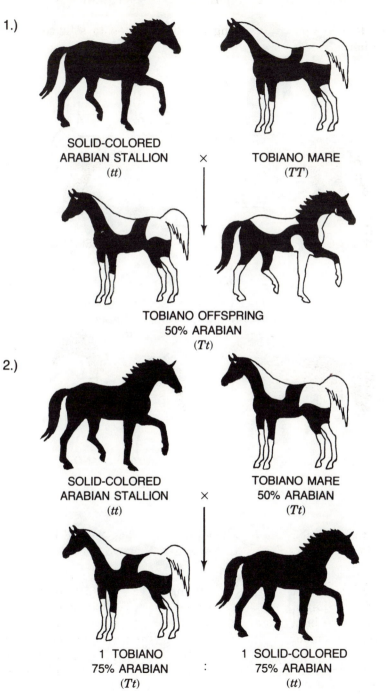

SOLID-COLORED
ARABIAN STALLION
(tt)

×

TOBIANO MARE
(TT)

TOBIANO OFFSPRING
50% ARABIAN
(Tt)

2.)

SOLID-COLORED
ARABIAN STALLION
(tt)

×

TOBIANO MARE
50% ARABIAN
(Tt)

1 TOBIANO
75% ARABIAN
(Tt)

:

1 SOLID-COLORED
75% ARABIAN
(tt)

3.)

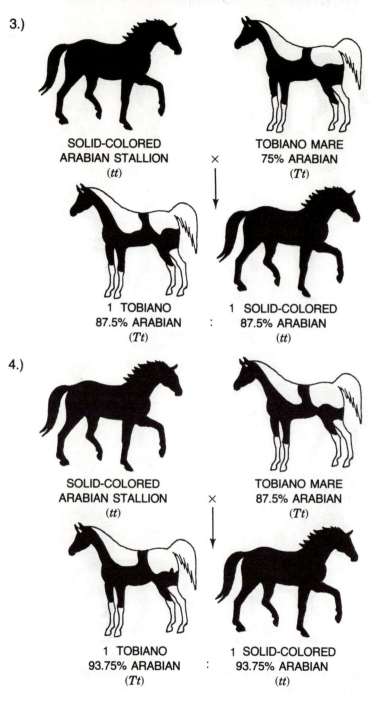

SOLID-COLORED
ARABIAN STALLION
(*tt*)
×

TOBIANO MARE
75% ARABIAN
(*Tt*)

1 TOBIANO
87.5% ARABIAN
(*Tt*)
:

1 SOLID-COLORED
87.5% ARABIAN
(*tt*)

4.)

SOLID-COLORED
ARABIAN STALLION
(*tt*)
×

TOBIANO MARE
87.5% ARABIAN
(*Tt*)

1 TOBIANO
93.75% ARABIAN
(*Tt*)
:

1 SOLID-COLORED
93.75% ARABIAN
(*tt*)

5.)

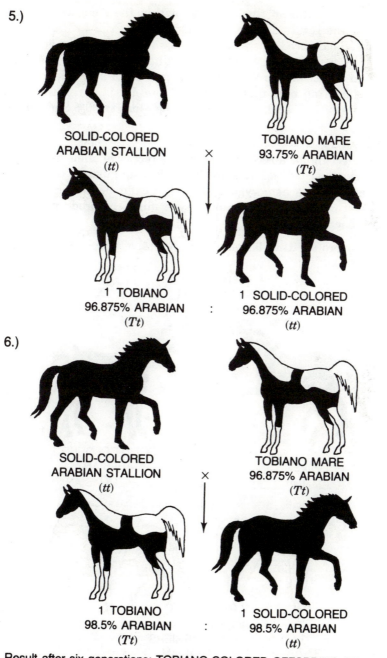

SOLID-COLORED
ARABIAN STALLION
(*tt*)

×

TOBIANO MARE
93.75% ARABIAN
(*Tt*)

1 TOBIANO
96.875% ARABIAN
(*Tt*)

:

1 SOLID-COLORED
96.875% ARABIAN
(*tt*)

6.)

SOLID-COLORED
ARABIAN STALLION
(*tt*)

×

TOBIANO MARE
96.875% ARABIAN
(*Tt*)

1 TOBIANO
98.5% ARABIAN
(*Tt*)

:

1 SOLID-COLORED
98.5% ARABIAN
(*tt*)

Result after six generations: TOBIANO-COLORED OFFSPRING OF ARA-BIAN TYPE.

Something similar to this occurred within the American Saddle-bred breed and resulted in spotted Saddlebreds. Prior to 1948, mares from Pinto-Saddlebred crosses could be used as broodmares, and their offspring by a registered Saddlebred stallion could be registered as Saddlebreds. Such a mare's offspring was bred to a Saddlebred, the resulting offspring bred to a Saddlebred, etc. In six generations, the spotted horses from such a cross would be almost 100% Saddlebred.

Introducing a Recessive Trait

When selecting for a recessive, since the recessive allele must be in the homozygous condition to show, the offspring must be intermated to fix the gene in a homozygous state. For example, horses with recessive overo spotting and Arabian type might be produced by the following backcross.

If overo spotting is completely recessive to solid color:

$$\text{overo} = oo$$
$$\text{solid-colored} = OO \text{ or } Oo$$

1.)

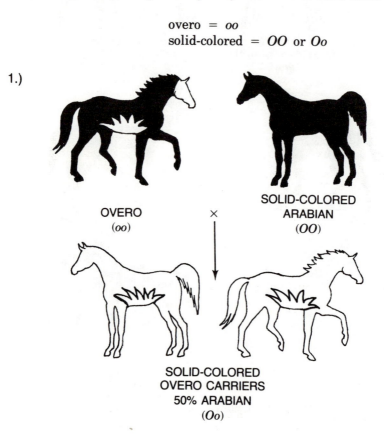

OVERO   ×   SOLID-COLORED
(oo)   ARABIAN
                             (OO)

SOLID-COLORED
OVERO CARRIERS
50% ARABIAN
(Oo)

2.)

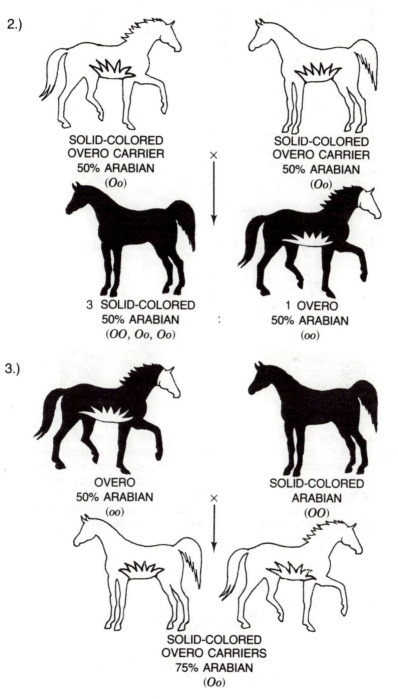

SOLID-COLORED
OVERO CARRIER
50% ARABIAN
(Oo)

×

SOLID-COLORED
OVERO CARRIER
50% ARABIAN
(Oo)

3 SOLID-COLORED
50% ARABIAN
(OO, Oo, Oo)

:

1 OVERO
50% ARABIAN
(oo)

3.)

OVERO
50% ARABIAN
(oo)

×

SOLID-COLORED
ARABIAN
(OO)

SOLID-COLORED
OVERO CARRIERS
75% ARABIAN
(Oo)

4.)

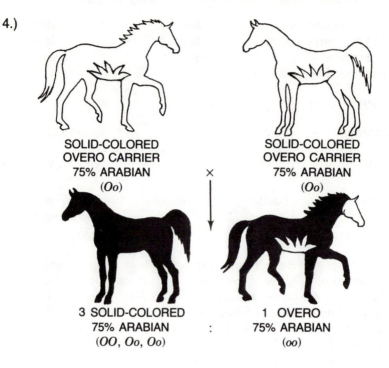

SOLID-COLORED
OVERO CARRIER
75% ARABIAN
(*Oo*)

×

SOLID-COLORED
OVERO CARRIER
75% ARABIAN
(*Oo*)

3 SOLID-COLORED
75% ARABIAN
(*OO, Oo, Oo*)

:

1 OVERO
75% ARABIAN
(*oo*)

5. The backcrossing pattern is repeated until the resulting offspring become a uniform type. In this case, the breeder hopes to obtain a herd of overo horses with strong Arabian type. Although each generation is labeled 50% Arabian, 75% Arabian, etc., some of the partbred offspring may have received very few genes from their purebred parent. (Others may closely resemble their Arabian parent.) For this reason, the designated descriptions are only approximated averages for each generation.

It is interesting to note that the half-breed registries require back-crossing to a purebred parent. This is done to insure that at least 50% of the partbred offspring's genes originate from the purebred side of the pedigree. For example:

PP	×	GG	=	PG
Purebred	×	Grade	=	Partbred (½ purebred genes)
PG	×	PG	=	1 PP : 2 PG : 1 GG
Partbred	×	Partbred	=	Offspring with anywhere from 0-100% of the original purebred alleles

Introducing a recessive trait takes twice as long to achieve the same amount of progress as with a dominant trait. It is much easier to select for a recessive already in the herd than to introduce one, since every homozygous recessive horse will show the desired characteristics and, when bred to other recessives, will produce only offspring that also exhibit the trait (unless a mutation occurs).

For example, two liver horses should produce only liver offspring when mated:

bb	×	bb	=	bb
liver	×	liver	=	all liver offspring

Actually, such factors as penetrance (how frequently animals that carry genes for a specific trait show the trait) and expressivity (the degree to which a gene is expressed in each individual) may affect selection for a recessive trait. The above example is correct for practical purposes (refer to **"Complicating Factors of Inheritance."**)

The goal of many breeders is fixation of a specific trait in the herd. Fixation refers to the expression of the characteristic throughout the herd. For example, the bay coat color gene is "fixed" in the Cleveland Bay breed, since every Cleveland Bay horse is bay in color. Fixation is most easily accomplished for recessive traits and is impossible for heterozygous traits (such as palomino color).

Selection Against A Trait

The progress that a breeder obtains when selecting against a trait depends on whether the trait is controlled by dominant or recessive genes. Undesirable traits that are controlled by dominant alleles are easy to detect (the presence of a dominant gene usually causes expression of the dominant trait) and, therefore, easy to remove from a breeding herd. Undesirable traits that are controlled by recessive alleles are often hidden by corresponding dominant alleles and are usually difficult to remove from a herd. For this reason, most undesirable traits that persist in breeding herds are recessive. (Also, geneticists believe that undesirable mutant genes are usually recessive to the original gene.)

Eliminating a dominant trait from a herd is usually fairly easy, since all carriers of a dominant gene will usually show the trait, whether homozygous dominant or heterozygous. For example, white face markings are believed to be dominant. If a breeder wishes each foal produced to have a solid face, he has only to cull all mares with white face markings and breed only to stallions without white markings. Sometimes it is impractical to cull all affected individuals in a herd, due to the many other good traits that they may carry. In such a case, the breeder may want to determine which horses are heterozygous (as indicated by their offspring) and continue using a good carrier (heterozygote) in the herd. As mentioned previously, a heterozygous animal crossed with a recessive will produce recessive offspring half of the time. For example, a breeder may not like white face markings, but he has an otherwise outstanding heterozygous stallion with a blaze. If the stallion is bred to mares without such markings, half of the offspring will have white markings and half will not:

$$Mm \quad \times \quad mm \quad = \quad 1\,Mm \quad : \quad 1\,mm$$

| heterozygous for white markings | × | homozygous - no markings | = | 1 with markings | : | 1 without markings |

In other words, 50% of the offspring have white face markings, and 50% do not. The breeder may then decide to retain only the solid-colored offspring and to cull those with white markings.

Selection against a trait caused by a recessive gene is more difficult, since the gene may be hidden by a dominant allele in a heterozygous animal. Merely culling the recessive horse that shows the trait will only reduce the incidence of the gene, not eliminate it. Therefore, in the future, recessive offspring may still appear, due to the mating of two heterozygous horses.

It is probable that many diseases are transmitted by recessive genes. It has been suggested that laryngeal paralysis (roaring) in the

Clydesdale breed is caused by an autosomal recessive. (In other breeds, roaring may also be caused by a recessive allele.) If it is absolutely certain that the trait is a simple recessive, the heterozygous parent carriers are positively identified when a foal with the trait is produced. The parent carriers can be eliminated from the breeding program at this time. It must be remembered that, if a carrier produces a normal foal, the foal still has a 50% chance of carrying the undesirable allele. This concept is very important when considering serious conditions, such as CID (Combined Immunodeficiency Disease) in the Arabian. ♞

10

CONTROVERSIAL THEORIES OF SELECTION

For years, breeders have searched for the ultimate breeding system, one that could guarantee optimum production results. Breeding systems were proposed and rejected throughout the history of breed development. Those systems that discriminated against certain families often resulted in the extinction of those lines. Many of the earlier breeding theories are now obviously fallacious; others are still practiced by breeders today (e.g., Bruce Lowe's theory).

There is no simple way to design the "perfect" breeding system. The variables involved in any production program are complex, and the goals of each breeder may differ considerably. Most of the theories included in the following discussion have been discredited due to their over-reliance on the performance of distant relatives. Today, breeding systems emphasize performance, conformation, and the achievements of close relatives.

Galton's Law of Hereditary Influence

Galton's Law is the basis for many breeding theories. It is not a statement of scientific fact, but a guide for determining the relative influence of each generation within a horse's pedigree. The work of Sir Francis Galton (1822-1911) established the foundation of eugenics, a branch of genetics that concentrates on the improvement of a breed or race by selecting for, or against, inherited traits. Although Galton's Law is now known to be fallacious, the application of modern eugenics has proven to be an important aid in breed improvement.

As a result of extensive research and statistical analysis, Sir Galton formulated the following law:

The offspring receives half his genetic material from his parents, one-fourth from his grandparents, one-eighth from his great-grandparents, one-sixteenth from his great-great-grandparents, etc.

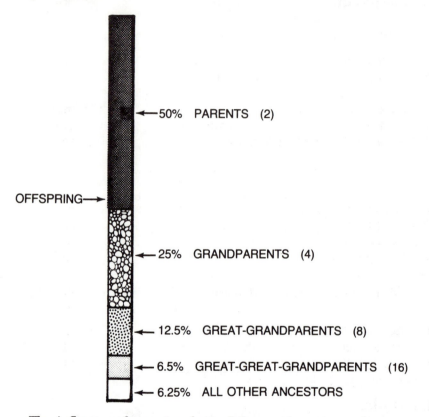

The influence of ancestors beyond the great-great-grandparents follows the same pattern. Therefore, the influence of each ancestor decreases with each generation removed.

The fallacy of this theory is that the influence of the ancestors approximates but never equals one. (The horse must receive all of his genetic material from his ancestors.) Sir Galton's estimate of mathematical proportions (e.g., 50%, 25%, etc.) was wrong, but his general principle of decreasing hereditary influence was correct.

Stamina Index

The Stamina Index estimates distances at which a prospective race horse should be able to compete (i.e., his stamina). Based on the performance of ancestors, this index has been applied primarily to Thoroughbred racing. Because this calculation does not consider the heritability of stamina, and because it involves the application of Galton's Law, its predictions are not accurate.

Determining the Stamina Index:

1) The longest winning distance for each ancestor (back to at least the great-great-grandparents) is determined.

2) The average distance for each generation is then determined. Example: If the sire's longest winning distance was 15 furlongs, and the dam's was 10 furlongs, the average for the first generation would be:

$$\frac{10 + 15}{2} = 12.5 \text{ furlongs}$$

3) The average distance for each generation is multiplied by the influence of that generation upon the prospective race horse's genotype (refer to Galton's Law). Example: If the average winning distance for the first generation was 12.5 furlongs, the following calculation would be made:

(12.5 furlongs) × (50% influence) = 6.25 furlongs contributed by the first generation

4) After step 3 has been completed for each generation, the resulting figures are added. Because Galton's Law does not account for 100% of the individual's inheritance, a correction factor must be used to calculate the final Stamina Index.

Example:

Generation	Greatest Winning Distance (average/generation)	% Genetic Influence	Winning Distance contributed by each generation
parents	12.5 furlongs	1/2	6.25 furlongs
grandparents	12.5 furlongs	1/4	3.125 furlongs
great-grandparents	12.0 furlongs	1/8	1.5 furlongs
great-great-grandparents	13.75 furlongs	1/16	0.86 furlongs
		15/16	11.735 furlongs

CORRECTION: Note that the total distance (11.735 furlongs) accounts for only 15/16 of the horse's genetic capacity. To determine the missing 1/16, multiply 11.735 furlongs by 1/16 and add the product to the original total:

1.	(11.735) × (1/16)	=	.733 furlongs (correction factor)
2.	15/16	:	11.73 furlongs
3.	15/16 + 1/16	:	11.73 furlongs + .733 furlongs
4.	1	:	12.468 furlongs

According to the Stamina Index, the prospective race horse should be able to compete successfully up to a distance of 12.468 furlongs.

Stamina of a Stallion's Progeny

The stamina of a stallion's progeny estimates the distance that each of his progeny should be able to run successfully. This theory is inaccurate, however, since it does not consider the dam's influence upon the racing ability of each of her foals.

The stamina estimate is calculated by averaging the distances of every race won by a specific stallion's progeny. (The races of two-year-old offspring are not included, due to their immaturity and to the greater distances normally run by older horses.) For example, suppose that a stallion sired six offspring with the following winning distances (furlongs):

	(1)	(2)	(3)	(4)	(5)	(6)
	10	9	9.5	11	12	10
	11	8.5	9.5	12	15	10

These distances are then added: $21+ \quad 17.5+ \quad 19.0+ \quad 23+ \quad 27+ \quad 20 = 127.5$

The total distance is then divided by the number of winning races to determine the average winning distance:

$$\frac{127.5 \text{ furlongs}}{12 \text{ races}} = 10.625 \text{ furlongs}$$

The resulting figure, 10.625 furlongs, is the distance that a specified stallion's offspring should be able to run successfully (according to this theory).

Vuillier Dosage System

Colonel J. Vuillier, a French cavalry officer, proposed a breeding theory during the early 1900's. (This theory is now known to be inaccurate.) On the basis of Galton's Law and an extensive Thoroughbred pedigree study, Vuillier determined that the same 16 horses appeared in the pedigrees of every superior Thoroughbred. He counted the number of times that each of these ancestors appeared in hundreds of pedigrees (back to the 12th generation), and decided how influential each was within the ancestry of superior Thoroughbreds. This influence was designated numerically as "standard dosages" (sometimes referred to as the ratio of ingredients necessary to produce superior offspring).

STANDARD DOSAGES

Ancestor	Standard Dosage
Pantaloon	140
Gladiator	95
Melbourne	184
Hermit	235
Isonomy	280
Voltaire	186
Birdcatcher	288
Newminster	295
Hampton	260
Bend Or	210
Touchstone	351
Bay Middleton	127
Stockwell	340
Galopin	405
St. Simon	420
Pocahontas (only mare)	313

According to Vuillier, the ideal race horse should have the "standard dosage" of each of the 16 ancestors within its pedigree. To determine the "dosage" that a specific horse carried, the following steps were taken:

1) **Pedigree Study**: Vuillier identified the 16 ancestors within a horse's pedigree and noted how many times and in what generations they appeared.

2) **Ancestral Influence**: Vuillier claimed that within the 12th generation (where the horse has 4096 ancestors) each ancestor contributed 1/4096 of the genetic influence. Each 11th generation ancestor

provided 2/4096, each 10th generation ancestor contributed 4/4096, and so on. Using these factors, Vuillier established a "dosage system." Ancestors that appeared in the 12th generation counted as 1; those that appeared in the 11th generation counted as 2; and the pattern continued:

Generation	Influence
12th	1
11th	2
10th	4
9th	8
8th	16
7th	32
6th	64
5th	128
4th	256
3rd	512
2nd	1,024
1st	2,048

3) **Pedigree "Dosage:"** The appearance of St. Simon within the 3rd, 5th and 6th generations of a horse's pedigree would have indicated to Vuillier that the horse had a St. Simon "dosage" of 704 (= 512 + 128 + 64).

4) **Standard "Dosage" Comparison:** The St. Simon pedigree "dosage" would have been compared with the standard "dosage" (ideal proportion) to determine whether the horse in question had an excess or deficiency of St. Simon influence:

Pedigree Dosage		Standard Dosage		Excess Dosage
704	–	420	=	284

5) **Application to Breeding**: To Vuillier, the horse's pedigree "dosage" would have indicated an excess of St. Simon influence and, therefore, a need to mate him to a horse with a corresponding St. Simon deficiency.

Because Vuiller's calculations were based on Galton's Law, the "standard dosages" are not considered accurate. Also, it should be remembered that a parent does not pass the exact same genotype to each of his offspring.

Bruce Lowe System

During the late 1800s, Bruce Lowe and William Allison concluded that every Thoroughbred mare traced back to one of about 100 foundation mares in the original English Thoroughbred Stud Book. A breeding theory (Bruce Lowe Theory) was designed by classifying every Thoroughbred into one of 43 families, according to tail female lines that traced back to the 43 foundation mares still found in modern pedigrees. (Tail female lines = dam − maternal granddam − maternal great-granddam, etc.)

Using the following guidelines, Lowe numbered each family 1-43; families with the lower numbers were considered most desirable.

1) Winners of the English Derby, the English Oaks, and the St. Leger races were identified within each family.

2) The family with the greatest number of winners was labeled No. 1. Family No. 2 had the second highest number of winners. Families 1-5 were designated as running families.

3) Families that contained no winners were classified according to Lowe's personal assessment. For example, families that included superior stallions were arbitrarily labeled No. 3, No. 8, No. 11, No. 13 and No. 14.

Every Thoroughbred foal was given the number of its dam's family. If the sire was from family No. 1, and the dam from family No. 3, the foal was labeled No. 3:

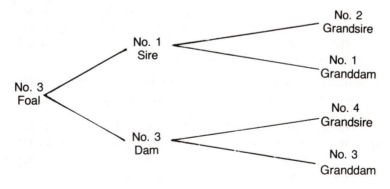

The classification of each foal, according to tail female lines, emphasized the dam's contribution to each of her foals and ignored the sire's genetic contribution. Many breeders accepted this system and concentrated on breeding horses from lower numbered families.

Consequently, many of the higher numbered families were discriminated against and eliminated from the breed.

Telegony

The theory of telegony claims that a pregnant mare can become "saturated" or "infected" with the characteristics that her foal received from his sire. Supposedly, the "infected" mare could pass the "absorbed" characteristics to future offspring (even if they were sired by a different stallion). At one time, telegony was a widely accepted theory. Many breeders even believed that more than one mating of a mare to the same stallion caused a buildup of the stallion's "blood" within the mare.

In the early 1800s, a 7/8 Arabian mare was bred to a Quagga stallion (extinct zebra species). The mare produced a Quagga-horse hybrid that carried the striping pattern of his sire. Later, the same mare was bred to an Arabian stallion and produced a bay foal with a dorsal stripe and zebra striping on the shoulder and legs. A second mating to the same Arabian stallion produced similar results. Both geneticists and breeders concluded that this incident proved the telegony theory. Today, it is obvious that the mare (and/or the Arabian stallion) carried the genetic material for dorsal and zebra striping. (Refer to **"Dilution"** under **"Coat Color and Texture."**)

The telegony concept is easily disproved by the fact that mares do not produce foals with donkey characteristics once they have produced a mule. In addition, mares that have been bred to zebra stallions will not continue to produce foals with zebra characteristics when they are bred to horses (unless they carry genes for zebra striping). Advanced knowledge of genetics and physiology shows that telegony is obviously erroneous. It deserves mention, however, since it has not been discarded by breeders in some parts of the world.

Throughout the history of breed development man has attempted to design the perfect breeding system, a system that would allow the breeder to produce optimum results in a minimum amount of time. Today, systems such as the selection index, minimum culling level and tandem methods try to accomplish similar goals. Breeding programs, such as linebreeding, outcrossing and crossbreeding, have also been designed to fit the needs of different breeders. Although the selection systems designed by Galton, Vuillier, and Bruce Lowe are not commonly employed by the modern horseman, they each contributed to the acceptance of an important genetic concept: "With each generation removed, the influence of an ancestor decreases — regardless of his outstanding quality." ♘

11

COAT COLOR AND TEXTURE

The great emphasis placed on the genetics of coat color and texture is due to three important factors: these coat characteristics are easily observed traits; they are controlled by relatively few gene pairs; and they are not usually affected by environmental influences. These attributes make color and texture an excellent starting point for the study of basic genetic principles. For this reason, the inheritance of coat color and texture is analyzed in detail within this text. This chapter introduces many new concepts, uses a variety of allelic symbols, and begins the study of detailed genetics. Although the discussion may seem complicated and at times confusing, once understood, the reader is well on the way toward understanding the full spectrum of practical equine genetics. In addition to its usefulness as a "stepping stone" to further knowledge, this study provides the horseman with considerable interesting and vital data. Examples are:

— The lethal nature of the roan coat color can contribute to lowered fertility on the breeding farm.
— Genes for excessive white in the overo are closely associated with fatal atresia coli.
— The life span of a grey horse will probably be slightly shorter than his nongrey brothers.
— High white leg markings do not necessarily indicate that the horse is a tobiano genetically.
— It is possible to produce nothing but palomino foals by crossing chestnut and cremello horses.

At this point, it might be helpful to review the general concepts presented in "**Basic Genetic Definitions**," and to note that a comprehensive glossary is included in the back of the text.

The inheritance of coat color in horses has been an area of controversy among both equine geneticists and horsemen for many years. One of the largest stumbling blocks to an understanding of equine coat color is the discrepancy in color definitions. Many horsemen and genetic authorities disagree on the exact definitions for coat colors. This leads to confusion when theories about those colors are discussed. For this reason, descriptions of specific colors throughout this section should be carefully noted.

The horse's hair helps to regulate body temperature (as it does with all mammals) by trapping a layer of air next to the skin. In cold weather, this minimizes heat loss; in hot weather, it insulates the horse against the heat of the environment. In addition, a horse uses the long and especially sensitive tactile hairs of his muzzle to feel objects.

The coat color of a horse is caused by pigment units (known as granules) within the hair. The color of the horse's hair was vital to his survival in the wild, since concealing colors protected him from predators. Cryptic (concealing) coloration, such as that seen in Przewalski's horse, makes the horse blend into his background. Disruptive coloration, like that of a zebra, breaks up the outline, making recognition by predators at a distance difficult. As man domesticated the horse, the need for protective coloring disappeared. Subsequently, when unusual coat colors appeared (i.e., the result of mutations), they were allowed to survive. Some of these color mutations were selected for by man and eventually spread throughout the equine population.

The Mechanics of Coat Color

We see color when light reflected off objects reaches our eyes. What specific color we see depends on the wave length of the light that reflects off the object being viewed. For example, the eyes of a blue-eyed horse appear that color because small protein particles (pigment granules) in the iris scatter or reflect light, particularly the blue and violet wave lengths. Therefore, we see the eyes as blue. Similarly, we see black when light falls on an object and is absorbed instead of reflected. Black hair in horses is seen as that color because it contains dark pigment that absorbs light. White is caused by the physical structure of the object reflecting and scattering all light that strikes it. White hair in horses contains air spaces that scatter light, and has no pigment to absorb light.

Coat color is influenced by the type of pigment present in the hair, in conjunction with the pigment's structure and environment within the horse's body. There are two major types of pigment in horses, both of which are forms of melanin: eumelanin, a black/brown pigment, and phaeomelanin, a yellow/red pigment. In the horse, all forms of melanin are found in granules (i.e., grain-like structures that develop from specialized cells, called melanocytes). The neural crest of the embryonic horse (which is the structure responsible for development of the nervous system, among other body systems) produces melanoblasts, which travel from the neural crest toward the skin of the horse, gradually turning into melanocytes. The number of melanocytes found in the skin of each individual horse varies, but is apparently unrelated to skin pigmentation. Therefore, differences in skin color seem to be due to differences in the activity of the melanocytes, not to differences in concentration.

Melanocytes reside in the hair follicles, which are the points of origin for the hairs of the horse's coat. There are two kinds of melanocytes present in the follicle: amelanotic and melanotic. Amelanotic melanocytes contain no melanin pigment, while melanotic melanocytes do contain melanin pigment. Most horses have both kinds of melanocytes, while dominant white horses lack them entirely. These two types of melanocytes can switch back and forth, changing from one type to the other as needed. Should a layer of pigmented skin be removed in an injury, some of the amelanotic melanocytes may begin producing melanin to repigment the skin. Sometimes a deep injury may damage the melanocytes along with the skin. In those cases, hair from a wound may grow back white since no pigment is manufactured.

The melanocytes gradually move from the neural crest toward the skin surface, multiplying along the way. They lose their ability to make pigment as they age, and eventually slough off with dead skin cells. The melanocytes that cause hair pigmentation are found in the bulb of the hair follicle. As the hair grows from the follicle, the melanocytes inject melanin granules into the hair shaft. (Incidentally, the type and degree of pigmentation in the hair can be different from that of the surrounding skin, as in the case of white hairs growing in from an injury on a dark-skinned horse.) In older grey horses, the greying gene somehow prevents pigment from entering the hair shaft, but concentrates it in the surrounding tissue.

Hair Follicle

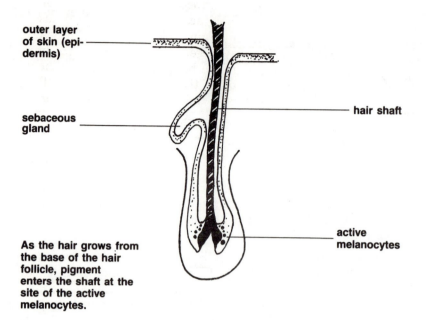

outer layer of skin (epi-dermis)

sebaceous gland

hair shaft

active melanocytes

As the hair grows from the base of the hair follicle, pigment enters the shaft at the site of the active melanocytes.

Pigment granules in the iris (the part of the eye surrounding the pupil) give the equine eye its color, and increase its efficiency by partially restricting the amount of light that enters the eye. This helps explain why blue-eyed horses are light sensitive. Since they lack much of the necessary pigment, light entering their eyes is much less restricted. It is interesting to note that the color of the retina may vary from green to blue to pink, and may be closely associated with the inheritance of eye and coat color.

Factors which affect the horse's coat color include the type of melanin granules present (including their size, shape, number, and arrangement in the hair), the melanocytes, and the type of hair (refer to the discussion on coat texture within this chapter). The intracellular environment of the melanocyte also affects pigmentation. In addition, the melanocytes would not be in a position to inject melanin granules into the hair if they had not successfully migrated from the neural crest where they originated. A wide variety of genes may affect any one or more of these factors, thus changing the horse's coat color.

Many genes have pleiotropic effects, meaning that they influence more than one trait. For example, the overo gene which causes white spotting can also cause the lethal "white colt" syndrome. In this situation, an all-white foal with defects such as atresia coli (an abnormality of the digestive tract) is born from overo parents. The foal is either stillborn or dies soon after birth.

Genes which affect the melanoblasts right after they leave the neural crest may prevent them from maturing into melanocytes, thus causing a white coat color. Genes acting at this point may reduce the number of melanoblasts, thereby reducing the eventual number of melanocytes and decreasing the amount of pigment in the coat. This results in diluted colors such as palomino.

Genes can also change coat color by somehow affecting the structure of the melanocyte in such a way that it cannot make and secrete pigment granules in the normal manner. Coat color is also modified if a gene causes the pigment granules to clump together, instead of allowing them to be evenly distributed in the hair. This close association of granules lightens hair color.

Pigment granules consist of melanin (either eumelanin or phaeomelanin) attached to a protein framework. If a gene changes the shape of the protein framework on which the melanin is deposited, the shape of the pigment granules is changed, and the coat color is altered. Eumelanin may be altered from black to liver, and perhaps phaeomelanin may be changed from red to yellow. The locus, or location on the chromosome, at which formation of the protein framework can be affected is the B locus (which will be discussed later).

Because the production of melanin is controlled by genes (as are all body functions), melanin production can be affected at any point in the process by a gene, resulting in a change in coat color. During melanin production, genes can switch production from eumelanin to phaeomelanin or vice versa. Two loci that affect this switch from one form of melanin to the other are the A locus and the E locus (both of which will be discussed later).

Coat color may also be affected by the structure of hair. If the hair follicle or framework is abnormal (e.g., hair that is freeze-branded will fall out and grow back white), pigment may not be produced, may not be able to enter the hair shaft, or it may be distributed in an abnormal way. For example, the brindle coat pattern in dogs and cattle is due to abnormal hair structure.

Much of what is known about coat color in horses has been derived from both actual reports of color in horses and from genetic research done in other mammals. Studies on different species of animals help to unravel coat color genetics, because they have similar sets of alleles, with similar relationships and functions. For example, the locus that causes a certain color in rodents, cats, dogs, etc., is probably the same one that causes an identical color in horses.

Black and Liver: The B Locus

As stated in the introduction, the basic color of a horse is due to the type of pigment present, its structure and its distribution throughout the hair. The B locus controls color by regulating the formation of the protein framework on which eumelanin, a dark pigment, is deposited.

The capital letter B is used to represent the dominant allele for black, a color which is caused by eumelanin. The B allele causes the formation of elongated eumelanin granules, which appear black. The small letter b is used to describe the recessive allele for brown pigment, which is also a form of eumelanin. The b allele causes the production of small, round, or oval pigment granules which appear brown rather than black, and is responsible for the liver coat color. Horses with either BB or Bb in their genotypes are usually born a mousey grey/brown and gradually darken to black. This color fades in the sun, giving reddish highlights to the horse's coat, so the black color caused by B is sometimes called "fading black" to distinguish it from the more rare, nonfading black due to the effects of another locus on B. This type of black will be discussed later.

. Previously, many authorities believed that b represented the allele for phaeomelanin, a golden-red pigment seen in the coat of chestnut or sorrel horses. However, in other species (such as the dog and rabbit), b actually represents a liver or chocolate color. Recent research indicates that this also holds true in horses, accounting for liver horses *(bb)* which, although mistakenly called liver chestnuts, are really more of a uniform brown than red. (Uniform brown means that each hair is the same brownish color, unlike the dark seal brown coat in which black hairs are mixed with red ones to give an overall brown appearance.) A liver horse, although uniform brown, may have reddish, golden, or purple highlights in his coat, perhaps due to the fading action of sunlight (just like the reddish highlights in the fading black coat). A nonfading liver color, due to the action of another locus on b, will be discussed later.

Chestnut or sorrel is probably due to gene action at another locus. Reasons why bb should logically represent liver instead of chestnut will be presented when the action of bay and dilution genes on black, liver, and chestnut is outlined later.

The following table lists the respective genotypes for the black and liver phenotypes.

PHENOTYPE	GENOTYPE
fading black	*BB, Bb*
nonfading black	*BB, Bb* plus a certain allele at another locus (described in later discussions)
fading liver	*bb*
nonfading liver	*bb* plus a certain allele at another locus (described in later discussions)

Bay and Seal Brown: The A Locus

Mutations of the genes at the A (agouti) locus are responsible for changing the primitive equine coat color into the various coat colors of the modern horse. Originally, the horse's environment necessitated drab, protective coat coloring. After the domestication of the horse, man's preferences began to influence equine breeding patterns, and selection for brighter, more distinctive colors occurred. These colors were possible because the original A^+ allele, which caused the wild-type coat pattern, mutated (changed).

The original wild type coat pattern is sometimes referred to as agouti coloration, because of its similarity to the coloration of the South American agouti, a relative of the guinea pig. The A^+ allele is dominant over any other allele found at this locus, but rarely, if ever, occurs in modern-day horses. It can, however, be seen in the coat color of the primitive Przewalski's horse, some of which still are found in the steppes of Mongolia. The markings of these horses are characteristic of the A^+ allele: they have dark points (mane, tail and legs) with a coat containing a mixture of eumelanin (pigment which causes the black color) and phaeomelanin (pigment which causes the red/yellow color) in the hair shaft. This mixture causes the coat to appear grey along the back and neck, but as it extends down to the abdominal region, the phaeomelanin pigment becomes prominent, which makes the animal appear to have more yellow hair in that area.

A mutation of the A^+ gene is the A allele, commonly seen today as the bay coat pattern. When A is present at one position of the A locus, and A^+ is present at the other position, A is recessive to A^+ (the A^+ will overpower A and produce primitive coloring). A is dominant to both the a^t and a alleles. The A allele creates the bay pattern by causing eumelanin to be restricted to the points, resulting in the characteristic black points. The rest of the body hair will still contain some eumelanin, and its abundance in relation to the phaeomelanin will determine the exact shade of the coat color (e.g., light bay, dark bay, liver bay, etc.).

The A allele affects both black and liver horses; because both have eumelanin in their coats, the A can restrict it. But the A allele will not create the bay pattern in chestnut or sorrel horses. This is because another gene (which will be discussed later) has already restricted their eumelanin to their skin and eyes, and they have none in their coats. The A allele may, however, have a lightening effect on the phaeomelanin in their coats and may be responsible for the variation in shades of chestnut and sorrel.

Another allele of this locus is the a^t, which has a modifying effect on the horse's coat color but does not cause as extreme a change as the A allele. The a^t is recessive to the A allele, and will not change the bay pattern, but when it is in the homozygous form $(a^t a^t)$, it will influence the color of the horse. When a^t is present in one of these two forms as the dominant allele, then a genotypically black horse (BB or Bb) will have the dark seal brown color (his genotype will be $B\_a^t a$ or $B\_a^t a^t$). A genotypically liver horse (bb) with the allelic combination $a^t a^t$ or $a^t a$ will be a light seal brown.

A dark seal brown horse has a mixture of red and black hairs in his coat, but more black are present, accounting for the dark brown color. Similarly, a light seal brown horse has a mixture of red and liver hairs in his coat. In both cases, some reddish hair will be visible on the flanks, muzzle, and abdomen, but the legs remain dark. Although the A allele will not produce the bay pattern in a chestnut horse, the a^t allele, when found in conjunction with the D dilution gene, will create a red dun (claybank) color. The important thing to remember is that this dilution gene (D) must be present for the dun color to be expressed (refer to the discussion on dun and grulla).

The a allele is recessive to all other alleles at this locus, and even in its homozygous form (aa), has no influence on the shade or distribution of color on the horse's coat. Therefore, the horse with this genotype (aa) will show his normal coloring without any restrictions.

PHENOTYPE	GENOTYPE
bay	$A\_\_B\_\_$
liver bay	$A\_\_bb$
dark seal brown	$a^t\_\_B\_\_$
light seal brown	$a^t\_\_bb$
wild type	$A^+ \_\_ \_\_$
black	$aa\ B\_\_$
liver	$aa\ bb$

Note: the blank (__) means that any allele or pair of alleles from that locus could be present without affecting the color. For example, A determines the bay pattern, so AA, Aa^t, and Aa horses are all bay. This is represented by $A\_\_$. Similarly, since a^t is dominant to a, both $a^t a^t$ and $a^t a$ horses are seal brown; this is represented by $a^t\_\_$.

Chestnut and Black: The E Locus

The E locus affects the extension of both eumelanin and phaeomelanin throughout the coat. If one of these pigments is extended, then the other is restricted, and vice versa. Thus, this locus sometimes is referred to as the "extension/restriction" locus.

The E locus contains three possible alleles and, of these, the E^D allele is dominant over the other two. This allele allows the full extension of the black $(B\_\_)$ and liver (bb) pigments throughout the coat. (There will be no phaeomelanin present.) Thus, a horse with the BB or Bb genotype with the E^D allele will be a jet black that does not fade in the sun. Unlike the fading blacks, which are born mousey grey/brown, these horses are born with the black spread throughout their coats and remain that way throughout life. Similarly, a liver horse (bb) will become a rare nonfading liver (sometimes called black chestnut). Because the horse is bb, it has brown pigment granules (instead of black) in its coat. The E^D allows these pigments to be extended throughout the coat. E^D can also completely mask the A locus (meaning E^D is epistatic to A, a^t, and a), so that a jet black or nonfading liver horse may carry the allele for bay (A) or brown $(a^t a^t, a^t a)$ and not express those colors.

Just as the dominant E^D permits extension of black and liver pigments, the recessive allele e restricts their expression (i.e., black and liver pigment granules are left only in the skin and eyes), leaving only the golden-red phaeomelanin in the coat. The result is a chestnut or sorrel horse. Some authorities have believed that chestnut, or sorrel, is due to the presence of bb at the B locus. Recent research, however, indicates that the colors result from ee, not bb. Reasoning used to substantiate this theory includes the following points.

1) In other species, bb is known to stimulate production of brown pigment granules and produce the same uniform brown color seen in liver horses (as mentioned under Black and Liver: The B Locus). It does not affect the phaeomelanin, the golden-red pigment found in chestnuts.

2) Also, in other species, ee restricts black and brown pigment to the skin and eyes, leaving only phaeomelanin in the hair and producing a red coat. Animals with this coloration include red foxes, red squirrels, and orangutans.

3) Chestnut or sorrel horses which carry the A (bay) allele are unaffected by it; they do not exhibit the characteristic bay pattern. Because, in other species, bb (liver) animals are affected by A in the same way as are BB and Bb (black) animals, chestnuts and sorrels must not be caused by (bb) or they would also be affected (and would have dark points).

4) If chestnut or sorrel horses resulted from ee, then logically they would remain unaffected by A. This is because A produces the bay pattern by restricting eumelanin to the points. Since in chestnuts or sorrels the eumelanin already is restricted to the eyes and skin, they have no eumelanin in their coats and cannot show the bay pattern. These coats contain only phaeomelanin; the A allele might lighten this pigment and cause the variation in chestnuts and sorrels (a lighter sorrel might carry A and a darker chestnut aa).

E is the original or "wild type" allele which allows normal expression of the black or liver pigment. A horse with a BB or Bb genotype and either EE or Ee is fading black, while a horse with bb and EE or Ee is liver. E is dominant to e, but not to E^D; an $E^D E\ B\_\_$ horse would be jet black, and a $E^D E\ bb$ would be nonfading liver. When E^D is present, the colors are nonfading.

A horse carrying the E allele can be affected by the bay (A) or seal brown (a^t) alleles at the A locus. (A and A^t mask the effects of E.) Some authorities have believed that E causes the bay coat pattern, and that e allows the expression of normal coloring. However, in nature, the normal, or wild type, allele is usually dominant, while the recessive allele is a mutation. This principle can be applied to the E locus. The dominant E allele is the wild type (i.e., it allows the expression of black and liver), and e is the recessive mutation which causes restriction of eumelanin and total extension of phaeomelanin (i.e., chestnut or sorrel coloration).

Some other species contain an intermediate allele, between e (which restricts eumelanin to the eyes and skin) and E (which allows expression of the eumelanin). This allele, e_j, produces a pattern like that of the bay allele *(A)*, which restricts eumelanin to the points. It has not been proven to be present in horses, however. Even if it does exist, it is a different allele than E, because it does not allow expression of the normal coat color, but instead causes the bay color.

The following table lists the possible genotypes for each of the colors discussed so far, including the A, B, and E loci.

PHENOTYPE	GENOTYPE
jet black	_B_E^D_
nonfading liver	_bb E^D_
fading black	aa B_ EE or aa B_ Ee
liver	aa bb EE or aa bb Ee
chestnut	_ _ ee
sorrel (light chestnut)	A_ _ ee
bay	A_ B_ EE or A_ B_ Ee
dark seal brown	a^t_ B_ EE or a^t_ B_ Ee
light seal brown	a^t_ bb EE or a^t_ bb Ee

Grey, Roan and Dominant White: The Epistatic Modifiers

Even though a horse's basic genotype may be chestnut, black, liver, bay, brown, or a dilute color, the horse may phenotypically appear grey, roan, or white. This is because all three of these genes (grey, roan, and dominant white) are epistatic to the basic and dilute colors. For example, just as the a^t allele at the A locus was epistatic to the B allele at the B locus, making a black horse become dark seal brown, the G allele at the G locus can make a dark seal brown horse turn grey as he ages.

GREY: THE G LOCUS

In the horse, progressive silvering that takes place after birth is known as greying. As the grey horse matures, white hairs replace colored hairs, eventually causing a white appearance. A grey "white" horse can be distinguished from a cremello (pseudoalbino) or a dominant white horse because the skin will always be dark and the eyes of a grey horse will usually be dark. (Of course, if the horse originally had white markings, the skin in those areas is pink, just like on a nongrey horse.)

A horse that will turn grey is born the color coded for by his genotype, but the color may appear darker than usual. For instance, a bay foal that will grey may be a darker bay at birth than a bay foal that will retain his color. As the foal matures, white hairs will begin to appear, first around the eyes and ears, then gradually over the rest of the head and body. Frequently, the legs will grey later than the rest of the coat. Some horses grey more quickly in the mane and tail, causing an iron grey appearance with a silver mane and tail, while others grey more quickly in the body, causing a white appearance with dark grey mane and tail.

All horses that are not grey carry the recessive pair of alleles gg at the G locus. Grey horses carry either Gg or GG, since greying is completely dominant to nongreying. Although there has been speculation that a homozygous dominant horse, GG, may turn white more rapidly than a heterozygous horse, Gg, there has been no research to prove this theory.

The greying gene causes silvering by preventing pigment from entering the hair shaft; instead, the pigment is retained in the body. (Unlike white hair due to the greying gene, white hair due to the dominant white gene totally lacks pigment granules.) In older grey horses (over 15 years of age) melanomas, or pigment tumors, become common. Tumors rising from the accumulation of pigment in the body usually form in the region below the ear (parotid), in the region of the dock (solid part of the tail), around the anus, or on the sheath. Melanomas sometimes metastasize, meaning that they spread through the body, and can affect many organs. If a melanoma in the parotid region spreads to a deeper group of lymph nodes (the retropharyngeal lymph nodes) the secondary growth at the deeper location is often much larger than the primary growth below the ear. The retropharyngeal lymph nodes may enlarge to such an extent that they obstruct the airway of the throat and can prove fatal. In other cases, pigment tumors can spread to lymph nodes in the gastrointestinal tract. Although a melanoma can be fatal, most grey horses with pigment tumors live to a good age; the tumors remain superficial and do not spread to vital structures. As a result, they are usually only of cosmetic importance. Nevertheless, some authorities (Tesio) refer to greying as a disease, perhaps because of this tendency for grey horses to develop tumors. One study of English Thoroughbred stallions indicated that the lifespan of greys was approximately two years less than that of their nongrey brothers.

Even though the greying gene stops pigment from entering the hair shaft, some horses eventually develop small "freckles" of colored hair, resulting in a "flea-bitten" grey appearance. Occasionally, grey horses will become dappled, forming rings of lighter hair around circles of darker hair. These dapples are due to differences in the rate of greying and are probably caused by a modifying gene, as yet unidentified. They may be the same as the dapples that become evident on the coat of a horse that is blooming with health due to superior nutrition, grooming, and management. A dappled grey horse will eventually lose the contrast which causes these markings as he continually lightens. He may become either completely grey and appear to have a pure white coat over black skin, or he may have the gene for the "flea-bitten" or "freckled" appearance.

Because grey coat color is completely dominant over nongrey, it is not possible to determine whether a horse is homozygous dominant *(GG)* or heterozygous *(Gg)* by simple visual appraisal. Therefore, even in breeds in which the grey color has been a selection characteristic, nongrey offspring are occasionally produced, since some of the grey parents are heterozygous. In the Thoroughbred, an examination of the English, French, and United States stud books revealed only three stallions that could definitely be considered homozygous grey because of an adequate number of offspring, all of which were grey. Through 1973, Dancing Dervish, Wise Exchange, and Al Hattab had respectively sired 145, 42, and 32 foals, all of which were grey. Also, both parents of each stallion were grey, another requirement for the homozygous grey genotype.

Bloody Shoulder Markings

Bloody shoulder markings are permanently restricted areas of dark hair (intensified dappling or roaning) seen only on the grey horse. These markings may also be present, but undetectable, on coat colors other than grey. Although they are called "shoulder" markings, the patches of either red or black hair commonly appear in the girth and saddle area; occasionally, these markings will extend to the shoulder.

Horses with bloody shoulder markings are a type of genetic "throwback." Throwback refers to a horse which resembles a relative from the distant past, one which may be removed by decades or centuries. Bloody shoulder markings are very rare and appear only in Arabians and in horses with Arab ancestry (e.g., it has been reported in the Thoroughbred). This suggests that the the trait originated as a mutation within the Arabian breed many years ago. The sporadic appearance of bloody shoulder markings indicates that the controlling allele is a recessive found in very few horses.

The possible genotypes for the grey and nongrey phenotypes are listed below.

PHENOTYPE	GENOTYPE
grey	Gg or GG
nongrey	gg
fading black that turns grey	$aa\ B\_\ CC\ dd\ E\_\ G$
dark seal brown that turns grey	$a^t\_\ B\_\ CC\ dd\ E\_\ G\_$
bay that turns grey	$A\_\ B\_\ CC\ dd\ E\_\ G\_$
liver that turns grey	$aa\ bb\ CC\ dd\ E\_\ G\_$
chestnut that turns grey	$\_\ \_\ CC\ dd\ ee\ G\_$
palomino that turns grey	$\_\ \_\ Cc^{cr}\ dd\ ee\ G\_$
buckskin/dun that turns grey	$A\_\ B\_\ Cc^{cr}\ dd\ E\_\ G\_$ $A\_\ B\_\ CC\ D\_\ E\_\ G\_$

ROAN: THE R LOCUS

Nonprogressive silvering, which is present at birth in the same proportions that it will be throughout the horse's life, is known as roan. A roan horse does not become completely white with age, unless he also carries the greying gene. Unlike grey horses, which develop white hairs first on the face, roan horses always show their basic color on the face and lower legs. The skin of a roan horse is dark, like that of a grey, except for areas under natural white markings (such as stockings or face markings). When the skin of a roan is injured, the hair will often grow back as the basic color. In nonroan horses, scarred areas are frequently mixtures of white and base-colored hairs.

Roan is epistatic to all coat colors, except grey and dominant white. Therefore, roaning may appear on a black, brown, bay, liver, chestnut, etc. It may also appear on diluted coat colors such as palomino, buckskin, and dun. Terms such as blue roan (roaning on black), red roan (roaning on bay) and strawberry roan (roaning on chestnut) are sometimes used to describe the roaning patterns. To avoid confusion, only the base color name, in conjunction with "roan," will be employed in this book. For instance, a horse that is basically black with the roaning pattern will be called a black roan, not a "blue" roan. A bay with the roaning pattern will be called a bay roan, not a "red" roan, etc.

Frequently, horses which are believed to be roans when young are actually in the early stages of greying. In fact, some registries mistakenly classify roan and grey together. As stated above, a roan horse is born roan, with white hairs distributed throughout the coat and a dark face and legs, while a grey horse is born a solid color (except for possible white markings) and progressively becomes white, starting at the face. It is possible, however, for a horse to be both roan and grey, since the genes that control these two color modifications are not mutually exclusive. For example, a black horse which carries both roaning and greying genes would be born a black roan and gradually become completely white. However since roan is a relatively uncommon color, this is unlikely to happen.

Roan (R) is dominant to nonroan (r). However, in the homozygous condition (RR), the roan allele is lethal to the embryo. Therefore, only heterozygous roan horses (Rr) are born. Because roan is lethal in the homozygous state, the foal crop is decreased when roan horses are mated to each other, but the percentage of live foals that are roan is increased.

Rr (roan) $\times$ Rr (roan) $=$ 1 RR (lethal), 2 Rr (roan), 1 rr (nonroan)

The roan parents will produce this 1:2:1 ratio, but the RR embryos will die, decreasing the foal crop or herd fertility by 25%. Of the foals that are born, there will be 2 roans to every solid color (2Rr:1rr) or a 67% roan foal crop.

Cc^{cr} (palomino) $\times$ Cc^{cr} (palomino) $=$ 1 CC (chestnut)
2 Cc^{cr} (palomino)
1 $c^{cr}c^{cr}$ (cremello)

The same 1:2:1 ratio is produced, but the homozygous dominant condition is not lethal, so all the foals are born alive (unless, of course, they are defective in some other way). This type of cross results in an approximately 50% palomino foal crop.

It is impossible to have a herd of true-breeding roan horses, just as it is impossible to have a herd of true-breeding palominos or any other heterozygous type of horse.

A mutation of the R allele which basically restricts the roan pattern to a frosting over the hips is probably responsible for the roan type of Appaloosa, and is represented by Rn^{Ap} (refer to "**Appaloosa Spotting**"). Another type of roaning, which is restricted to the flank area, is probably controlled by another set of genes (perhaps additional alleles of the R locus or a nearby locus) that are yet unidentified.

Because the grey and roan genes are epistatic to those controlling the horse's base coat color (except for dominant white), they are considered undesirable in many of the color breeds (e.g., Appaloosa, Pinto, Paint, etc.). The greying gene will cause total fading of the base color in time, thus hiding any pattern, while roaning will cause a permanent, unchanging faded appearance.

The following table lists the possible genotypes of roan and nonroan horses.

PHENOTYPE	GENOTYPE
roan	*Rr*
nonroan	*rr*
lethal roan	*RR*
nongrey roan	*Rr gg*
grey/roan	*Rr G__*
black roan	*aa B__ CC dd E__ gg Rr*
bay roan	*A__ B__ CC dd E__ gg Rr*
chestnut roan	*__ __ CC dd ee gg Rr*
sorrel (light chestnut) roan	*A__ __ CC dd ee gg Rr*
dark seal brown roan	*a^t__ B__ CC dd E__ gg Rr*
light seal brown roan	*a^t__ bb CC dd E__ gg Rr*
liver roan	*aa bb CC dd E__ gg Rr*
palomino roan	*__ __ Cc^cr dd ee gg Rr*
buckskin/dun roan	*A__ B__ Cc^cr dd E__ gg Rr*
	A__ B__ CC D__ E__ gg Rr

DOMINANT WHITE: THE W LOCUS

Dominant white is the term used to describe a horse that is born completely white, with pink skin and colored eyes (blue, brown, amber, or hazel). A dominant white horse is truly white, not cream-colored like a cremello (pseudoalbino). Therefore, white markings do not show up on a dominant white horse. This type of white differs from white caused by grey; a dominant white horse has pink skin, while a grey horse that has gone completely white has dark skin. In addition, the dominant white horse is born white, unlike the grey "gone white" animal.

Dominant white is epistatic to all other colors, including grey and roan, so a horse could have the basic genotype of a black, bay, roan, etc., and appear completely white if carrying the dominant white gene. Like roan, dominant white is lethal in the homozygous state (WW) in the early stages of gestation, so only heterozygous horses (Ww) are born white. Homozygous recessive horses (ww) are colored. The offspring of two dominant white horses would result in the ratio of 2 white foals (Ww) to 1 colored foal (ww). The WW embryos die during gestation.

$$Ww \text{ (white)} \times Ww \text{ (white)} = 1\ WW \text{ (lethal)}$$
$$2\ Ww \text{ (white)}$$
$$1\ ww \text{ (colored)}$$

Since dominant white is epistatic to all other colors, it would be possible for a dominant white horse to carry the genes for black, bay, brown, etc. These might be revealed in a mating to either a colored horse or another dominant white horse. Suppose that a dominant white stallion was mated for three consecutive years to a chestnut mare. The matings result in two white foals and one black that turns grey. Since the mare does not carry the gene for grey (G), it had to be passed on by the stallion. In addition, the foal would not have been black at birth unless it had an E allele in its genotype $(E$ is necessary for the expression of black). The chestnut mare could not have carried the E $(ee$ is necessary for chestnut color), so the stallion must have been E.

Occasionally, a dominant white horse with pale blue eyes will appear pink-eyed when viewed in strong sunlight at certain angles. Regardless of this, the dominant white horse is not an albino; a true albino horse would have pink eyes when viewed from any angle. True albino horses, which would have absolutely no pigment, are unknown. In other species, the gene for albino is not usually lethal in the homozygous state, so there are true-breeding lines of albinos, such as white mice and rabbits. True albino is a recessive, not a dominant.

Recently, a white coat color has appeared in the Thoroughbred. Like dominant white horses, these horses have pink skin, dark eyes, an occasional small dark spot, and have produced white offspring. Their phenotype is probably caused by the dominant white allele *(W)*, or by another allele at the dominant white locus.

The following lists some of the numerous possible genotypes a dominant white horse may have; they all have in common the presence of a *W* allele.

PHENOTYPE	GENOTYPE
dominant white	*Ww*
dominant white with the black genotype	*aa B__ CC dd E__ gg Ww rr*
dominant white with both black and grey genes	*aa B__ CC dd E G__ Ww rr*
dominant white with the chestnut genotype	*__ __ CC dd ee gg Ww rr*
dominant white with both the chestnut and roan genes	*__ __ CC dd ee gg Ww Rr*

Diluted Coat Colors:
The C and D Loci

Diluted coat colors in the horse (such as palomino and buckskin) are believed to be controlled by two gene pairs. These genes cause the restriction of pigment to one side of the hair shaft, leaving the other side transparent. In some cases, the pigment is almost completely restricted from the hair shaft (as in the very dilute cremello and perlino colors). Dilution is not caused by a mixture of white and colored hairs, as are the grey and roan coat colors.

Before identifying the dilution genes and examining their effects upon the various base colors, two coat colors should be examined. One thing to note, when considering dilution genes and their effects, is that the description of dun and buckskin are not always agreed upon by geneticists and breed registries. For this reason, the following discussion will emphasize the different patterns and shades of color, and suggest a possible name for each.

THE C LOCUS

The C locus affects the degree of pigmentation within the hair shaft. Geneticists believe that the C allele is necessary for the expression of either eumelanin or phaeomelanin. In some species, a mutation of C to another allele, c, is associated with albinism, because the c allele results in the complete absence of all pigment. Although c is present in many species, it is non-existent in the horse. Hence, there are no true albino horses. True albino (pink skin, pink eyes, and completely white coat) should not be confused with pseudo-albino (pink or grey skin, blue or hazel eyes, and almost white coat) which is caused by the allele c^{cr} at the C locus.

The c^{cr} allele (another mutation of C) causes dilution of the horse's red pigment (phaeomelanin). For this reason, the presence of c^{cr} causes the dilution of chestnut, bay, and the lighter area of the seal brown coat colors. The c^{cr} allele has no noticeable effect on black, liver, or the darker areas of the seal brown coat. (Note: Dilution is present at birth. "Dilution" of a color refers to the effect of c^{cr} upon a genetic base color, during embryonic development.)

Palomino and Cremello

When the c^{cr} allele is found in the heterozygous state with the C allele, the red pigment is restricted to one side of the hair shaft. The overall effect is partial dilution of what would otherwise have been red or reddish brown hair. A base color of chestnut, for example, is diluted to palomino (golden body color with very pale mane and tail) by the presence of a single c^{cr} allele.

$$\text{CHESTNUT} + C\ c^{cr} = \text{PALOMINO}$$

When the c^{cr} is homozygous, the red pigment is almost completely restricted from the hair shaft. (Refer to the discussion on codominance within "**Complicating Factors of Inheritance**.") This results in the dilution of a red or reddish brown coat to a creamy offwhite, known as cremello. The cremello coat color might easily be confused with dominant white (an absence of color caused by the W locus). The two can be differentiated, however, by the fact that white markings, such as a star or stocking, are apparent on the cremello coat, but never appear on the dominant white horse. The cremello's skin appears pink, due to the sparse pigmentation, and his eyes are amber or blue. Again, the mane and tail are extremely pale.

CHESTNUT + $c^{cr}\, c^{cr}$ = CREMELLO

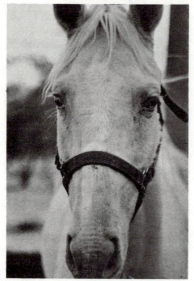

Double c^{cr} dilution of the chestnut base color results in the creamy off-white color of the cremello. Note the contrast between the coat color and the white face marking.

Because the cremello horse must carry two c^{cr} alleles, while the chestnut carries two C alleles at the same locus, mating a chestnut with a cremello always results in palomino dilution $(C\, c^{cr})$.

$$CC \qquad \times \qquad c^{cr}c^{cr} \qquad = \qquad Cc^{cr}$$

CHESTNUT × CREMELLO = PALOMINO

Palominos vary in shade from a true golden coat with white mane and tail to a light tan coat with off-white mane and tail. This continuous variation is caused by two important factors:

1) The effects of c^{cr} vary according to the shade of the base chestnut. The dilution of a uniform darker chestnut, for example, would be slightly darker than the dilution of a lighter chestnut with a flaxen mane and tail.

2) Other modifying genes may cause similar diluting effects, so that the term "palomino" might be used to describe several different genotypes. Although the c^{cr} allele is responsible for the typical palomino phenotype (golden body with a flaxen mane and tail), a dilution allele at the D locus can produce similar results (refer to the discussion on dun and grulla). The silver dappling gene modifies chestnut to a light golden brown with a pale (white or off-white) mane and tail. This color pattern in the Shetland Pony might also be confused with palomino.

To explain why chestnut-chestnut crosses have produced palomino foals, when the dilution allele should have been expressed in at least one parent, a genetic theory was proposed. This theory states that the restriction allele A (the allele that turns black and liver to bay) is necessary for the expression of the palomino color. According to this hypothesis, the horse with the chestnut genotype and a c^{cr} dilution allele, but without the A restriction allele, is not diluted to palomino:

$$\text{CHESTNUT} + c^{cr} - A = \text{CHESTNUT}$$

A simpler, more consistent explanation, however, is that the reports of chestnut-chestnut crosses producing palomino foals may not have interpreted coat color accurately. Liver, a true brown color sometimes mistaken for dark chestnut, cannot be diluted by c^{cr} (c^{cr} dilutes only phaeomelanin; nonfading liver contains no phaeomelanin and fading liver is minimally diluted). Therefore, liver horses may carry the dilution gene without being diluted, and produce palomino foals when mated to chestnuts (or heterozygous bays).

Diluted Seal Browns

Although the c^{cr} dilution allele has no effect upon black hairs, the few reddish hairs around the muzzle and flanks of the dark seal brown coat can be diluted. In a similar manner the brown hairs of the light seal brown coat color (which is derived from liver) is also unaffected, although the reddish muzzle and flank hairs can be diluted. These dilutions are usually unnoticed, especially in the case of dark seal brown. Although both types of seal brown horses must carry the a^t allele at the A locus, and the a^t allele is associated with dorsal stripes and zebra shoulder or leg stripes, the markings are not apparent when superimposed on the seal brown coat color.

Bucksin/Dun: Dilution of Eumelanin

Because the bay coat is a mixture of red and black hairs, the action of a single c^{cr} allele on the bay pattern results in a diluted coat with a black mane and tail. The actual shade of the diluted coat depends upon the degree of black hairs intermingled with red hairs. The dark bay, for example, is diluted to a sooty yellow (mixture of black and yellow hairs) with black points (buckskin). On the other hand, the bright red bay is diluted to a clear yellow with dark points (dun). There are many shades of diluted bay coats, ranging from off-white to sooty tan, making it difficult to form a clear-cut definition of buckskin or dun. The black dorsal stripe (back stripe), zebra shoulder stripes, and zebra leg stripes are often found with the buckskin/dun patterns. (The inheritance of these markings has been associated with the a^t allele at the A locus.)

Perlino

When c^{cr} is homozygous, its effect upon the bay color is doubled. The coat is diluted to near white, while the dark points either remain dark or become a light rusty or chocolate shade (depending on the ratio of black to red pigment granules). The darker dorsal stripe or zebra markings (sometimes present) have been associated with the a^t allele at the A locus. The skin will have minimal pigmentation and may appear pink or grey. This color pattern is often referred to as perlino.

Red Dun

Red dun (claybank dun) horses have a light red body with dark red points and often a dorsal stripe. These horses sometimes have zebra markings on the legs and/or shoulders. Recent studies indicate that the c^{cr} allele is responsible for the dilution of chestnut to red dun. In this case, however, the a^t allele probably must be present. This restriction allele (which also changes black to seal brown) controls the inheritance of the dorsal stripe, explaining why the red dun is frequently characterized by this feature. The inheritance of darker points might also be caused by the presence of a^t.

$$\text{CHESTNUT} + a^t \text{ and } C\,c^{cr} = \text{RED DUN}$$

The possible c^{cr} dilution genotypes are listed below:

PHENOTYPES	GENOTYPES
Palomino	$\_\_\ Cc^{\,cr}ee$
Cremello	$\_\_\ c^{cr}c^{cr}\ ee$
Red Dun	$a^t\_\_\ Cc^{cr}\ ee$
Buckskin/Dun (dorsal stripe)	$Aa^t\_\ Cc^{cr}\ E\_$
Buckskin/Dun (no dorsal stripe)	$A\_\_\ Cc^{cr}\ E\_$
Perlino (dorsal stripe)	$Aa^t\_\ c^{cr}c^{cr}\ E\_$
Perlino (no dorsal stripe)	$A\_\_\ c^{cr}c^{cr}\ E\_$
Dark seal brown/	$a^t\_\ B\_\ Cc^{cr}\ E\_$
yellow highlights on muzzle and flanks	$a^t\_\ B\_\ c^{cr}c^{cr}\ E\_$
Light seal brown/	$a^t\_\ bb\ Cc^{cr}\ E\_$
yellow highlights on muzzle and flanks	$a^t\_\ bb\ c^{cr}c^{cr}\ E$

THE D LOCUS

Like the C locus, the D locus also controls the degree of pigmentation along the hair shaft. But the D locus differs in that it affects both phaeomelanin and eumelanin. There are two known alleles at the D locus:

1) The dominant D allele causes both eumelanin and phaeomelanin to be restricted to one side of the hair shaft; the overall result is dilution of both red and black hairs. (The reddish hairs may be diluted to a lighter shade than the black ones.)

2) The d allele causes no dilution. This allele is completely recessive to the D allele.

Because D is completely dominant to d, the effects of D in the homozygous state (DD) are exactly the same as in the heterozygous state (Dd). When d is homozygous (dd), the D locus has no control over the degree or location of pigment along the hair shaft. For example, the presence of d in the homozygous state has no effect upon the chestnut base color:

$$\text{CHESTNUT} + dd = \text{CHESTNUT}$$

Isabella

The presence of either homozygous D (DD) or heterozygous D (Dd) causes the dilution of chestnut to a uniform yellow with a yellowish mane and tail:

$$\text{CHESTNUT} + DD \text{ (or } Dd) = \text{YELLOW DUN/YELLOW DUN}$$
$$\text{MANE AND TAIL}$$

Dilution of a chestnut with a flaxen mane and tail could result in a yellow body with a pale flaxen mane and tail. This pattern is referred to as Isabella. Isabella might easily be confused with the c^{cr} dilution of chestnut, which forms the palomino color pattern. It is interesting to note that the homozygous Isabella (DD) could theoretically breed true, unlike the palomino (Cc^{cr}).

Grulla

Many sources state that the presence of one or two D dilution alleles (i.e., DD or $Dd)$ causes the dilution of black to a smokey blue with black points (grulla). The black dorsal stripe, shoulder stripes, and leg stripes are often found on the grulla pattern. Reviewing the description of D dilution, an important question might arise: if the D allele causes uniform dilution of both eumelanin and phaeomelanin, why does the blue grulla have dark points? This question has not yet been answered by geneticists. Because the seal brown allele (a^t) causes the dorsal stripe (and perhaps dark points), grulla could be a dilution of dark seal brown $(a^t\_\ B\_\ E\_)$, rather than black $(aa\ B\_\ E\_$ or aa $B\_\ E^D\_)$. Because the true (jet) black horse is very rare, its uniform dilution is very unlikely. The dilution of true black by the D locus probably causes a uniform smokey blue or mouse color.

Dilution of liver by the D allele causes a brownish shade of grulla (with or without dark points, dorsal stripe, and/or zebra markings). Fading liver is probably diluted to a uniform, light brown grulla. On the other hand, the dilution of nonfading liver might result in a smokey brown grulla with slightly darker points and a dorsal stripe, with or without zebra markings. The dilution of a liver bay (yellowish liver with liver mane and tail) could cause a light brown grulla with dark liver points and, possibly, a dorsal stripe with or without the zebra markings. Dilution of a light seal brown horse $(a^t\_\_ bb\ E\_\_)$ probably results in a lighter grulla than that resulting from the dilution of dark seal brown. $(a^t\_\_ B\_\_ E)$.

Buckskin/Dun: Dilution of Eumelanin and Phaeomelanin

Because the D allele dilutes both the red and the black pigment, it dilutes bay to a uniform yellow with darker points (dun or buckskin). This color (along with most of the other D dilutes) is often accompanied by the dorsal stripe, indicating that the D allele and the a^t restriction allele (which is thought to cause the dorsal striping) may be closely situated upon the same chromosome. (Closely linked genes are passed together from parent to offspring. Refer to the discussion on linkage under **CYTOGENETICS AND PROBABILITY.**) These horses may be very similar in appearance to the Cc^{cr} buckskin/duns.

COMBINED EFFECTS OF c^{cr} AND D DILUTION

When the D allele is present in conjunction with c^{cr}, dilution is not doubled, but there are a few phenotypic differences. When both c^{cr} and D are found with the basic chestnut genotype, for example, the resulting phenotype would probably be described as a pale palomino (yellowish tan with flaxen mane and tail). The combined effects of the c^{cr} and D alleles upon the bay coat color may account for the varying shades of buckskin and dun and the confusion between these two colors.

The effects of the two dilution alleles are not completely understood at this time, but their effects upon the various base colors seem far-reaching. Perhaps much of the confusion associated with labeling the diluted colors and detecting their paths of inheritance is caused by varying shades for each base color, and by the interaction of the dilution alleles with other modifying alleles (e.g., a^t and A).

Dilution genotypes:

PHENOTYPE	GENOTYPE
yellow body/yellow mane and tail	_ _ _ CC D_ ee
yellow body/pale mane and tail	_ _ _ CC D_ ee ff
uniform blue grulla	aa B_ _ _ D_ E^D_
uniform brown grulla	aa bb _ D_ E^D_
blue grulla/dark points and a dorsal stripe	a^t_ B_ CC D_ E_
brown grulla/slightly darker points and a dorsal stripe	a^t_ bb CC D_ E_
light brown grulla/dark points and dorsal stripe if a^t is present	A_ bb CC D_ E_
yellow dun/buckskin and dorsal stripe if a^t is present	A_ B_ CC D_ E_
buckskin/dun with darker points and a dorsal stripe if a^t is present	A_ B_ Cc^{cr} dd E_
paler buckskin/dun	A_ B_ Cc^{cr} D_ E_
pale palomino	_ _ Cc^{cr} D_ ee
red dun/dorsal stripe	a^t_ _ _ CC D_ ee

Silver Dapple: The S Locus

The term silver dapple is frequently used to describe horses with an almost black to light milk-chocolate coat color and a silver mane and tail, frequently accompanied by a dark "mask" on the face. The silver dapple gene may also cause dappling on chestnuts, sorrels, etc., but is more commonly associated with the liver and black coat colors, which become light milk-chocolate to almost black with silver dapples. Silver dapple is due to a dominant mutation within the Shetland Pony breed, first seen in 1886. The mutation occurs only in Shetland Ponies and in horses with Shetland ancestry.

The silver dapple gene (represented by S) somehow lightens coat pigment, resulting in a dappled appearance. The contrast between the body and the silver mane and tail (also caused by S) is much more striking in a heterozygous horse (Ss) than in a homozygous (SS), since SS lightens the coat more than Ss. The silver dapple gene lightens both the red and black pigments, but the contrast between the silver mane and tail and the dappled body is much greater with dark coat colors than with chestnut, or dilute colors such as palomino. Ponies that are basically chestnut or sorrel, and are homozygous for the silver dapple gene (SS), are sometimes mistakenly referred to as "palomino," since the phaeomelanin may be diluted to a golden cream or light sorrel color, with a pale flaxen (white or nearly white) mane and tail. The milk-chocolate color which most breeders desire is due to the presence of black pigment (eumelanin) in the coat, diluted by the silver dapple gene. Since these ponies have silver manes and tails, and are light chocolate in appearance, they are occasionally referred to as "silver chestnuts," even though they do not have the genotype of a chestnut.

The silver dapple gene will produce the most striking appearance and greatest contrast on a pony which is jet black (E^D). Shetland ponies are one of the few breeds other than the Tennessee Walker in which this true-black allele of the E locus is commonly found. (Since this allele can mask A, there are few bay Shetland ponies.) Breeders who wish to produce this dapple color frequently cross jet-black ponies with ones that carry the silver dapple gene (including some that are mistakenly called palomino), resulting in silver dapple and jet-black offspring. In conjunction with the greying gene (G), S causes a color known as "grey/white," in which ponies carrying both alleles $(S\_G\_)$ turn grey much more quickly than horses or ponies with only $G\_$. Some foals with this genotype are even born grey.

Because of its diluting action on pigment, the silver dapple gene S may, in some way, be related to the dominant dilution gene D, which causes both eumelanin and phaeomelanin to be restricted to one side of the hair shaft.

The following table lists possible genotypes for silver dapple Shetland ponies, using only the A, B, C, D, E, S, and G, loci.

PHENOTYPE	GENOTYPE
silver dapple (chocolate w/ great contrast)	$\_ B\_ CC\ dd\ E^D\_ Ss$
silver dapple (chocolate w/ contrast)	$aa\ B\_ CC\ dd\ E\_ Ss$
silver dapple (dark cream w/ less contrast)	$aa\ B\_ CC\ dd\ E\_ SS$
	$\_ B\_ CC\ dd\ E^D\ SS$
light red w/ white mane & tail	$\_ \_ CC\ dd\ ee\ S\_$
lighter red (may appear palomino) w/ white mane and tail	$\_ \_ CC\ dd\ ee\ SS$
palomino	$\_ \_ Cc^{cr}\ dd\ ee\ ss$
chestnut (dark)	$\_ \_ CC\ dd\ ee\ ss$
chestnut (light)	$A\_ \_ CC\ dd\ ee\ ss$
grey/white	$\_ \_ \_ \_ \_ \_ S\_ G\_$

Flaxen Mane and Tail

All true palominos and some chestnuts (sorrels) and livers have a flaxen (almost white) mane and tail. Some authorities believe that this trait is due to the action of one of the dilution genes. However, if this explanation were valid, only palomino horses (which carry a dilution gene) would show the flaxen mane and tail. An alternate theory is that this trait is caused by a recessive pair of genes; but, if this were so, not all palominos would have a flaxen mane and tail, because not all chestnuts and sorrels used to produce palominos have the trait.

A few experts have combined these two theories and believe that a recessive pair of genes, which must be accompanied by a dilution gene, results in a flaxen mane and tail. Here again, chestnuts or sorrels would never show the trait, because they would appear palomino (not chestnut or sorrel) if a dilution gene were present. In addition, black horses could show the trait: since black is unaffected by c^{cr}, a black horse, with a c^{cr} allele and the recessive pair of genes could theoretically have a flaxen mane and tail. Since bay, brown, and black horses (which have both B and E in their genotypes) have never been observed to have a flaxen mane and tail, this trait must act only in the absence of either B or E. (Since liver horses have E with b, they can show the trait. Since chestnut or sorrel horses are always ee, they can also show the trait.) Thus, liver and chestnut or sorrel horses could have a flaxen mane and tail due to a recessive pair of genes, represented by ff. (Nonflaxen mane and tail would be $F\_\_$.) The light mane and tail seen in palomino and cremello horses could be an effect of c^{cr} that gives the same mane and tail phenotype as the flaxen mane and tail of liver and chestnut or sorrel horses that carry ff.

The following table lists possible genotypes for chestnut or sorrel, liver, and palomino horses with a flaxen mane and tail. It should be noted that some horses have only a flaxen mane or only a flaxen tail, so two closely linked gene pairs could be involved.

PHENOTYPE	GENOTYPE
chestnut w/ flaxen mane and tail	__ __ CC dd ee ff
chestnut w/o flaxen mane and tail	__ __ CC dd ee Ff
light chestnut or sorrel w/ flaxen mane and tail	A__ __ CC dd ee ff
light chestnut or sorrel w/o flaxen mane and tail	A__ __ CC dd ee Ff
liver w/ flaxen mane and tail	aa bb CC dd E__ ff
liver w/o flaxen mane and tail	aa bb CC dd E__ Ff
palomino w/ flaxen mane and tail	__ __ Cc^{cr} dd ee __

Pinto/Paint: Tobiano and Overo Spotting

"Pinto" and "paint" were originally two terms used to describe horses that had white spotting different from that of the Appaloosa. Now, however, these terms are used to describe specific breeds, one the Pinto horse and the other the Paint horse, both having this distinctive coloration. Two other terms sometimes used to describe a particolored horse are piebald (a black horse spotted with white) and skewbald (a non-black horse spotted with white). In addition, there are two patterns of white spotting, tobiano and overo.

In both tobiano and overo horses, the amount and extent of the white spotting is due to the action of a number of modifying genes which are, as of yet, unidentified. These modifiers cause variations in spotting. The modifiers which increase the amount of spotting of a tobiano or overo horse probably work through additive gene action, meaning that the amount and extent of white increases as the number of modifying genes present increases. A horse with modifiers, but lacking the allele for tobiano or overo, is still a solid color. It may be possible for a horse with the allele for tobiano or overo, but lacking modifiers, to be a solid color also. This phenomenon has occurred in mice and rats. In both overo and tobiano horses, there may be a small band of "roaning" around each spot, where the white hairs of the marking mix with the solid-colored base hairs.

TOBIANO: THE T LOCUS

A tobiano horse is basically solid-colored, with four white legs. The face of a tobiano is marked similarly to that of a solid horse, with a dark head and perhaps a white face marking (such as a star or strip). The white spots of a tobiano are regular and rounded, with even borders, and tend to be dorsally located (meaning that they cross over the back of the horse). Frequently, a tobiano will have a large "shield-like" area of solid color on the chest and underside of the neck.

Tobiano spotting is dominant, and a tobiano horse crossed with a solid-colored horse can produce a tobiano offspring. The capital letter T is sometimes used to represent the dominant spotting, while t is used to represent normal solid color. However, in other species, such as rodents, similar spotting is symbolized by P (for piebald) or S (for spotting). Tobiano spotting in horses could be due to one of these two genes, or it may be the result of gene action at another locus that causes the same phenotype as P and S in rodents.

A tobiano horse can be either homozygous *(TT)* or heterozygous *(Tt)*. The base color can be black, bay, chestnut or sorrel, liver, brown, palomino, buckskin, etc. (This form of white spotting occurs with all colors except dominant white.) It is even possible for a horse to be grey and tobiano (or roan and tobiano) at the same time. For example, a horse could be born black and white, then eventually grey until it was completely white. A close look at the skin of such a horse would distinguish it from a nontobiano that had gone grey. This is because the skin beneath the white markings of both tobiano and overo horses is pink, while that under solid-colored areas (even if the hair has become white due to the greying gene) is dark.

OVERO: THE O LOCUS

Unlike the tobiano, the overo horse usually has four solid colored legs (except for normal white markings). In rare cases, extensive white leg markings give the overo several completely white legs. The overo may have a bonnet (a marking that covers the ears and poll and encircles the neck at the throat), or may be bald-faced (a wide blaze that covers both eyes), or apron-faced (white marking that extends beyond the ears and under the chin). The eyes of an overo that has extensive white facial markings are frequently blue or white (i.e., "china eyes"). Overo spotting is usually located ventrally (meaning that the white comes up from the belly) and rarely crosses the back. These white areas are usually irregular with uneven edges. The amount of white may vary, according to the number of modifying genes present. The white horse with colored ears is known as a "medicine hat" overo.

As with tobiano, overo can occur on any color background; however, this type of white spotting is recessive, not dominant, so it is less frequently seen. Because of its recessive nature, it is possible for two solid colored horses that carry the overo allele to produce an overo offspring. The capital letter O has been used to represent normal solid color, while o is used to designate recessive overo coloring.

Overo spotting may be caused by an allele of the W locus, the one that causes dominant white horses (Ww = dominant white, ww = solid color, and WW = lethal), but there has been no research to prove this theory. In overos, total white spotting results in the "white foal syndrome," a condition in which an overo foal is born completely white. These foals are either born dead or die soon after birth, and usually have atresia coli and brain defects (refer to **INHERITED ABNORMALITIES**). Lethal effects associated with white spotting also occur in other species, such as the cat and rodent. Mating overo horses with a great number of white modifiers (meaning that the parents have a high percentage of white spotting, such as with the medicine hat pattern), probably increases the chances of a lethal white offspring.

The following table lists the probable genotypes for tobiano and overo horses.

PHENOTYPE	GENOTYPE
tobiano	$T\_\_$
non-tobiano	tt
overo	oo
non-overo	$O\_\_$
non-tobiano, non-overo	$tt\ O\_\_$
tobiano/overo	$T\_\_\ oo$
black tobiano	$aa\ B\_\_\ C\_\_\ dd\ E\_\_\ gg\ rr\ T\_\_\ O\_\_$
black overo	$aa\ B\_\_\ C\_\_\ dd\ E\_\_\ gg\ rr\ tt\ oo$
dark seal brown tobiano	$a^t\_\_\ B\_\_\ C\_\_\ dd\ E\_\_\ gg\ rr\ T\_\_\ O\_\_$
dark seal brown overo	$a^t\_\_\ B\_\_\ C\_\_\ dd\ E\_\_\ gg\ rr\ tt\ oo$
bay tobiano	$A\_\_\ B\_\_\ CC\ dd\ E\_\_\ gg\ rr\ T\_\_\ O\_\_$
bay overo	$A\_\_\ B\_\_\ CC\ dd\ E\_\_\ gg\ rr\ tt\ oo$
chestnut tobiano	$\_\_\ \_\_\ CC\ dd\ ee\ gg\ rr\ T\_\_\ O$
chestnut overo	$\_\_\ \_\_\ CC\ dd\ ee\ gg\ rr\ tt\ oo$
light chestnut or sorrel tobiano	$A\_\_\ \_\_\ CC\ dd\ ee\ gg\ rr\ T\_\_\ O\_\_$
light chestnut or sorrel overo	$A\_\_\ \_\_\ CC\ dd\ ee\ gg\ rr\ tt\ oo$
palomino tobiano	$\_\_\ \_\_\ Cc^{cr}\ dd\ ee\ gg\ rr\ T\_\_\ O\_\_$
palomino overo	$\_\_\ \_\_\ Cc^{cr}\ dd\ ee\ gg\ rr\ tt\ oo$
buckskin/dun tobiano	$A\_\_B\_\_Cc^{cr}dd\ E\_\_\ gg\ rr\ T\_\_\ O\_\_$
	$A\_\_B\_\_CC\ D\_\_\ E\_\_\ gg\ rr\ T\_\_\ O\_\_$
buckskin/dun overo	$A\_\_\ B\_\_\ Cc^{cr}\ dd\ E\_\_\ gg\ rr\ tt\ oo$
	$A\_\_\ B\_\_\ CC\ D\_\_\ E\_\_\ gg\ rr\ tt\ oo$
grey black tobiano	$aa\ B\_\_\ CC\ dd\ E\_\_\ G\_\_\ rr\ T\_\_\ O\_\_$
grey black overo	$aa\ B\_\_\ CC\ dd\ E\_\_\ G\_\_\ rr\ tt\ oo$
roan black tobiano	$aa\ B\_\_\ CC\ dd\ E\_\_\ gg\ Rr\ T\_\_\ O\_\_$
roan black overo	$aa\ B\_\_\ CC\ dd\ E\_\_\ gg\ Rr\ tt\ oo$
grey roan black tobiano	$aa\ B\_\_\ CC\ dd\ E\_\_\ G\_\_\ Rr\ T\_\_\ O\_\_$
grey palomino overo	$\_\_\ \_\_\ Cc^{cr}\ dd\ ee\ G\_\_\ rr\ tt\ oo$

As is evident from the above list, the combinations are great in number, and many possible genotypes are not listed.

Appaloosa Spotting

The breed known as the Appaloosa shows a variety of unusual coat patterns accompanied by several distinct physical characteristics. Patterns seen in Appaloosas include small white spots on a darker background, an entirely white horse with dark spots, a white blanket with or without spots over the hip area, and any pattern in-between.

In addition to their unusual coat markings, Appaloosas also have several other unique breed characteristics. These include mottling of the skin on the face and genitalia, a visible white scelera around the eye, and narrow light and dark vertical striping on the hoofs. Another characteristic frequently associated with Appaloosa coloring is varnish marks, which are dark marks on the cheeks, elbows, and stifle areas. In addition, many Appaloosas have a noticeably sparse mane and tail, although this is becoming less common in modern Appaloosas. A kinked tail is also thought to be somehow related to intensive Appaloosa coloration.

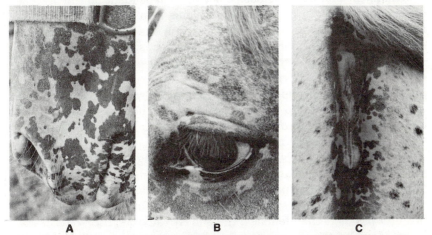

A   **B**   **C**

The Appaloosa coat pattern is usually characterized by mottled skin on the muzzle (A), around the eye (B) or around the anus (C).

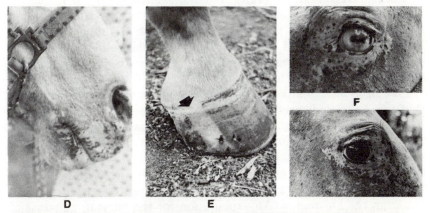

D   **E**

(D) Mottled skin typical of the pinky syndrome should be distinguished from the mottling associated with the Appaloosa coat pattern. (E) Characteristic striping on the Appaloosa hoof varies from one to several stripes. (F) Appaloosa with heterochromic eyes: The right eye (above) is blue, and the left eye (below) is brown.

Appaloosa markings may change in appearance throughout the horse's life, perhaps because the melanocytes migrate to different areas. The inheritance of the Appaloosa patterns is not completely understood and, since a variety of markings may result from a mating, an unborn foal's markings are difficult to predict. Although a number of theories have been proposed to explain Appaloosa coat coloring, they all tend to agree on the basic points which will be discussed here.

BASIC APPALOOSA GENE

The appearance of Appaloosa coloring is dependent on the presence of the dominant Ap gene. (This gene has also been labeled W^{ap} by some geneticists.) The Ap gene is located at the Ap locus and is not a mutation of the dominant white (W) locus. Solid-colored, non-Appaloosa horses are $apap$, while all Appaloosas are either homozygous $ApAp$ or heterozygous $Apap$. In addition, an Appaloosa horse possesses a number of modifying genes which produce the blanket and spots. A horse with Ap is still a solid color if he lacks the necessary modifier genes. On the other hand, a solid color horse without the Ap gene could carry many modifiers.

The genes responsible for Appaloosa coloring may perhaps be sex influenced, meaning that the sex hormones (such as testosterone) may affect the color and/or pattern. In other species, the effects of sex on coloration are readily seen. For example, in birds, the males have brighter plumage than females. In male Appaloosas, the patterns may be more varied than in the female.

WHITE BLANKET GENE

The designation W has sometimes been used to represent the white blanket locus. Due to possible confusion with the dominant white locus, Wb will be used here. Wb is the normal, solid-colored allele, while the recessive form, wb, is responsible for the appearance of a white blanket over the hip and loin area. Solid-colored horses may be $WbWb$ or $Wbwb$; blanket Appaloosas are $wbwb$. It should be noted that the Ap allele is necessary for the expression of the white blanket locus. A horse with the $wbwb$ alleles but, without the Ap allele, will not have a white blanket. Obviously, a horse with Ap and without $wbwb$ will also lack the blanket.

The wb allele acts in conjunction with Ap to prevent pigment from forming in the affected areas. The resulting white areas tend to radiate out from the hip region. An explanation for the varying amounts of possible white has been that there are actually a number of pairs of alleles which affect this trait $(wb_1 wb_1, wb_2 wb_2,$ etc). These act additively, so a horse with the genotype $wb_1 wb_1, wb_2 wb_2,$ for example, would have a larger blanket than one with the genotype $wb_1 wb_1$, $Wb_2 Wb_2$. The exact number of pairs that affect this pattern is unknown. Leopard horses probably have many of these recessive allelic pairs, since their white base color is the result of an extensive blanket. Like the Ap gene, genes at the Wb locus may be sex influenced. If this is the case, the male hormone testosterone may cause a heterozygous male *(Wbwb)* to have the white blanket, while a heterozygous female would not express this trait.

SPOTTING GENE

The presence of wb produces the white blanket, but any spots on the blanket are the result of a different locus, Sp. (This gene is sometimes referred to as S but, due to possible confusion with the silver dapple gene, it will be called Sp here). Sp represents the dominant, solid-colored allele, while sp in the homozygous state causes spots on the white blanket. As with the wb gene, sp may be sex influenced so that a male heterozygote *(Spsp)* may also be spotted.

Spotting occurs because the sp somehow interferes with the action of the wb alleles, allowing pigment to form in certain areas of the blanket. As with wb, there are actually a number of modifying pairs that affect spotting, represented by $sp_1 sp_1$, $sp_2 sp_2$, etc. A leopard Appaloosa will have a greater number of recessive sp modifiers (such as $sp_1 sp_1$, $sp_2 sp_2$) than an Appaloosa with spots only over a white blanket $sp_1 sp_1 Sp_2 Sp_2$).

Remember that, without the basic Ap gene, no spots will form. Therefore, a leopard Appaloosa horse probably has a genotype such as the following:

$$Ap\_\_\ wb_1 wb_1\ \ wb_2 wb_2\ \ldots\ sp_1 sp_1\ \ sp_2 sp_2\ \ldots$$

An Appaloosa with a white blanket and spots might be:

$$Ap\_\_\ wb_1 wb_1\ \ Wb_2 Wb_2\ \ldots\ sp_1 sp_1\ \ Sp_2 Sp_2\ \ldots$$

LEOPARD AND ROANING PATTERNS

Another theory about Appaloosa coloring attributes the leopard pattern to the presence of the Sl^{ap} gene at the Sl locus. According to this theory, $Sl^{ap} Sl^{ap}$ with modifiers would cause the leopard pattern. $Sl^{ap} sl^{ap}$ would result in either a roan blanket over the hips or overall roaning, depending on the number of modifiers present. Sl^{ap} in the heterozygous state may be responsible for the varnish roan pattern. A nonleopard horse would be $sl^{ap} sl^{ap}$ or $Sl^{ap} Sl^{ap}$ without modifiers.

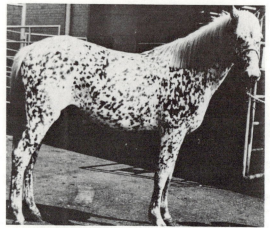

This unusual color pattern
is a striking example of the
leopard Appaloosa.

The roaning effect attributed to the Sl^{ap} allele implies a relationship between this locus and the roan locus R. Some geneticists believe that these may be alleles of the same locus. In addition, Appaloosas also appear to have a special roaning gene, represented by Rn^{ap}. This gene, which may be a mutation of R, is not lethal when homozygous $(Rn^{ap}Rn^{ap})$, unlike the roan gene R. As with the other Appaloosa genes, Rn^{ap} is considered sex influenced. In this case, female Appaloosas with Appaloosa roaning are $Rn^{ap}Rn^{ap}$, while males are either $Rn^{ap}Rn^{ap}$ or $Rn^{ap}rn^{ap}$. In the male, $Rn^{ap}Rn^{ap}$ may cause a small band of roaning around each spot in the blanket. Occasionally, females will also show the roaning around each spot. Perhaps a heterozygous female with few modifiers shows this small amount of roaning instead of the overall Appaloosa roan pattern.

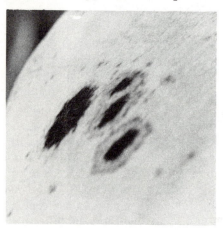

The presence of an Rn^{ap} allele may
cause a band of roaning around dark
spots located on a white blanket.

DIMINISHING CONTRAST

If an Appaloosa horse carries the greying gene, *G,* his coat pattern will gradually fade away as he ages. As a result, the greying gene is undesirable in Appaloosas. In the same way, the roan gene *R* is also undesirable, since it diminishes the contrast on the Appaloosa pattern. (The interspersion of white hairs causes an overall faded appearance that is present at birth.)

Other genes that interfere with contrast in the Appaloosa coat patterns include the dilution alleles; c^{cr} and *D.* Both of these genes dilute the base color, reducing the contrast between the blanket, spots, and basic coat color. For example, a white blanket might not show the desired contrast on a palomino horse and would barely be apparent on the cremello. To preserve the most intensive contrast between the Appaloosa markings, breeders should use only dark base-colored horses.

It is possible for a horse to carry both tobiano or overo and Appaloosa genes, and show a white blanket on top of tobiano or overo markings, or small Appaloosa spots inside of larger white tobiano or overo markings. This type of mixed pattern is considered undesirable and unregisterable by the Appaloosa, Paint, and Pinto registries.

The following list includes possible genotypes and phenotypes for Appaloosa-colored horses:

MALE

PHENOTYPE	GENOTYPE
no Appaloosa pattern	*ap ap*
pattern allowed to appear	*Ap ap* or *Ap Ap*
no white blanket areas	*Wb Wb*
white blanket areas	*Wb wb* or *wb wb*
no spotting	*Sp Sp*
spotting	*Sp sp* or *sp sp*

example genotype: *ApAp, WbWb, SpSp* = no white blanket or spotting in the male

FEMALE

PHENOTYPE	GENOTYPE
no Appaloosa pattern	*ap ap*
pattern allowed to appear	*Ap ap* or *Ap Ap*
no white blanket areas	*Wb Wb* or *Wb wb*
white blanket areas	*wb wb*
no spotting	*Sp Sp* or *Sp sp*
spotting	*sp sp*

example genotype: *ApAp, wbwb, spsp* = white blanket with spotting in the female

black Appaloosa stallion with white blanket and spots	aa B__ CC dd E__ rr gg ss ww Ap__ wb__ sp__
black Appaloosa mare with white blanket and spots	aa B__ CC dd E__ rr gg ss ww Ap__ wbwb spsp
grey/chestnut leopard stallion (contrast lost with age)	CC dd ee rr G__ ss ww Ap__ wb$_1$wb$_1$ wb$_2$Wb$_2$ sp$_1$sp$_1$ sp$_2$Sp$_2$
bay/roan mare with white blanket, no spots (contrast less than desired)	A__ B__ CC dd E__ Rr gg ss ww Ap__ wbwb SpSp

These examples show only a few of the many possible genotypes for
Appaloosa horses.

White Markings: The Face and Legs

The term white markings refers to areas of white hair found on the horse's head and/or legs. The underlying skin is pink, and a band of roaning may occur around the edges of the marking. White markings are frequently used as a means of identification, since they do not usually change with age. The expression of these markings is highly variable. A star may be faint or extensive; it may or may not be connected to a strip. A star, strip, and snip may form one continuous marking on the horse's face, forming the blaze or bald patterns. The numerous shapes, sizes, and possible combinations for white markings provide extensive variation and, therefore, contribute to accurate identification.

Geneticists agree that white markings are controlled by several alleles which act as modifiers upon the horse's coat color. The exact number of modifiers and their genetic "roles" (i.e., whether they are dominant, recessive, epistatic, etc.) are not clearly understood, however.

In the past, several researchers suggested that white markings on the face and legs were inherited as simple recessives (meaning that the genotype must have two recessive alleles for the trait before it can be expressed). Recent studies indicate that this may not be true in all cases. Chin spotting, for example, may not be a simple recessive as previously thought. Some geneticists now believe that the star, strip, and snip are dominant traits. If *St* designates the dominant allele for a star, and *st* designates the allele for no star, a horse with a star would be *StSt* or *Stst*, while a horse without the marking would be *stst*.

star = *StSt* or *Stst*
no star = *stst*

Light chestnut with light chestnut mane and tail.
Genotype: A_eeF_

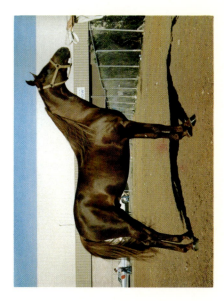

Medium chestnut with chestnut mane and tail.
Genotype: eeF_

Chestnut with flaxen mane and tail.
Genotype: eeff

Chestnut with flaxen mane and chestnut tail.
Genotype: eeff

276

Red Chestnut.
Genotype: eeF_

Liver (nonfading) with flaxen mane.
Genotype: bbE^D_ff

Dark chestnut with chestnut mane and tail.
Genotype: eeF_

Liver (nonfading).
Genotype: bbE^D_F_

Dark bay.
Genotype: Aa^tB_E_

Red bay: A_B_E_

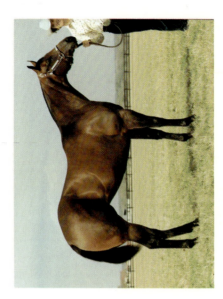

Black (nonfading).
Genotype: B_ED_

Black (fading).
Genotype: aaB_Ee or aaB_EE

278

Light bay
Genotype: A_B_E_

Medium seal brown.
Genotype: a^t_B_E_

Dark seal brown.
Genotype: a^t_B_E_

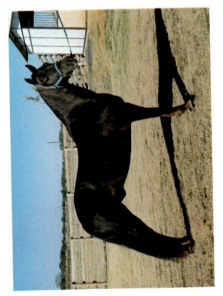

Light seal brown.
Genotype: a^t_bbE_

Black roan (extreme roaning).
Genotype: aaB_E_Rr

Chestnut with flaxen mane and tail (flank roaning).
Genotype: eeffRr

Chestnut roan (average roaning).
Genotype: eeF_Rr

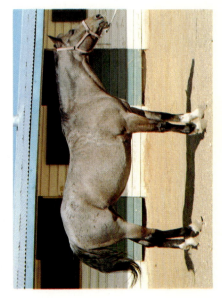

Bay roan (average roaning).
Genotype: A_B_E_Rr

Bay mare and buckskin foal.
Genotypes: A_B_E_ and A_B_E_Cc^{cr}

Bay foal losing baby coat and going grey.
Genotype: A_B_E_G_

Dapple grey mare and chestnut foal.
Genotypes: G_ and eeF_

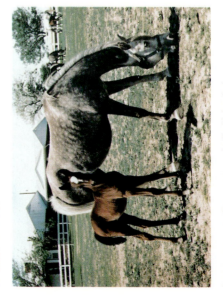

Dapple grey mare (flaxen mane and tail) with bay foal.
Genotypes: G_ff and A_B_E_

Dark seal brown going grey.
Genotype: a^t_B_E_G_

Rose grey (note: light face = grey).
Genotype: eeFfG_

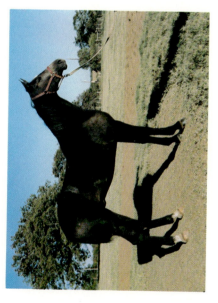

Liver (fading), flaxen mane and tail, flank roaning (dark face = roan)
Genotype: bbE_ffRr

Dapple grey (black mane and tail)
Genotype: G_ with gene for dappling

282

Grey - flea-bitten front/dappled rump.
Genotype: G_ with genes for dappling, flea-bitten

Grey gone white.
Genotype: G_

Flea-bitten grey.
Genotype: G_ with gene for flea-bitten

Silver dapple Shetland pony.
Genotype: aaB_E_Ss

283

Palomino.
Genotype: CcCᶜʳee

Cremello.
Genotype: eecᶜʳcᶜʳ

Chestnut (red).
Genotype: eeF_

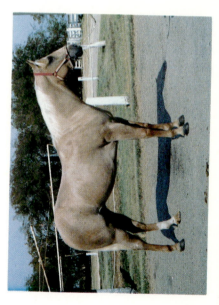

Palomino (roan)
Genotype: CcCᶜʳeeRr

284

Buckskin/Dun with dapples.
Genotype: A_Cc^{cr}E_ or A_D_E_ with gene for dappling

Buckskin/Dun (uniform yellow).
Genotype: A_Cc^{cr}E_ or A_D_E_

Buckskin/Dun with dorsal stripe.
Genotype: Aa^tCc^{cr}E_ or Aa^tD_E_

Perlino.
Genotype: A_c^{cr}c^{cr}E_

Light red dun.
Genotype: Aa^{t}Cc^{cr}ee or Aa^{t}D_ee

Dark red dun.
Genotype: a^t_Cc^{cr}ee or a^t_D_ee

Light grulla with zebra marks and dorsal stripe.
Genotype: a^t_B_CCD_E_

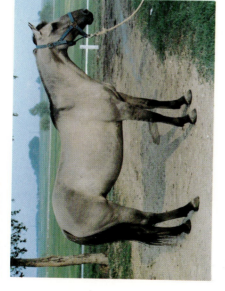

Grulla with zebra marks and dorsal stripe. .
Genotype: a^t_B_CCD_E_

286

Overo (chestnut).
Genotype: eeF_oott

Overo (chestnut).
Genotype: eeF_oott

Overo (bay).
Genotype: A_B_E_oott

Overo (palomino).
Genotype: Cc^{Cr}eeoott

Tobiano (bay).
Genotype: A_B_E_T_O_

Tobiano (note, loss of contrast due to greying gene).
Genotype: G_T_O_

Tobiano (chestnut).
Genotype: eeF_T_O_

Tobiano (black).
Genotype: aaB_E_T_O_

288

Dun Appaloosa with extended blanket, no spots, little contrast.
Genotype: Ap_A_B_Cc^{cr}DdE_wb$_1$wb$_1$wb$_2$wb$_2$Wb$_3$_Sp$_1$_

Chestnut Appaloosa with extended blanket and spots.
Genotype: Ap_eeF_wb$_1$wb$_1$wb$_2$wb$_2$Wb$_3$_sp$_1$sp$_1$Sp$_2$_

Chestnut Appaloosa with blanket, few spots.
Genotype: Ap_eeF_wb$_1$wb$_1$Wb$_2$_sp$_1$sp$_1$Sp$_2$_

Chestnut Appaloosa with spotted blanket, roaning around spots.
Genotype: Ap_eeF_wb$_1$wb$_1$Wb$_2$_sp$_1$sp$_1$Sp$_2$_

289

Chestnut Appaloosa with small white spots over body.
Genotype: Ap_Wb$_1$_sp$_1$sp$_1$Sp$_2$—

Blanket Appaloosa (grey, contrast lost).
Genotype: Ap_wb$_1$wb$_1$Wb$_2$_Sp$_1$_G_

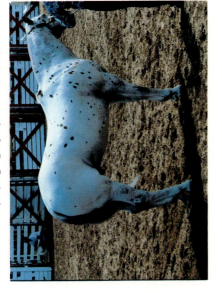

Leopard Appaloosa.
Genotype: Ap_wb$_1$wb$_1$wb$_2$wb$_2$wb$_3$wb$_3$sp$_1$sp$_1$sp$_2$sp$_2$Sp$_3$—

Blanket Appaloosa with greying gene and extensive spotting.
Genotype: Ap_wb$_1$wb$_1$wb$_2$wb$_2$Wb$_3$_sp$_1$sp$_1$sp$_2$sp$_2$Sp$_3$_G_

Cremello.
Genotype: $c^{cr}c^{cr}ee$

Grey gone white.
Genotype: G_

Appaloosa gone grey.
Genotype: Ap_G_

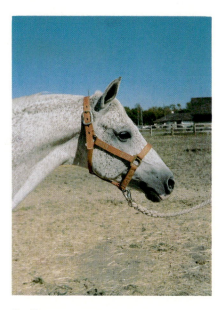

Flea-bitten grey.
Genotype: G_ with gene for flea-bitten

Left: Small star connected to a narrow strip that extends halfway down the face. Right: Offset blaze and a white muzzle. Notice that the horse has one dark eye and one heterochromic eye.

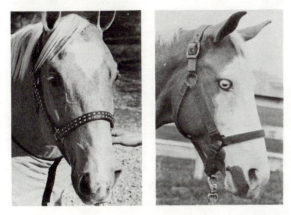

Left: Large triangular star and a snip. Right: Bald face with white lower lip and dark upper lip.

The horse with a homozygous star *(StSt)* should always produce offspring with stars on their foreheads. This is because the horse will always transmit the dominant *St* to his offspring, causing the expression of the star regardless of the corresponding allele (allele that is transmitted by the other parent). Similarly, the strip and snip are thought to be dominant traits:

strip = *SrSr* or *Srsr*
no strip = *srsr*
snip = *SnSn* or *Snsn*
no snip = *snsn*
strip/snip = *SrSr* (or *Srsr*) and *SnSn* (or *Snsn*).

Small star.

Snip.

Star, strip and an
unconnected snip.

Connected star, strip and
snip.

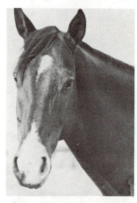

Star with white muzzle.

Blaze.

Forked blaze.

White underlip.

Apron face.

In the Arabian, small, white markings located on the belly or side of the horse are occasionally seen. These spots, known as "body marks," are frequently associated with white face and/or leg markings. The close association between high white leg markings and the tobiano pattern, or bald face markings and the overo pattern, indicates that the controlling genes may be located on the same chromosome (refer to the discussion on linkage under **CYTOGENETICS AND PROBABILITY**). It should be noted, however, that the presence of high white leg markings, white face markings, and a small body mark does not necessarily indicate the presence of tobiano or overo genes. Instead, such white markings might indicate that the horse carries a number of white modifiers, but lacks the tobiano or overo genes (refer to the discussion on tobiano and overo spotting).

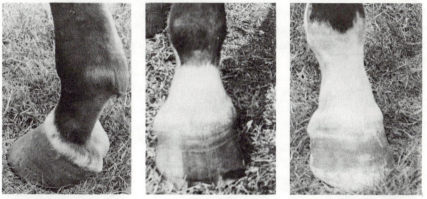

Coronet.   Pastern.   Fetlock.

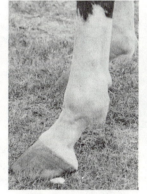

Sock (half stocking).   Three-quarter stocking.   Full stocking.

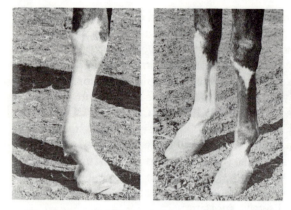

Left: Stocking plus.
Right: Stocking plus (right foreleg) and a pastern with a white patch below the knee (left foreleg). Knee patches without additional white on the legs are also seen.

The inheritance of shorter stockings (socks), white pasterns, white coronets, and a white muzzle has not yet been determined. The genes which control leg markings on a particular leg are thought to be unique to that leg (i.e., markings on each leg are controlled by separate genes).

Distal leg spots are small dark markings just above the hoof. These spots can appear only on white legs and are believed to be caused by a dominant gene. If the markings include the coronary band, a dark stripe will appear in the hoof below the coronary band.

This distal leg spot (arrow) is a spot of color found near the coronary band on a white leg marking. Note that the hoof is pigmented below the distal leg spot.

Distal leg spots:	*D1D1* and the gene for white legs
	D1d1 and the gene for white legs
No distal leg spots:	*D1D1* and no gene for white legs
	D1d1 and no gene for white legs
	d1d1

Spots may also occur within face markings, and leg markings are sometimes broken by a patch of intervening color (e.g., a 3/4 white stocking, a band of color, and a white knee patch). The inheritance pathway of these deviations within white markings is not yet known.

Blaze with two colored spots in the star area.

Colors That Always Breed True

PHENOTYPE	GENOTYPE
chestnut or sorrel	_ _ CC dd ee gg rr tt OO ww
isabella	_ _ CC DD ee gg rr tt OO ww ff
cremello	_ _ $c^{cr}c^{cr}$ dd ee gg rr tt OO ww
overo	_ _ _ _ _ _ _ _ _ oo ww*

Colors That Never Breed True

PHENOTYPE	GENOTYPE
palomino	_ _ Cc^{cr} dd ee gg rr tt OO ww
buckskin	A_ B_ Cc^{cr} dd E_ gg rr tt OO ww
roan	_ _ _ _ _ _ Rr _ _ ww
dominant white	_ _ _ _ _ _ _ _ _ _ Ww

Colors That Breed True When Homozygous

PHENOTYPE	GENOTYPE
bay	AA BB CC dd EE gg rr tt OO ww
dark seal brown	$a^{t}a^{t}$ BB CC dd EE gg rr tt OO ww
light seal brown	$a^{t}a^{t}$ bb CC dd EE gg rr tt OO ww
fading black	aa BB CC dd EE gg rr tt OO ww
non-fading black	aa BB CC dd $E^{D}E^{D}$ gg rr tt OO ww
fading liver	aa bb CC dd EE gg rr tt OO ww
non-fading liver	aa bb CC dd $E^{D}E^{D}$ gg rr tt OO ww
grey	_ _ _ _ _ GG _ _ _ ww
perlino	AA BB $c^{cr}c^{cr}$ dd EE gg rr tt OO ww

tobiano	_ _ _ _ _ _ _ _ TT _ ww*
buckskin/dun	AA BB CC DD EE gg rr tt OO ww
yellow dun/yellow dun mane and tail	_ _ CC DD ee gg rr tt OO ww
brown grulla	$a^t a^t$ bb CC DD EE gg rr tt OO ww
blue grulla	$a^t a^t$ BB CC DD EE gg rr tt OO ww

*NOTE: tobiano and overo horses, just like all other colored horses, may turn grey with age and eventually lose their color pattern. Nevertheless, homozygous tobiano and overo horses do breed true.

Coat Texture

The hair of the horse's coat consists of several different types. First, there are the short, smooth hairs which comprise the basic coat. Secondly, long, coarse hairs form the mane and tail. Thirdly, hair that is intermediate between these two extremes may cause feathering around the fetlocks, pasterns, and cannons. Besides these variations in the hair type, there are differences between breeds. For example, the mane and tail of an Arabian may be much finer than the feathering of a draft horse.

During the winter, the horse grows a longer coat of hair which provides added insulation and protection from the weather. Although the winter coat tends to be coarser and thicker than the summer coat, there are still definite breed differences. For example, the winter coat of a Thoroughbred would not be as thick as that of a Shetland pony.

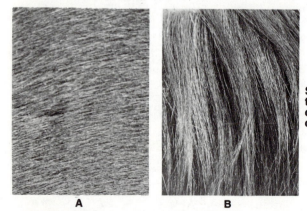

Short smooth hair (A) of the horse's coat compared with long, coarse mane hair (B).

    A   **B**

Differences in coat texture, however, consist of more than simple variations in thickness, length, or fineness. Feathering, curly hair, Appaloosa spots, and whorls are all conditions which affect coat texture.

FEATHERING

Feathering is the term used to describe the long hair that is found on the cannon, fetlock, and pastern area of some horses. It may range from a small tuft of hair on the back of the fetlock to an abundance of long hair from the knee (carpus) or hock down (as in the Shire breed).

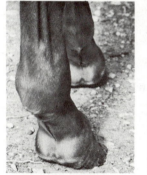

Absence of feathering.   **Moderate feathering at the fetlocks.**   **Heavy feathering on the forelegs of a draft horse.**

At one time, draft horse breeders believed that abundant feathering was associated with good bone structure. Although feathering might conceal poor bone structure, a genetic relationship between extensive feathering and good bones has not been established. Horses with heavy feathering easily pick up mud, dirt, etc., and require extra grooming. Because feathering holds dirt and grime, and because of the extra washing needed to keep these areas clean, horses with feathering tend to develop grease heel, scratches, and other skin conditions that affect the heel and pastern area.

Gradually, as feathering fell into disfavor, breeders began selecting for more clean-legged horses. Feathering is a quantitative trait since there is a continuous variation in the amount of feathering a horse could have — from none at all to abundant. Therefore, it is probably the result of additive gene action, which means that there are undoubtedly several pairs of genes which increase feathering. A horse with none of these genes would be clean-legged, while a Shire or Clydesdale would probably carry many of the genes.

Additive Gene Action
(*Fe* = feathering; *fe* = no feathering)

fe fe fe fe = no feathering
Fe Fe Fe Fe = abundant feathering
Fe Fe fe fe = moderate feathering

CURLY HAIR

Most horses have straight hair on the body and in the mane and tail. A cross-section of a strand of straight hair appears round.

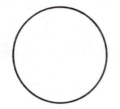

Occasionally, horses will have curly hair, which is flat in cross-section.

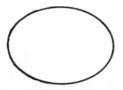

Curly hair may occur only in the mane and tail, or it may be all over the body. The coats of some breeds tend to be curly at birth, then gradually shed out to a normal straight coat. The degree of curl may vary, from slight waving to tight marcelled waving or "poodle" curls. A very curly-haired horse frequently has a mane and/or tail that is in corkscrew curls.

A   B

(A) Genetically influenced curly coat. (B) Close-up showing texture of the poodle-like curls.

Curly hair in Percheron horses has been shown to be due to a simple recessive. If curly hair is represented by *cu,* and straight hair by *Cu,* then straight-haired Percherons are *CuCu* or *Cucu,* while curly-haired Percherons are *cucu.* Curly coat has also been seen occasionally in the Missouri Fox Trotting horse and other breeds.

Bashkir, the name for the curly horses raised by the Bashkir people of Russia, is also the name adopted for an American breed of small horses that possess a curly coat as a breed characteristic. These horses are reputed not to cause an allergic reaction in people who normally are allergic to horses. (It is usually dander which causes an allergic reaction; i.e., the small scales of dead skin and debris that flake off the skin and hair. Perhaps the dander of a curly-coated horse is different in some way so that it does not cause an allergy.) The mode of inheritance of the curly coat in Bashkir horses is unclear, but it is probably some type of recessive inheritance, similar to that seen in Percherons.

Long, wavy hair is also seen in aged horses with pituitary tumors. The tumor causes excessive water drinking and urinating, along with the unusual hair coat. In addition, these horses are frequently thin and in poor condition. Old horses with pituitary tumors should not be mistaken for horses with naturally curly coats.

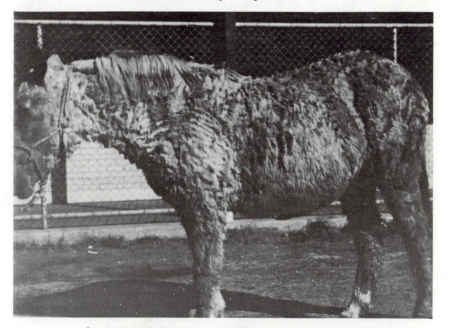

Long, wavy hair due to a pituitary tumor should not be confused with wavy or curly hair caused by heredity.

Wavy mane.

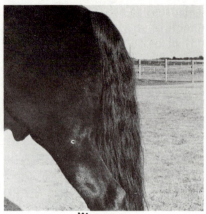
Wavy tail.

WHORLS (TRICHOGLYPHS)

Whorls, or trichoglyphs, are the swirls of hair caused by the hair at one specific spot changing its direction of growth relative to the surrounding hair. They commonly occur along the midline of the neck or the forehead. As with chestnuts, no two whorls are absolutely identical in appearance or location. This permits them to be used for identification. The exact cause of this change in the direction of hair growth is unknown. It is probable, however, that it is genetic in origin as it is in other species. Whorls in swine, for example, are known to be caused by two dominant genes.

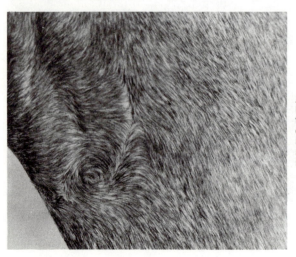

A swirl of hair, or whorl, located on the midline of the neck. Whorls are often used as part of an identification system, since no two are identical.

APPALOOSA SPOTS

The dark spots on an Appaloosa horse are of a different texture than the surrounding hair, unlike those of a tobiano or overo horse, which are of the same texture as the overall coat. Dark Appaloosa spots have a coarser feel than the smoother background coat.

The reason for this phenomenon is unknown, but hair structure is known to influence color. Perhaps the texture of the hair at a certain area allows a spot to develop at that point. In any case, this change in hair texture is believed to be inherited in association with Appaloosa spotting.

12

HEREDITY AND GAITS

The walk (a four-beat gait), the trot (a two-beat diagonal gait), and the canter (a three-beat gait) are usually considered the three basic gaits of the horse, and almost all horses perform them. However, some horses are bred to perform specialized gaits and, in these instances, the absence of a true trot or even a canter may be considered desirable. Theoretically, there are 104 possible gaits for a four-legged animal, 55 of which a horse might be physically capable of performing. In earlier times, before the advent of superior roads and the introduction of harness, most horses were used for riding. These horses were selected for a comfortable, ambling gait (a broken pace). When roads improved and wheeled transportation became prevalent, trotting horses were in greater demand. As a result, horses with ambling gaits were bred only in such areas as Spain, Iceland, South America, and the Caribbean, areas which lacked comprehensive road systems. Now that horses are more of a luxury item than a necessity, "gaited" horses have regained some of their popularity. The gaits performed by the wide variety of gaited horses include the running walk, rack, fox trot, pace and paso. In some cases, the horse performs one or more of these movements naturally from birth, while in others this native ability is improved by training.

The walk is identified by its regular four-beat pattern; each hoof strikes the ground successionally — right hind, right fore, left hind, left fore, etc. This sequence illustrates the characteristic pattern. Note that there is no moment of suspension during this gait; at least two feet are on the ground at all times.

Walk

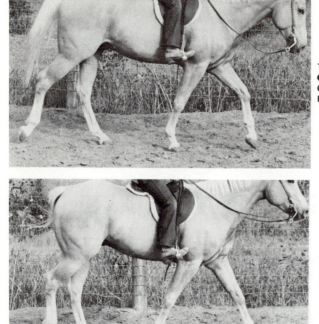

1. Left forefoot extends prior to contact, while the right hind flexes.

2. Left forefoot strikes the ground; right hind is pulled forward.

3. Extension of the right hind occurs as the right fore leaves the ground.

Walk

4. The lateral pattern of the walk is well-illustrated at this point: the right hind strikes the ground as the right fore prepares to do the same.

5. Right forefoot flexes as the body is pushed forward by the hindlimbs.

6. Right forefoot moves forward.

Walk

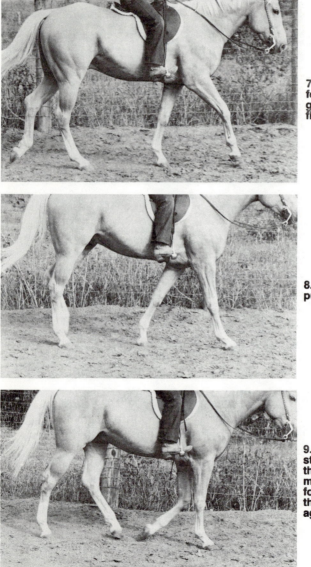

7. As the right fore strikes the ground, the left hind flexes.

8. The left hind is pulled forward.

9. The left hind strikes the ground as the left fore begins its movement up, forward, down, and the sequence begins again (see step 1).

Trot

The trot is characterized by the paired motion of diagonal front and hind limbs: e.g., the left forelimb and the right hind-limb strike the ground simultaneously. A "true" trot is an even two-beat gait. There may be a slight period of suspension, if two diagonal feet leave the ground just before the other pair strikes.

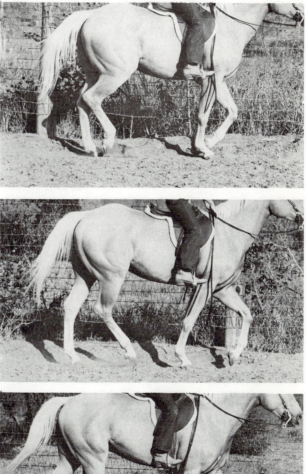

1. The right hind and left forelimb move forward.

2. The right hind and left fore extend simultaneously.

3. As the right hind and left fore strike the ground, the opposite diagonal pair leaves the surface.

Trot

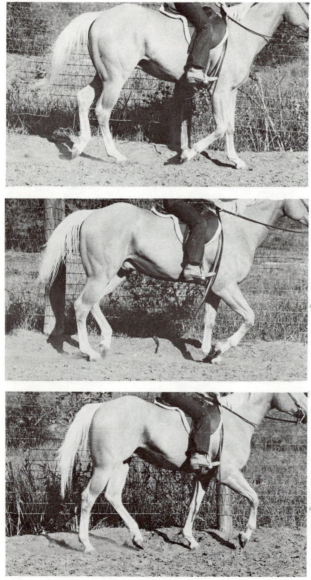

4. The left hind and right forelimb flex at the same time.

5. Left hind and right fore move forward.

6. The body is propelled forward by the right hind and left fore, as the left hind and right fore move toward the ground.

Trot

7. The left hind and right fore strike the ground, as the opposite pair begin their movement up and forward (see step 1).

The canter (a restrained gallop) is a three-beat gait: 1) the beat of the leading front foot, 2) the simultaneous beats of the opposite front foot and its diagonal hind foot and 3) the beat of the leading hind foot. Like the gallop, the canter places extra stress on the leading forelimb and hindlimb. (Changing leads relieves the stressed limbs.)

Canter

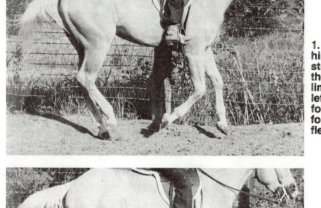

1. The leading hindlimb (left hind) strikes the ground, as the paired diagonal limbs (right hind and left fore) move forward; the leading forelimb (right fore) is flexed.

2. The diagonal pair move toward the ground. Note that, at this point, the body is supported only by the leading hindlimb.

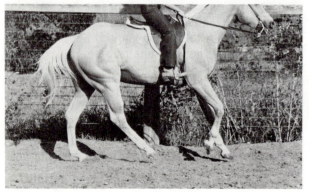

3. The diagonal pair strike the ground, as the leading forelimb (right fore) moves forward.

Canter

4. The body is supported by the diagonal pair, while the leading forelimb begins its descent, and the leading hindlimb leaves the ground.

5. The leading forelimb is extended prior to contact with the ground.

6. After the leading forelimb strikes the ground, the leading hindlimb moves forward, and the diagonal pair begin to leave the ground. Note the stress on the leading forelimb (arrows in figures 6, 7 and 8) as it receives the impact of forward movement coupled with the horse's body weight.

Canter

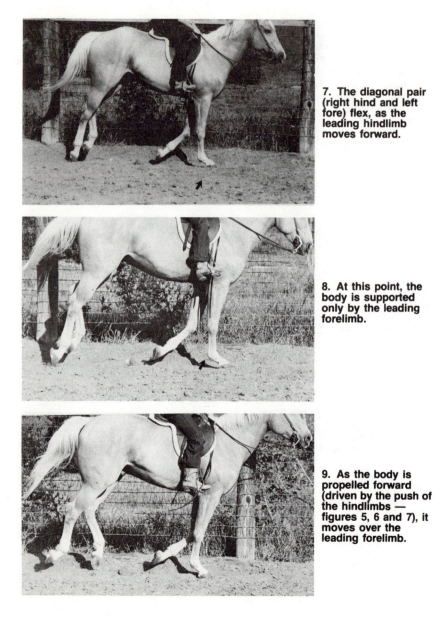

7. The diagonal pair (right hind and left fore) flex, as the leading hindlimb moves forward.

8. At this point, the body is supported only by the leading forelimb.

9. As the body is propelled forward (driven by the push of the hindlimbs — figures 5, 6 and 7), it moves over the leading forelimb.

Canter

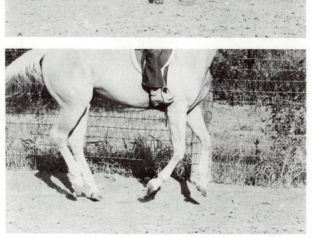

10. Moment of suspension. As the leading forelimb leaves the ground, the body is temporarily suspended in air.

11. The leading hindlimb (left hind) extends just prior to contact with the ground, completing the canter sequence (see step 1).

Backing is a two-beat diagonal gait. The following sequence illustrates how backing resembles a reversed trotting pattern.

Backing

1. The left foreleg and right hindleg leave the ground.

2. The left fore and right hindlegs move backwards simultaneously.

3. As the left fore and right hindlegs strike the ground, the opposite diagonal (right fore and left hind) prepare to leave the surface.

Backing

4. The right fore and left hindlegs begin their movement backwards.

5. The right fore and left hind are on the ground as the opposite diagonal limbs continue the pattern (see step 1).

Running Walk

The running walk (a four-beat gait) is the characteristic "slow gait" of the Tennessee Walking Horse, and is sometimes found in lesser-known breeds, such as the Galiceno and the Icelandic. (In this instance, running walk does not refer to the fast walk of the Quarter Horse.) During the running walk, the horse oversteps the print of the front foot with the hind foot. The length of overstepping may vary from a few inches to a few feet, and the speed attained can reach that of a canter. The horse nods his head vigorously during this smooth, gliding movement. The conformation of the Tennessee Walking Horse, with its characteristic sloping shoulder, short back and set-back rear legs, frequently accompanied by slightly sickled hocks, probably facilitates execution of the running walk. This gait may be dominant, since it is sometimes performed by half-bred Tennessee Walking Horses.

Rack

The rack is a fast, high-stepping, four-beat gait which is extremely smooth for the rider, and is commonly performed by the five-gaited American Saddlebred. During the rack, each foot strikes the ground separately at equal intervals, but the hind foot does not overstep the front foot as much as in the running walk. Because the rack involves a great deal of knee and hock flexion, it requires considerable exertion on the part of the horse. Although this gait is usually classified as man-made (artificial), some horses are born with a natural ability to perform the movement with little training. Some trainers believe that a horse with a strong trot can be taught to rack more easily than a horse with a strong rack can be trained to trot. Since training has such a great influence on a horse's ability to rack, the role of inheritance in this gait is unclear.

Fox Trot

The fox trot is a broken trot, in which the diagonal feet do not strike the ground at the same time; the hind foot reaches the ground slightly before the front foot of the diagonal pair. A tendency to perform this gait is probably inherited, although some trainers feel most horses can be taught to do it by forcing them to shift their weight from side-to-side while trotting. There are two registries for fox trotting horses.

Pace

The pace is one of the two racing gaits of the Standardbred (the other being the trot). The pace is a two-beat lateral gait in which the two legs on the same side contact the ground at the same time. This gait is faster than the trot and, consequently, more Standardbreds race at this gait. Although there are some natural "free-legged" pacers, most wear pacing hopples during a race to prevent them from breaking gait. Studies to determine whether or not pacing is dominant or recessive to trotting have had conflicting results, so the mode of inheritance is unknown.

Paso Gaits

The basic paso gait is similar to a broken pace, in which the lateral feet do not strike the ground at the same time. Instead, the hind foot reaches the ground a moment before the front foot, greatly increasing the smoothness of the ride. There are slight variations between the paso gaits, depending on speed and animation.

A number of breeds of horses, all of Andalusian origin, perform the same basic gait. In the United States, the two most popular breeds that perform this gait are the Paso Fino and the Peruvian Paso. The major difference between the two is that the Peruvian Paso also has "termino," which is an outward swinging of the front legs in an arc before the foot touches the ground. The sloping hindquarters and low-set tail of the Paso horse probably contribute to the performance of this gait. The Paso registries state that all Paso horses perform the gait, so it is undoubtedly an inherited trait. In addition, half-bred Paso horses occasionally perform some form of this movement, although they usually do not perform a pure paso gait. This implies that the gait is not due to a simple recessive allele.

The running walk is a fast four-beat gait (intermediate between the walk and the rack). This gait is characterized by overstepping, where the hind foot reaches past the print of the front foot. During the running walk the hind foot strikes the ground just before the front foot of the same side contacts the surface (broken pace). This similarity between the running walk and the pace is illustrated by the following sequence.

Running Walk

1. As the left hind foot leaves the ground, the right fore moves forward. Note the position of the left fore in comparison to the paper in the foreground (future reference point).

2. The right forefoot strikes the ground, as the left hind moves forward. The left forefoot leaves the surface.

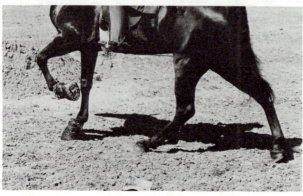

3. The left hind foot strikes the ground, as the left fore prepares to do the same. This broken lateral movement is characteristic of both the walk and the running walk. Compare the position of the left forefoot in step 1. This overstep gives the horse a gliding motion for a smooth comfortable ride.

Running Walk

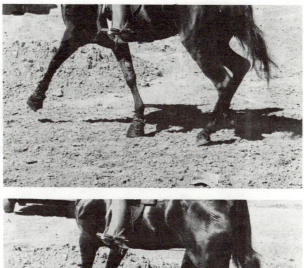

4. As the left forefoot moves toward the ground, the right hind is lifted from the surface.

5. The right hindlimb extends, as the right forelimb flexes. The sequence is then repeated (see step 1).

This sequence illustrates a variation of the trot as performed by an American Saddlebred. This trot is characterized by higher action.

Collected Trot

1. The right fore and left hindlimbs move toward the ground, as the opposite diagonal prepares to leave the surface.

2. The right fore and left hindlegs support the body while the left fore and right hind move upward.

3. The high action of the trot is illustrated by the flexed positions of the right hind and left forelimbs.

Collected Trot

4. The right hind and left forelimbs extend as the opposite diagonal pair leave the surface.

5. The right fore and left hind move upward, and the pattern is repeated.

The rack is an exaggerated four-beat gait, characterized by flashy high-stepping action and smooth gliding movement. Although the rack involves less overstep than the running walk, it is much harder on the horse.

Rack

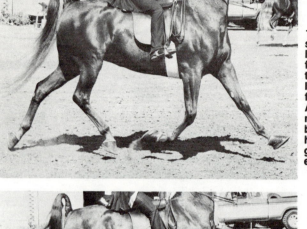

1. A moment of suspension: the right hind foot and the left forefoot move toward the ground, while the left hind and right forelimbs flex. Although the movement appears to be diagonal, each hoof strikes the ground separately at even intervals.

2. The right hind strikes the ground well under the body, while the left fore reaches forward. The opposite diagonal pair move upward.

3. The left fore strikes the ground, and the body is pushed forward by the right hindlimb.

Rack

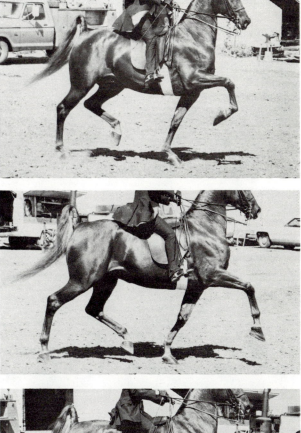

4. Extension of the left hindlimb is accompanied by the high action of the right forelimb.

5. The left hindlimb moves toward the surface, as the right fore reaches forward.

6. Seconds after the left hind makes contact with the surface, the right fore strikes the ground. Again, the high action of the diagonal limbs occurs as the body is supported over the opposite diagonal pair. The foot-fall pattern is repeated: right hind, left fore, left hind, right fore, etc.

13

INHERITED
ABNORMALITIES

It has been estimated that approximately one horse out of every five is born with some type of abnormality; a small number of these anomalies are severe or lethal. Considerable economic loss may be caused by a related decrease in breed quality (e.g., the production of physically handicapped horses or an increase in the number of stillbirths and abortions). A knowledge of genetic influence upon undesirable traits increases the breeder's chances of producing a full, healthy foal crop.

Abnormalities may be the result of improper body function (physiological abnormality), an imperfect body structure (anatomical abnormality), or a "cause and effect" interaction between structure and function (e.g., abnormal organ or gland structure resulting in improper function). These defects can be caused by:

1. heredity (e.g., hemophilia)
2. environment (e.g., limb fracture)
3. an interaction between heredity and environment (e.g., inherited calf-knees increase the horse's susceptibility to carpal chips)

Frequently, the relative importance of heredity versus environment as the cause of a defect is difficult to determine. In fact, the environment may cause a condition which mimics an inherited one. These environment-related abnormalities are referred to by geneticists as phenocopies. For example, epiphysitis (inflammation of the growth plates in the long bones), caused by a nutritional imbalance and/or trauma, may be indistinguishable from a similar condition caused by an inherited growth pattern or conformation weakness. This chapter considers both defects caused directly by heredity and those due to interaction between heredity and environment.

A genetic predisposition to a disorder is an inherited susceptibility to develop the disorder when stressed by certain environmental conditions. Genetic predisposition in a herd is suspected when only a few closely related horses develop the disorder although the entire herd is subjected to the same environment. If the frequency of a condition increases within a breeding herd at a rate that cannot be justified by environment alone, genetic control must be considered.

Because equine genetic research has been somewhat limited, the exact role of heredity (dominant, epistatic, etc.) is unknown for many inherited abnormalities. This chapter considers defects and disorders that have been established as hereditary as well as abnormalities that are thought to be genetically influenced; any supporting evidence is cited. The fact that a defect is limited to a specific breed, for example, suggests that it is hereditary. (Related horses have a greater number of similar gene types than horses of different breeds. An undesirable recessive trait, caused by the homozygous recessive condition, will occur more frequently in the offspring of related or linebred horses than in the offspring of unrelated horses. Refer to **"Inbreeding and Outbreeding."**)

Many of the abnormalities covered in this chapter are congenital, meaning that they are apparent at birth. This abnormal development during gestation could be the result of environmental factors (e.g., nutritional imbalances in the dam, improper medication during pregnancy, maternal illness, etc.), inherited factors, or both.

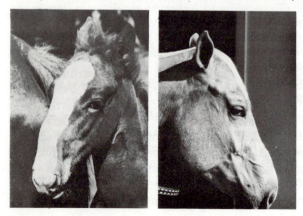

Left: Congenital face deformity known as bent nose.

Right: Another congenital face deformity, camel nose.

Only those congenital defects which are believed to be inherited will be included within this chapter. Because many congenital defects of unknown cause could be the result of both environmental and genetic factors, a list of inborn abnormalities that have occurred in the horse is provided within the appendix.

If an inherited abnormality causes the death of the affected horse, it is referred to as a lethal trait. Because lethal genes do not always cause immediate death, varying degrees of lethality are recognized. Lethal genes are classified into the following categories:

True Lethals: True lethal genes cause death before birth. The death often occurs during early gestation, and results in absorption of the fetus. True lethals are an important cause of decreased foaling percentages. If the true lethal acts immediately after conception, it causes an apparent loss of fertility. (The mare is believed to have never been in foal.)

Delayed Lethals: Delayed lethal genes cause death sometime after birth. Inherited heart defects, which cause heart failure in the mature horse, are classified as delayed lethals.

Partial Lethals: Partial lethal genes cause death only in conjunction with certain environmental conditions. As a case in point, hemophilia (an inherited abnormality of the clotting mechanism) results in death only when accompanied by physical injury. Otherwise, the horse's health is unaffected by the defect.

Detrimental Genes: Genes which do not usually cause death, but frequently interfere with the horse's performance (reducing athletic ability, vigor, or life span), are referred to as detrimental genes. The genes which control many conformation defects fit this category. A horse with short upright pasterns, for example, is highly susceptible to breakdown when physically stressed. Inherited weaknesses can result in the need to euthanize a severely injured horse.

Prior to domestication, the horse constantly developed in a direction that enabled it to survive in its environment. Both detrimental and helpful mutations occurred; the desirable traits often remained within the population, while lethal and detrimental traits were sometimes eliminated. The role of heredity is explained as follows:

1. Harmful recessive mutations usually remained in the population, since heterozygous carriers transmit the recessive alleles to about half of their offspring. (The effect of a harmful recessive allele is overridden by the effects of a corresponding dominant allele.)

2. Mutations which caused either delayed or partial lethals were passed from one generation to the next only if the affected horse

lived to maturity and reproduced.

3. When completely dominant, the true lethal mutations were removed from the population, since carriers always died during the gestation period.

4. When incompletely dominant, the true lethal mutations could be carried by healthy heterozygotes. (The allele which causes the roan coat pattern, for example, is an incompletely dominant, true lethal mutation. In the heterozygous form, the allele causes the roaning pattern. In the homozygous form, the allele causes the death of the fetus. Refer to "Coat Color and Texture.")

Today, most detrimental and lethal genes are either recessive or incompletely dominant.

If the overall incidence of undesirable recessive, or incompletely dominant, alleles is low, they are usually revealed only by inbreeding. (These harmful alleles are expressed in the homozygous form; inbreeding causes an increase in homozygous gene pairs. Refer to "Inbreeding and Outbreeding".) Unfortunately, as all breeds of pedigreed horses are inbred (to some degree), there is a constant danger of these harmful recessives being revealed. In the more highly inbred breeds, the incidence of harmful recessive and incomplete dominant abnormalities may be high. For example, the occurrence of CID (combined immunodeficiency disease) in approximately 2% of all Arabian foals indicates that as many as 25% of all Arabian horses carry the controlling recessive allele.

A breeder must decide the relative importance of any abnormality within his breeding stock. For example, the dominant white coat may be so important to his goals that it is worth a decreased foal crop. (White coat is controlled by an incompletely dominant, true lethal allele. Refer to **"Coat Color And Texture."**) A racing stallion, known to pass on a minor conformation defect to his offspring may also pass on speed and a strong will to run. Therefore, he may still be valuable to a breeding program.

The breeder should also determine the breeding program philosophy most feasible and applicable to his own circumstances. An outbreeding program will limit the appearance of detrimental and lethal alleles, but also limits the breeder's ability to isolate and eliminate the undesirable alleles from his herd. Inbreeding, on the other hand, increases the chance of exposing undesirable alleles but, by the same token, provides the breeder with an opportunity to cull affected horses and eliminate the harmful alleles from his herd. (Refer to **"Inbreeding and Outbreeding."**)

The breeder can use the following guidelines, in conjunction with the information contained in this section, to help reduce the incidence of inherited abnormalities in his herd:

1. *Recognize when a disease is inherited.* Generally, if there is a marked familial tendency (if it "runs in families or lines"), or if a non-infectious disorder tends to occur even under good management, genetic factors should be suspected as the cause.

2. *Make an accurate diagnosis* when an inherited disorder appears. The signs of some inherited abnormalities are obvious, while others can be confused with conditions caused by the environment.

3. *Identify the carrier animals.* If an inherited recessive defect is seen even in only one offspring, both the sire and dam must be carriers of the detrimental gene. The breeder might choose to eliminate one or both parents from his breeding herd.

4. *Know the mode of inheritance involved* (whether the inherited abnormality is dominant, recessive, controlled by many genes (polygenetic), etc.). Because these factors limit selection progress, they influence the breeder's ability to eliminate a trait from his herd. (Refer to "Breed Improvement Through Applied Genetics.")

This book will help the breeder employ these guidelines in making intelligent decisions about the future of his selection and breeding program. This chapter provides a list of genetically influenced abnormalities that should be considered throughout the selection process.

Conformation

The importance of a thorough soundness examination prior to a horse's acquisition and the close relationship between conformation and function have been emphasized in previous chapters. In this discussion, conformation defects which might be passed from parent to offspring will be examined.

Due to limited research in the field of equine genetics, it is not possible to classify all conformation abnormalities as either inherited or environmental. Whether a physical defect is inborn (congenital) or apparent sometime after birth (acquired), it could be the result of environmental and/or genetic factors. For this reason, all conformation defects and breakdowns that may have been caused by inherited weakness should be carefully considered when selecting a horse for breeding purposes. Although there are also many environment-related conformation defects (e.g., due to injury or poor nutrition) that affect the horse's movement or ability to withstand stress, a complete analysis of their causes and effects is beyond the scope of this discussion.

HEAD AND NECK
Parrot Mouth

Parrot mouth is an inherited deformity involving either the overgrowth of the upper jaw (superior prognathia) or an underdeveloped lower jaw (inferior brachygnathia). Normally, the horse's teeth grow continuously; correct alignment between the upper and lower incisors and molars is necessary for even wear and successful grazing. Parrot mouth interrupts this normal arrangement and allows the premolar and molar teeth to form sharp, uneven edges. For example, unchecked growth of the sixth cheek tooth in the lower jaw permits the formation of a sharp hook. This hook can cause a deep wound in the gum of the upper jaw (at the angle of the jaw behind the sixth cheek tooth). In severe cases, overgrowth of the lower incisors may irritate the gum and palate above. Malalignment of the lower incisors is not necessarily accompanied by malalignment of the molar teeth.

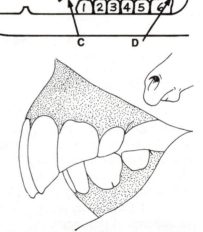

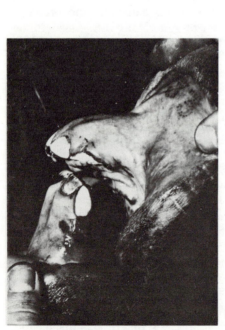

Parrot Mouth. Although malalignment of the incisors (A) is not necessarily accompanied by malalignment of the molars (B), parrot mouth may cause both. Due to the formation of hooks on poorly aligned front and back molar teeth (C & D), malalignment of the molars causes oral pain more frequently than poor alignment of the incisors.

Parrot Mouth.

Some authorities believe that parrot mouth is caused by a dominant allele. Therefore, at least half of an affected horse's offspring should (theoretically) inherit parrot mouth:

$$\text{Parrot Mouth} = p^D p^d \text{ or } p^D p^D$$
$$\text{Normal} = p^d p^d$$

1)

Heterozygous
Parrot Mouth   X   Normal

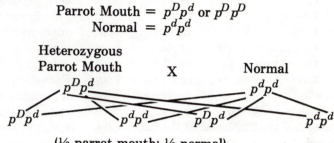

(½ parrot mouth; ½ normal)

2)

Homozygous
Parrot Mouth   X   Normal

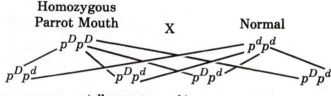

(all parrot mouth)

Although parrot mouth might be considered undesirable, it does not interfere with athletic ability, unless oral pain is caused by a persistent ulcer of the mouth or digestive problems are caused by improper chewing. Proper dental care can alleviate discomfort due to sharp uneven teeth. It should be noted that poor alignment of the incisor teeth interferes with prehension (grasping of food), and poor alignment of the molars may interfere with mastication (chewing), possibly resulting in digestive problems. From a medical viewpoint, poor alignment of the molar teeth is more likely to cause trouble than poor alignment of the incisors.

Bulldog Mouth

Bulldog mouth is a rare congenital condition, involving an underdeveloped upper jaw or an overgrowth of the lower jaw. This deformity positions the upper incisors and molars behind corresponding teeth on the lower jaw, resulting in unrestrained growth, uneven wear, and sharp edges on the teeth. In this case, there is a tendency for a hook to grow on the sixth cheek tooth of the upper jaw. As with parrot mouth, regular dental care prevents damage to the gums.

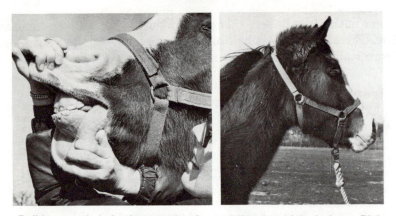

Bulldog mouth. Left, close-up showing malalignment of the incisors. Right, head profile of the same horse.

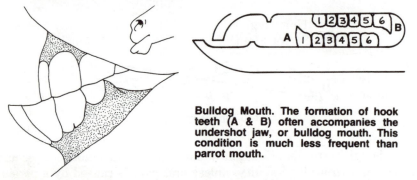

Bulldog Mouth. The formation of hook teeth (A & B) often accompanies the undershot jaw, or bulldog mouth. This condition is much less frequent than parrot mouth.

The occurrence of undershot jaw in several pony lines indicates heredity as a causative factor. The exact mode of inheritance has not yet been indicated, however.

Predisposition to Periodontitis

Periodontal disease involves inflammation and erosion of certain areas of the gum (usually between two slightly separated teeth where pieces of food might be lodged). If the erosion is not stopped, the cavity harbors food and continues to decay the supportive tissue until the tooth is lost.

The initial development of periodontitis is aggravated by the absence of normal "shearing" (wearing of the teeth caused by a side-to-side grinding motion). Normal wear is essential for the maintenance of healthy (decay resistant) teeth and supportive tissue. Inherited abnormalities of tooth eruption (crooked teeth) or inherited deformities such as parrot mouth interfere with proper prehension, mastication, and tooth wear ("shearing"), and increase the horse's susceptibility to periodontitis.

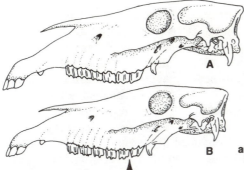

Normal (A) vs. irregular spacing (B) of the cheek teeth. Separation of the teeth (arrow) predisposes the gums and supportive tissue to decay.

According to some sources, several other defects of the teeth which may have hereditary basis are:

1) Supernumerary teeth (extra teeth)
2) Dental atresia (missing or small teeth)
3) Enamel defects (pitting)
4) Malposition of teeth (sheer mouth, wave mouth, etc.)
5) Dentigerous cysts (tooth bud just below the ear)

Big Head (Equine Osteomalacia)

Big head is a rare disorder in bone metabolism (i.e., bone formation and maintenance) which results in a swollen, fragile bone structure. Signs in the horse are an enlarged head, lowered croup, flat sunken sides, and occasional lameness.

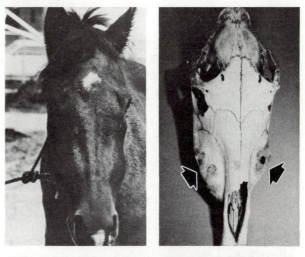

This horse suffers from osteomalacia, a disease also known as "big head" (left). The skull (right) shows the bulging facial bones characteristic of osteomalacia.

The strength of bone is determined by the levels of calcium, phosphorus, and vitamin D in the blood. Defective calcification of growing bone (e.g., rickets) can be caused by insufficient dietary calcium, excess dietary phosphorus, abnormal calcium-phosphorus ratio, or a vitamin D deficiency. The ratio of serum calcium to serum phosphorus should range from 1:1 to about 2:1 if the ratio in the feed is 1:1 to 1.5:1. Depletion of calcium from mature bone (i.e., osteomalacia) results from a calcium-phosphorus imbalance in the blood; calcium is removed from the bone to support other physical processes, such as lactation.

In the early 1900's, big head was reported in several half-sib horses. The possibility of an inherited predisposition to big head was suggested, but complementary research has not been carried out to confirm this suggestion. Since all horses were on the same premises, it is possible that this occurrence may have been due to a diet high in phosphorus (e.g., bran).

Ewe Neck

Ewe neck is a conformation deformity involving a somewhat "upside-down" appearance of the neck. A concave outline from withers to poll, a thick underline, and a depression just in front of the withers are characteristic of ewe neck. Problems associated with this condition are decreased athletic ability, caused by limited flexibility of the neck, and unsightly appearance. Ewe neck is believed to be an inherited defect, but studies have not identified a controlling genetic system.

Cresty Neck

The crest, or convex shape, of the neck topline is a sex-influenced trait (controlled by sex hormones). A slight crest is a desirable conformation trait in some breeds, but an excessive crest on a stallion, or a thick neck on a mare, is usually undesirable. An extreme crest interferes with the neck's flexibility and the horse's maneuverability. On the mare, a cresty neck indicates a lack of femininity. Cresty-necked mares are usually difficult, and sometimes impossible, to get in foal (e.g., hypothyroid mare).

The fact that cresty neck is under hormonal control suggests that it is inherited. (Hormone production is genetically controlled.) The inheritance pathway has not been determined, however.

LIMBS

Abnormal Leg Set

Because genes control the growth and differentiation of body tissues, it is generally accepted that abnormal leg structure and resulting lameness are also genetically controlled. The inheritance pathways have not been established, but the appearance of a congenital conformation defect is most frequent when both the sire and dam express that trait. Because the limbs determine the usefulness of a horse, selection, for either athletic or breeding purposes, should always include a careful comparison of good vs. undesirable leg structure. The most common conformation deviations of the limbs are listed below:

Base Narrow	Camped Out in Front
Base Wide	Short Upright Pasterns
Toe In	Long Upright Pasterns
Toe Out	Long Pasterns and
Knock Knees	Excessive Slope
Calf Knees	Short Pasterns and
Offset Knees	Excessive Slope
Buck Knees	Straight Stifle
Bowlegs	Straight Shoulder
Cut Out Under the Knees	Cow Hock
Tied in Behind the Knee	Sickle Hock
Standing Under Behind	Camped Out Behind
Standing Under in Front	

(Refer to "**Important Selection Characteristics.**")

Many of these traits initiate problems such as splints, sidebone, and ringbone. Cow hock and sickle hock conformation predispose the affected horse to bone spavin (a painful enlargement of the joint on the lower inside of the hock). In a young horse, excessive stress upon a weak leg often results in inflammation of the epiphyseal plates (the growth area of long bones). Navicular disease is closely associated with poor conformation (e.g., upright pasterns, small feet, etc.). Angular deviations of the legs, such as base narrow or base wide, cause excessive strain on certain parts of the leg, increasing the possibility of tendon and ligament breakdown.

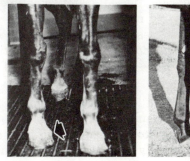

Left: Toe-in conformation (arrow), one of the many inherited deviations in leg set. Right: Calf knees.

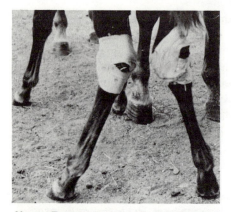

Above: Extreme angular leg deformity in the foal. This base-wide, toe-out conformation places excessive stress upon the hind limbs.

Right: An inherited weakness in conformation, coupled with stress, may result in an unsoundness such as bowed tendons (arrow).

Below: Radiograph showing ossification of lateral cartilages (arrows), a condition known as sidebone. Conformation weakness, such as short upright pasterns, predispose the horse to sidebones.

Right: Rafter hips (arrow).

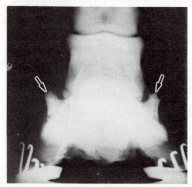

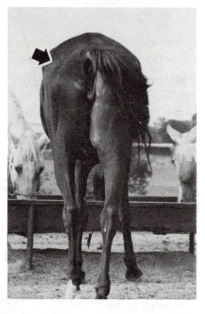

Epiphysitis

Epiphysitis is a lesion in the region of lengthwise growth in long bones. Normally, organized cartilagenous tissue located toward each end of the long bone lays down new bone for a gradual increase in length. As the bone reaches its mature length, the cartilagenous areas (epiphyseal plates) begin to close. If these areas are damaged or stressed before complete closure, the response is inflammation and disorganized growth (epiphysitis). This condition often appears as a swelling just above the knee ("big or open knees"), just above the pastern, or at the upper end of the long pastern bone. (Epiphysitis is less apparent in the hind leg.) Occasionally, the swelling is hot and painful, and slight lameness or stiffness is present.

Epiphysitis occurs in the young horse, especially when heavy training precedes epiphyseal closure (e.g., race horses). Overfed horses are highly susceptible to epiphyseal lesions, since excessive weight places additional stress upon the epiphysis (e.g., young, fat halter horses from early maturing lines). Environmental causes also include calcium or phosphorus excesses and protein deficiencies. An inherited predisposition to epiphysitis occurs with the base narrow-toe in conformation. This type of leg set places excessive strain on the outside of the various growing points. Large body size with light leg structure also places extreme pressure on the epiphysis. Any inherited lameness (or acquired abnormality) that causes the horse to place extra weight on one limb predisposes the respective epiphyseal plates to injury.

Contracted Digital Flexor Tendons

Contracted digital flexor tendons cause an abnormal flexion of the affected limb, varying in degree from slightly raised heels to extreme flexion at the fetlock. The flexor tendons connect the pastern bones and coffin bone (small bone within the hoof) to the muscles of the upper limb. Normal muscular contraction causes the fetlock to bend, but shortening of the tendons causes permanent flexion of the joint. Mild cases will often correct themselves; others require braces, casts, or surgery (tenectomy). In severe cases, the affected foal will walk on the front of the fetlock. Because these foals cannot recover, they are usually euthanized.

Congenital contraction of the deep digital flexor tendons (A). Severe congenital contraction of the digital flexor tendon: unilateral (B) and bilateral (C).

A   B   C

Acquired cases can affect either the forelimbs or the hind limbs, and are often caused by an injury or a nutritional deficiency. When the condition is present at birth, it usually occurs in both forelegs. Congenital cases can be caused by genetic factors, intrauterine crowding, or a prenatal nutrition deficiency. Genetic systems involved in this condition have not been identified.

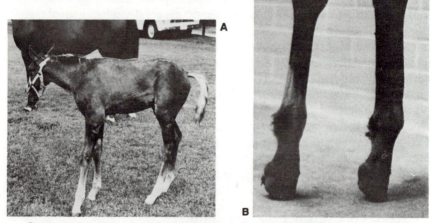

Contracted tendons in a foal (A), and a close-up of forelimbs in a severely affected horse (B), showing how weight is carried on the toe.

Upward Fixation of the Patella

Stress upon the various leg joints is limited by several specialized structures; the hoof, angles of the fetlock, and angles of the stifle absorb much of the pressure upon the hind leg. When a horse inherits an unusually straight hind leg, as viewed from the side, his joints are subjected to greater stress. This stress aggravates the normal movement of the patella, and increases the horse's susceptibility to upward fixation of the patella (i.e., locking of the knee cap above the femoral ridge). Once this occurs, stretched ligaments encourage its reappearance.

The signs of this abnormality include:
1) stiff hind leg (stifle and hock cannot flex);
2) altered movement to drag the affected leg forward; and
3) a slight catching (clicking) of the patella in less severe cases as the horse moves forward.

Although upward fixation of the patella might also be the result of injury or stress on weak legs, its close association with the straight stifle indicates a possible genetic predisposition. The problem is accentuated in young horses in poor condition.

Lateral Luxation of the Patella

Lateral luxation of the patella is a rare defect involving displacement of the patella (knee cap) to the outside of the stifle joint (knee). With flexion and extension of the stifle joint, the patella normally slides along a groove in the femur so that a "pulley-like" motion of the knee cap over the knee is established. Foals affected with lateral luxation of the patella are born with an underdeveloped lateral femoral ridge. This deformity allows the patella to be repositioned out of its groove and slide toward the outside of the knee where abnormal "pulley-like" motion causes painful irritation and inflammation of the joint. Lateral luxation of the patella appears as a severe lameness in the affected hind leg.

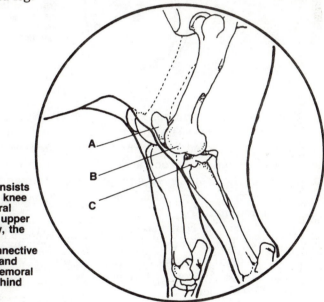

The stifle joint consists of the patella - or knee cap (A), the femoral ridge (B) and the upper tibia (C). Normally, the patella (which is positioned by connective tissue) slides up and down within the femoral groove when the hind limb is in motion.

Although the signs of this defect appear sometime after birth, the actual deformity is present at birth. Studies suggest that this congenital abnormality is an inherited defect, but genetic control has not been identified.

Contracted Heels

Shock absorption within the limbs depends upon the coordinated action of several specialized structures. Perhaps the most important of these is the hoof; initially, concussion is distributed over the hoof wall, bars, and frog. The hoof wall should be strong enough to avoid excessive wear, the heel should expand with pressure, and the bars should be far enough apart to allow expansion of the frog. The frog should be positioned upon the ground where pressure causes it to compress the digital cushion (an important shock absorber). Failure of this mechanism increases the shock upon other specialized structures, subjecting the limb to abnormal stress and possible injury.

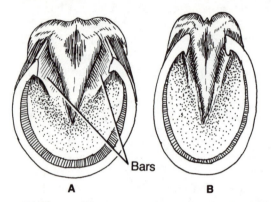

(A) Adequate distance between the hoof bars allows proper expansion of the frog. (B) The contracted heel limits expansion of the frog and interrupts the normal shock absorbing function of the digital cushion.

Bars

A   B

If, due to improper trimming or shoeing, the frog loses its contact with the ground, it serves no purpose and eventually atrophies. This is one cause of contracted heels. Similarly, a lameness which prevents the use of a foot can cause contraction. On the other hand, a congenital contraction of the heel prevents proper frog action and subsequently causes lameness. Drying and hardening of the hoof tissue also cause contraction and limit frog action.

Congenital forms of contracted heels usually occur on one forelimb, while acquired forms (caused by improper trimming, dryness, etc.) usually occur on both forelimbs. (Contraction is uncommon on the hind limb.) Congenital contraction is believed to be inherited, possibly controlled by two loci:

$$Ch^1 \ ch^1 \qquad \text{and} \qquad Ch^2 \ ch^2$$

Theoretically, either homozygous recessive form ($ch^1 \ ch^1$ or $ch^2 \ ch^2$) causes contracted heels regardless of the other locus.

Flat Feet

The healthy hoof, one which is capable of withstanding concussion, should have a round toe and wide heel. The horse's weight should be distributed over the hoof wall, bars, and frog, not upon the inner sole. Because the sole is not a weight-bearing structure and because it is highly sensitive, it should be well-arched to avoid contact with the ground. A lack of concavity (flat foot) increases the horse's susceptibility to bruised sole and possible lameness. Flat-footed horses often walk on their heels to relieve some of the pressure on their soles. Special shoes can alleviate much of the discomfort, but the condition has no known remedy.

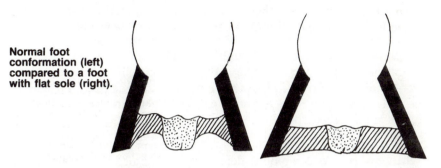

Normal foot conformation (left) compared to a foot with flat sole (right).

Although flat feet are not common within the lighter breeds, draft horses have long suffered from this defect. Extreme shock upon the hooves of heavy work horses necessitates the presence of proportionately large hooves with thick walls, deep wide heels, and well-arched soles. Flat feet in drafters is an inherited defect which appears, primarily, in the forelimbs.

Osteochondritis Dissecans

Osteochondritis dissecans is a developmental problem seen in young, rapidly growing horses. Characteristically, a fragment of articulating cartilage will break loose within the shoulder joint capsule. This free-floating fragment, or "joint mouse," results in pain and lameness. The condition can be corrected by surgical removal of the offending fragment.

Osteochondritis dissecans has been observed in the hock joint of Standardbreds. It has been identified as hereditary in certain species (e.g., dogs), but its classification in horses has not been established. Genetic control in some species suggests, however, that the condition may also be inherited in horses.

Hip Dysplasia

Hip dysplasia, a rare developmental abnormality, is a congenital deformity of the hip joint. The head of the femur (thigh bone) is flattened and inserted into a shallow socket on the pelvic bone. This weak ball and socket connection causes wearing of the joint (i.e., areas of missing cartilage, missing ligaments, erosion of exposed bone, etc.). Horses with hip dysplasia show reduced flexibility in the stifle and hock, since any movement of the hip joint is painful. The horse takes short steps with his affected hind leg and frequently rotates over the outside of the respective toe. Atrophy of the croup muscles reveals the outline of the affected hip joint. The deformed side of the pelvis is usually lower than the normal side.

Surgical correction of hip dysplasia has not been successful in horses. Due to pain associated with the condition, affected horses are usually euthanized. The heritability of this hip deformity has been established in the dog, but the trait can be classified only as a suspected hereditary trait in horses. This condition has been reported in the Dole breed, the Standardbred, and the Shetland Pony, indicating that it is not breed specific.

Stiff Joints

Stiff joints is a congenital deformity of one or both forelegs. Affected foals have crooked limbs, stiff joints, and underdeveloped hooves. When both forelegs are affected, the foal cannot stand and will not usually survive.

In the 1930's, several stiff joint foals were produced by one stallion. The high incidence of stiff joint among the stallion's half-sib offspring coupled with no apparent nutrient deficiency suggests that the condition is inherited. The controlling genetic systems have not been identified, however.

Fibular Enlargement

The tibia is the long bone between the hock and the stifle. The fibula was once an important supplement to the tibia but has, over the years, become reduced to a small splint-like accessory.

Fibular enlargement is an inherited defect found within some Shetland Pony families. It involves the presence of a complete fibular shaft whose length does not correspond with that of the tibia. This results in splayed (turned out) hind legs from the hock down, and subsequent stress upon the entire leg. Its appearance within certain pony families suggests that the condition is controlled by an unidentified genetic system and that it is probably related to dwarfism.

Polydactyly

Polydactyly is the presence of an extra hoof and associated structures, usually on the inside of the forelimb below the carpus (knee). In some species, this gross deformity is believed to be caused simply by one allelic mutation. In others, its expression depends on many complicating factors and is not clearly understood. The condition is rare in horses and has not been studied extensively.

The polydactylous horse has an enlarged 3rd metacarpal (cannon) bone. Just below this bone, an extra set of digits (1st, 2nd, and 3rd phalanx) form the secondary pastern and hoof. The normal weight-bearing appendage is slanted outward and sometimes carries an extra ergot (remnant of a primitive foot pad) just above the normal one. Surgical correction has been successful.

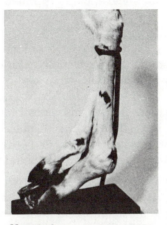
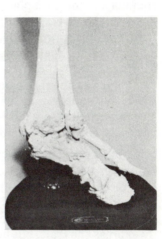

Mounted preparations of the polydactyly limb. Note the size differences between digits.

Abrachia

Abrachia refers to complete absence of one or more limbs in the newborn foal. Severe congenital deformities, such as abrachia, are usually closely associated with other inborn problems. The hereditary mechanisms involved in this disorder are unknown.

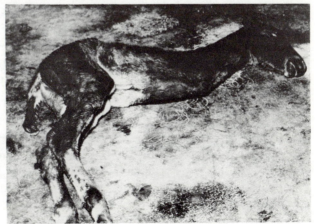

This severely deformed foal lacked both forelegs (abrachia) and eyes (anophthalmia).

BODY

Umbilical Hernia

Hernias in the newborn foal are usually caused by the failure of a prenatal opening (e.g., inguinal ring, umbilical opening, etc.) to close prior to birth. Refer to the discussion on scrotal hernia under "Inherited Abnormalities." If the umbilical opening is unusually large, or if it fails to close properly, pieces of intestinal tissue or a section of intestine may become displaced, forming a subcutaneous (below the skin) lump in the navel area. The hernia may not become apparent until food intake and gut movement begin. Usually, the condition is self correcting; the foal grows in size faster than his intestine grows in length, allowing the gut to tighten and move back into the abdominal cavity. If the opening contracts before this repositioning takes place, the blood supply may be cut off from the displaced area (i.e., gut strangulation). If gut strangulation occurs, or if the opening is not naturally corrected within one year, surgery is usually required.

Umbilical hernias can be caused by improper treatment of the umbilical cord at birth (e.g., cutting too short), intrauterine crowding, or by a recessive gene. Because most umbilical hernias are inherited, an affected horse will usually produce some carriers of the controlling recessive allele. The sire and dam of an affected foal will not necessarily express the trait, but both must be carriers of the controlling allele:

UU	$\times$	uu	$=$	$Uu + Uu + Uu + Uu$
Normal	$\times$	Hernia at Birth	$=$	All Carriers
Uu	$\times$	Uu	$=$	$UU + Uu + Uu + uu$
Carrier	$\times$	Carrier	$=$	One Normal Two Carriers One Hernia at Birth
uu	$\times$	uu	$=$	$uu + uu + uu + uu$
Hernia at Birth		Hernia at Birth	$=$	All Hernia at Birth

Umbilical hernia. The protrusion (arrow) is caused by the displacement of a portion of the intestine through an enlarged opening in the abdominal wall musculature. Normally this opening constricts shortly after birth.

Lordosis (Sway Back)

Lordosis, or sway back, is closely associated with flat croups, long backs, age, or underdeveloped vertebrae that cannot support weight properly. Normally, contact between the vertebrae (via intervertebral processes), and the presence of a muscular suspension bridge, counteract the pull of gravity and any additional weight upon the horse's back. Vertebral support at the shoulder and pelvis is provided by the limbs, while the mid-back is suspended between these two points by intervertebral muscles and ligaments. The angle of these muscles is very important since it counteracts the force of gravity. A flat croup decreases this angle and increases the horse's susceptibility to back strain (perhaps explaining why the frequency of sway back in Saddlebred horses is relatively high). A long back increases the distance requiring support, adding to the strain upon the supportive tissue.

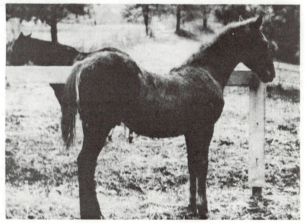

Foal with congenital lordosis, or sway back.

If the boney processes, which provide interlocking contact between the vertebrae, do not form completely during embryonic development, the resulting horse will probably suffer from overextension (concavity) of the vertebral column. Weight upon this type of back must be sup-

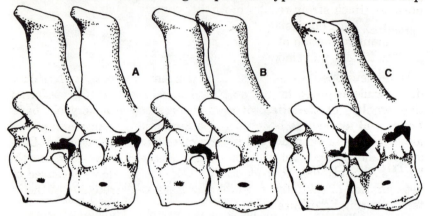

Intervertebral processes: normal (A), normal extension or bending of the back (B), hypoplastic intervertebral processes (arrow) cause over extension or sway back (C).

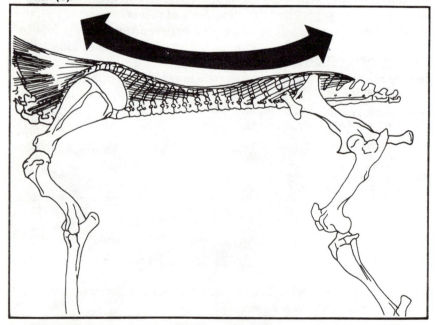

The vertebral column is supported by a specialized system of muscle and connective tissue. A long back places abnormal stress upon the supportive tissue and predisposes the horse to sway back.

ported by ligaments which often become stressed under the additional pressure. Back strain results in further complications:

1) sway back (lordosis)
2) impulsive action of the hind limbs
3) inefficient movement (e.g., reduced speed)

Underdeveloped vertebrae result in congenital (present at birth) lordosis, while acquired forms can be a delayed reaction to a conformation problem or the effect of normal stress upon a very old back. Congenital lordosis is believed to be inherited; instances of several lordotic foals being produced by one mare have been reported. Back length and croup level are inherited traits which can increase the horse's susceptibility to back strain. "Old age" lordosis is caused by deterioration of the intervertebral processes and stretching of the interconnecting ligaments.

A sway back appearance in elderly horses may be caused by genetic predisposition, loss of muscle tone, relaxation of the connective tissue and/or a long broodmare career. This grey mare is 26 years old and has produced 16 foals.

Note: Several conditions are often mistaken for sway back:
1) high neck carriage

2) short dorsal spinous processes (vertical extensions from each vertebra)

3) dipped back (low behind the withers and high at the croup) which is sometimes seen in race horses

Kyphosis (Roach Back)

Kyphosis, or "roach back," is an unsightly conformation abnormality involving a short, arched back. This deformity causes inflexibility of the spine, shortness of stride, and interference (contact between forelimbs and hind limbs) during fast gaits. Kyphosis is considered a serious weakness since a rider's weight is placed directly upon, and supported by, the arched back bone. (Such a horse is highly susceptible to back injury.)

Kyphosis is a congenital defect which may be hereditary. Because the condition is rare, information regarding possible genetic control is limited.

Scoliosis (Lateral Curvature of the Spine)

Scoliosis is a lateral deviation (bending to one side) of the spine caused by varying degrees of vertebral instability. Extreme scoliosis is a characteristic of the contracted foal (foal that is stillborn with gross deformities of the limbs and back). Occasionally, foals affected with mild scoliosis are viable and able to adapt to their unusual anatomy. Although the congenital nature of this condition suggests that it may be hereditary, research has not identified a controlling genetic system.

Multiple Exostosis

Multiple exostosis is an inherited abnormality involving the growth of boney projections from the ribs, pelvis, and/or long bones. These growths appear in various shapes and sizes, and usually stabilize as the horse reaches maturity. (One exception is the rare occurrence of malignant growths.) There is no known remedy for multiple exostosis, but athletic ability is not usually affected by its presence.

The inheritance of multiple exostosis in horses is very similar to its inheritance in humans. Its transmission from parent to offspring suggests that it is controlled by a single dominant allele. In the absence of any previous family history, its sudden appearance might be the result of a mutation.

The Eye

There are a number of congenital abnormalities that may affect the equine eye. As with all congenital defects, when one such abnormality is present, other defects may also be present in the same animal. Research has indicated that some of these eye abnormalities are inherited.

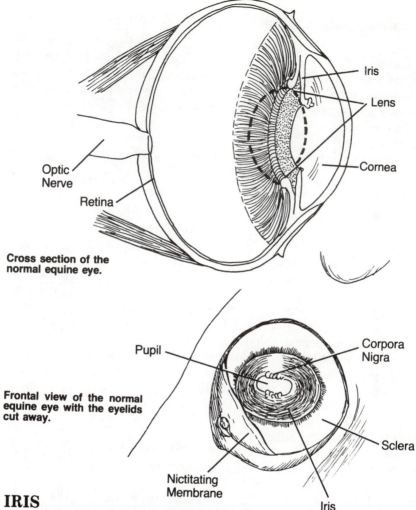

Cross section of the normal equine eye.

Frontal view of the normal equine eye with the eyelids cut away.

IRIS

The iris is the diaphragm of the eye, which controls the size of the pupil and regulates the amount of light that enters. Heterochromia, hyperplastic corpora nigra, aniridia, and persistent pupillary membranes are a few of the congenital abnormalities that may affect the iris.

Heterochromia

The color of the eye is actually the color of the iris. Most horses have dark brown eyes, but other colors include amber, blue, and white. Iridial heterochromia is a condition involving eyes of two different colors or two or more colors in the iris of one eye.

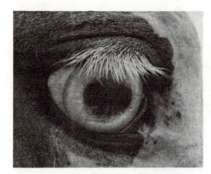 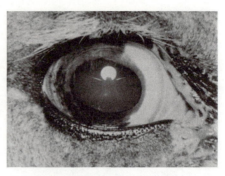

The blue eye ("white eye") is frequently seen with white face markings.

A heterochromic eye; note that the iris is
two different colors.

American Cream draft horses have amber eyes as one of their breed characteristics, while many Paint horses will have blue or white eyes. A horse may also have a combination of blue, white, amber and/or brown in either one or both eyes. If part of the iris is white, the condition is known as "albinismus partialis" (wall eye). If the iris is almost completely white, with only the corpora nigra pigmented, it is known as "albinismus totalis" (glass eye). Heterochromia occurs most often in white, spotted, palomino, or cremello horses. Fortunately, no disease is associated with this condition. However, horses with a scarcely pigmented iris will often be sensitive to bright sunlight, just as blue-eyed people are sometimes light-sensitive. One of the foundation horses of the Tennessee Walking Horse breed had a "glass eye." Because of the sensitivity to bright light which these horses exhibit, breeders have often selected against this condition.

Heterochromia is frequently found with white face markings, such as in a bald-faced horse; these markings are dominant. Horses with blue eyes and white face markings are more prone to develop squamous cell carcinoma. This is a type of skin cancer involving the eyelids and nictitating membrane (the third eyelid) when these structures lack pigment. (Refer to the discussion on skin abnormalities within this chapter.)

Aniridia

Aniridia is an inherited absence of most of the iris. This condition has only been reported in Belgian horses. Usually, the iris is almost completely absent, and vision is poor or absent. Aniridia is caused by an autosomal dominant gene. Frequently, affected Belgian foals develop secondary cataracts at approximately two months of age. If cataracts develop, the horse will be blind.

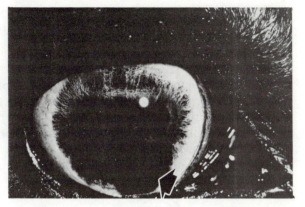

Coloboma iridis: a defect
of the eye in which part
of the iris is missing
(arrow). (In aniridia, all of
the iris would be absent.)

Coloboma Iridis

Coloboma iridis is a more frequently seen congenital condition involving a hole in the iris. These holes are often located near the six or twelve o'clock position. Although more light is permitted to reach the retina, the condition usually has little effect on vision.

Iris Cysts

Cysts are hollow structures that may contain a gas, fluid, or solid matter. They may occur in the body of the iris or in the corpora nigra, in either one or both eyes. Cysts may be congenital and, due to their occurrence in ponies with heterochromia, they may be caused by a developmental defect of the pigment epithelium. Congenital iris cysts often cause no problem, although they persist for the life of the animal. In some cases, however, the cysts may rupture, and the iris may atrophy. Iris cysts may be surgically removed.

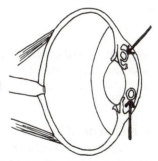

Iris cysts may be attached
(top arrow) or free-floating
(bottom arrow) in front of
the lens and pupil.

In one study, three blue-eyed ponies had similar black cysts in both eyes. Researchers suggested that lightly pigmented eyes (e.g., blue, heterochromic, etc.) may be more susceptible to cyst formation. If thi͏ hypothesis is true, cysts of the eye may be, to some extent, inherited. It should also be noted that the black cysts may just be more apparent on the lightly pigmented iris.

Hyperplasia of the Corpora Nigra
A horse may be born with overgrown (hyperplastic) corpora nigra. The corpora nigra are the "black bodies" seen in the eye which help to reduce the amount of light that enters. Hyperplastic corpora nigra may be so large that they interfere with the horse's vision. If necessary, they can be surgically removed. As with other congenital defects, hyperplasia of the corpora nigra may be inherited.

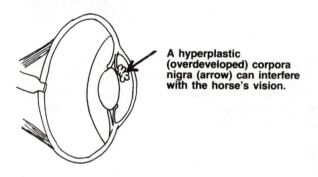

A hyperplastic (overdeveloped) corpora nigra (arrow) can interfere with the horse's vision.

CORNEA
The cornea is the transparent window to the front of the eyeball. There are a number of congenital abnormalities that may affect the cornea. Frequently, the defects occur in association with other ocular abnormalities. Corneal defects may be due to genetic defects in the developmental process.

Microcornea
Microcornea (abnormally small cornea) is often present with microphthalmia (abnormally small eyeball). Cataracts are often present with microcornea and microphthalmia. Vision will be absent in small eyes and slightly impaired in almost normal-sized eyes.

Congenital Keratopathy
Congenital keratopathy is a rare condition that affects both eyes of a foal. It is a painless condition involving corneal opacity of the deeper layers of the cornea (a normally transparent membrane around the eyeball). This condition causes blindness and cannot be surgically corrected, except by a corneal graft (i.e., replacing corneal tissue).

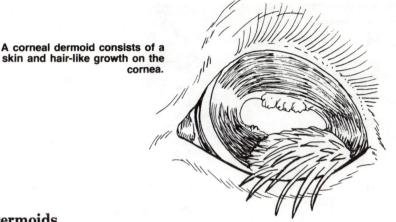

A corneal dermoid consists of a skin and hair-like growth on the cornea.

Dermoids

A horse may be born with a dermoid on the cornea. This growth, which is the same composition as the eyelids, may be covered with long hairs. Corneal dermoids may be surgically removed, since only the superficial layers of the cornea are involved, but scarring may result. The genetic basis for this condition is unknown.

Melanosis

Abnormal deposition of black or brown pigment (melanin) in the eye is a condition known as melanosis. This occasionally affects the cornea, but it does not usually interfere with vision, unless it is located directly in front of the pupil.

LENS

The lens is the transparent body lying between the iris and the vitreous body (inner eye). It is a biconvex body that is surrounded by a fine membrane. There are several congenital abnormalities of the lens, including cataracts and periodic ophthalmia.

Cataracts

A cataract is a loss of transparency in the lens of the eye; a dense, white opacity fills the entire pupil when the cataract is mature. Congenital cataracts are frequently bilateral, and may be associated with other eye problems such as congenital microphthalmia (very small eye). Three types of cataracts occur in the horse:

1) "Y-type" cataracts appear in both the anterior and posterior cortices of the lens. They may result from an increase in the ground substance of the lens, or from a shortness of the lens fibers. These cataracts are usually congenital and early surgical removal may be successful.

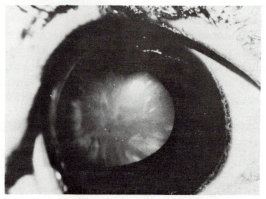

Above: A "Y-type" congenital cataract of the foal. The suture lines that form the "Y" are clearly visible. Right: A mature cataract in an adult horse. Note the cloudiness of the lens.

2) Nuclear cataracts involve only the central part of the lens and appear as irregular, concentric rings. There is a sharp line of demarcation between the nucleus and the normal lens cortex. This type of cataract only partially obstructs vision and may tend to clear with age.

3) Complete mature cortical cataracts cause blindness, while the above types may cause either partial or complete loss of vision. These cataracts are generally associated with aging but are occasionally found in younger horses. They may also occur secondary to trauma or diseases such as periodic ophthalmia.

Cataracts due to an autosomal dominant gene have been reported in Belgian horses, in association with aniridia (missing iris). In other breeds, cataracts may also be caused by an autosomal recessive gene. Since matings between affected animals have not been carried out to test this theory, repeated breeding of parents which have produced offspring with congenital cataracts may confirm the mode of inheritance. Surgical removal of cataracts has been successful only in foals, not in mature horses.

Lens Luxation

The lens can be either totally dislocated (luxated) or partially luxated (subluxation). When luxation occurs, the lens may fall forward into the anterior chamber of the eye, or backward into the posterior chamber. Luxation is a rare disorder which can result in glaucoma. It may be a congenital defect possibly of genetic origin. The mode of inheritance for this condition, if any, is not known.

Periodic Ophthalmia

Periodic ophthalmia, also known as recurrent uveitis or "moon blindness," is a recurring inflammation of the eye. Signs of this condition include excessive lacrimation (tearing), acute photophobia (painful reaction to light), pain, conjunctivitis, capillary invasion of the sclera (the part of the eye known in humans as the "white of the eye") in which blood vessels appear in the eye where there were none before, loss of clarity at the margin of the pupil, cloudy lens, yellowing of the iris, and an ocular discharge. If allowed to continue without treatment, adhesions may develop between the iris and the lens capsule (posterior synechiae). An episode of periodic ophthalmia may last anywhere from a few days to as long as a month.

Several factors indicate an inherited susceptibility to develop periodic ophthalmia, if not a direct inheritance of the disease. For example, Great Britain almost eradicated the condition within 40 years through a law which prevented the use of stallions with cataracts (an end result of periodic ophthalmia). Admittedly, other factors could also have changed in such a period of time (1918-1958). In addition, periodic ophthalmia appeared in South Africa following the importation of American horses. The disease then disappeared over the next decade, suggesting that it was selected against.

One early study (1942) indicated that a complex genetic theory would be required to explain periodic ophthalmia. In this study, one stallion produced several foals which developed periodic ophthalmia; several mares with periodic ophthalmia also produced foals which developed the condition. However, many foals which developed the eye defect had normal parents.

The exact cause of periodic ophthalmia is unknown, but inflammatory diseases are not usually inherited. The hereditary factor in periodic ophthalmia may be an inherited susceptibility to ocular inflammation in response to mild trauma or stress, or it could be an inherited predisposition of the ocular structures to be sensitized by penetrating antigens (foreign proteins). Some dysfunctions of immunizing systems (e.g., the reticuloendothelial system) prevent an animal from removing, or neutralizing, foreign proteins produced by parasites, such as *leptospira*. This dysfunction may result in allergies which could cause recurrent inflammation of the eye. (Tendencies to develop allergies are often heritable.) For example, inflammation of the iris (iridocyclitis) is known to be the result of a severe allergic reaction (anaphylactic hypersensivity) in some species.

There is a regional variation in the incidence of this disease, with a higher incidence in low-lying areas. This implies that heredity is not the only factor involved.

Persistent Hyaloid Vessel

The hyaloid artery supplies blood to the eye during gestation, then gradually degenerates during the first few weeks after birth. Occasionally, the vessel will persist for an extended period of time.

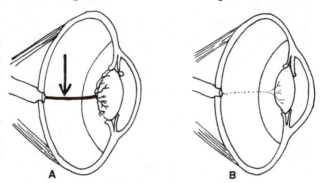

Normally, the hyaloid artery (arrow) is present only in the fetal eye (A) and gradually atrophies after birth (B).

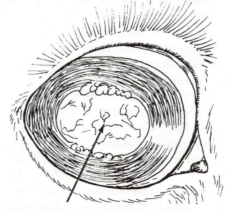

Frontal view of the equine eye, showing a persistent hyaloid artery (arrow).

Remnants of the hyaloid artery may appear as fine strands immediately behind the posterior lens capsule, or as filaments in the vitreous fluid between the lens and the optic disc. A persistent hyaloid vessel may be difficult to distinguish from a cataract. There will be some impairment of vision, with the degree of impairment depending on the size of the opacity in the vitreous fluid. While this defect is congenital, it is not known whether it is also hereditary in origin.

RETINA

The retina is the area within the eye which receives visual rays and directs the message to the brain via the optic nerve. The retina is also subject to various congenital abnormalities.

Absence of the Retina

A complete absence of the retina may occur in horses. This rare condition results in total blindness. Whether or not this defect is inherited is unknown.

Glaucoma

Glaucoma is a condition characterized by increased intraocular pressure. It may be a congenital abnormality, or due to trauma (such as glaucoma associated with a luxated lens). Signs include enlargement of the eyeball, dilation of the pupil (mydriasis), corneal edema (opacity of the cornea), and shutting of the eyelids (blepharospasm). Glaucoma should be differentiated from inflammation of the iris (iridocyclitis).

Untreated glaucoma will cause blindness due to atrophy of the retina and cupping of the optic disc. Glaucoma in the horse is rare compared to the incidence of this inherited condition in man and in the dog (especially in the Cocker Spaniel).

Congenital Night Blindness

Congenital night blindness causes the loss of night vision (scotopic, or dark-adapted, vision) due to a deficiency of rods in the retinal layers. Occasionally, there may be some impairment of the animal's day vision (photopic, or light-adapted, vision). Signs of congenital night blindness include bumping into objects, confusion, stumbling, standing alone in the pasture at night, and training difficulty. No ocular lesions are visible, and the retina (interior part of the eyeball seen through an ophthalmoscope) is normal.

Electroretinography (ERG) can be used to diagnose congenital night blindness. By this process, electrical impulses from the eye are monitored and traced upon special paper. The tracing is in a continuous up and down pattern of varying amplitude (height) and can be interpreted by an eye specialist. Under dark conditions, the amplitude of the ERG for a horse with congenital night blindness will be higher than that of a normal horse. Under light conditions, the ERG will appear normal, except for a decrease in amplitude.

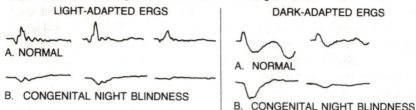

The ERGs (electroretinograms) of horses affected with congenital night blindness (A) as compared to that of normal horses (B).

According to several reports, only Appaloosa horses have been affected by congenital night blindness. Many of these affected horses have normal day vision. Night blindness in Appaloosas appears to be the same condition as congenital stationary night blindness (CSNB) in humans. CSNB with a normal retina is known to be heritable in man, but its exact mode of inheritance is not clear. (It has been classified as due to an autosomal dominant, autosomal recessive, or X-linked recessive.) In horses, the defect may be controlled by a recessive allele. A sire-daughter mating between two normal Appaloosa horses has resulted in offspring with the disorder. This implies that the sire was a carrier (heterozygote) that gave the recessive allele to his daughter, who then became a carrier. When the sire was mated to his daughter, the recessive allele was transmitted from both to the resulting offspring.

Optic Nerve Hypoplasia

Underdevelopment of the optic nerve is known as optic nerve hypoplasia, and may occur in conjunction with other congenital defects, such as absence of an eyeball (anophthalmia), abnormally small eyeball (microphthalmia), cataracts, and detached retina.

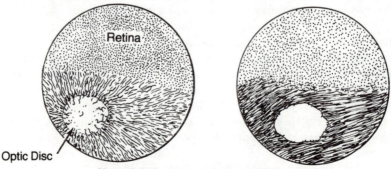

Normal optic nerve and disc (left) compared to a hypoplastic optic nerve and disc (right).

Blindness caused by a hypoplastic optic nerve may not be evident until weaning time, since the foal follows the mare closely while nursing. In this rare condition, the optic disc will be small and white with few, if any, retinal vessels surrounding it. The affected foal will have a decreased or absent pupillary response to light. Optic nerve hypoplasia may affect either one or both eyes.

The cause of optic nerve hypoplasia in the horse is unknown, but in the Basle waltzing strain of mice, it is due to an autosomal recessive gene.

OTHER DEFECTS

Microphthalmia

Microphthalmia refers to an abnormally small eyeball. Either one or both eyes may be affected. In severe cases, the eye may be completely missing (anophthalmia). A foal with microphthalmia is prone to develop cataracts. This unattractive condition usually causes blindness, except in cases where the eye is only slightly reduced in size. This defect has occurred in all breeds of horses; like so many other congenital defects which may or may not be inherited, microphthalmia does not yet have an established genetic basis.

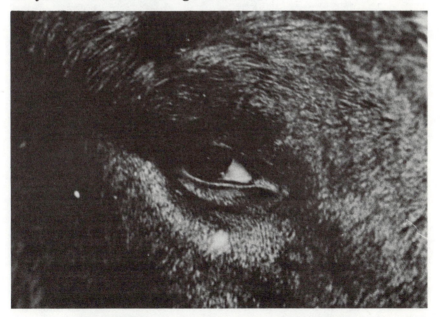

Microphthalmia (or small eye) usually causes blindness and may be accompanied by other congenital defects.

Entropion and Ectropion

Two congenital conditions affect the lower eyelid. Entropion is the inward rolling of the lower eyelid, while ectropion is the outward rolling, or drooping, of the lower lid. Entropion may cause the small eyelashes to scratch the cornea. A scratch could develop into a corneal ulcer which, if left untreated, could cause the cornea to rupture, resulting in blindness. Following corneal rupture, the globe (eyeball) shrinks in size. Corneal ulcers which heal, instead of rupturing, leave scars. The scar's effect on vision depends on its size, location, and density.

Ectropion may result in a dry eye. This is because tears cannot properly bathe the cornea to keep it moist. The dry condition that results is called keratitis sicca. Both of these defects can be surgically corrected, but the correction should be done at an early age.

The possible inheritance of these conditions has not been studied, but entropion is probably most frequently encountered in the Thoroughbred.

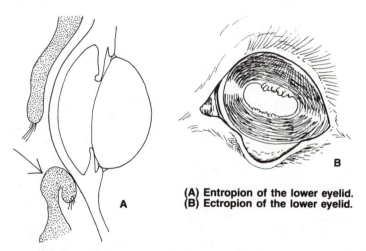

(A) Entropion of the lower eyelid.
(B) Ectropion of the lower eyelid.

Atresia of the Nasolacrimal Duct

The tear duct (nasolacrimal duct) runs from the inner corner of the eye to the line of union between the skin and the mucous membrane just inside the nostril. This duct carries tears away from the eye and discharges them into the nasal cavity. Usually, there is a single opening within the nostril through which tears are conveyed, but there may be double or rarely, triple openings. If the opening is missing, meaning that the duct is nonpatent, or closed, the foal will show an ocular discharge, since the tears cannot drain down the normal channel. This condition is known as atresia of the nasal punctum (opening) of the nasolacrimal duct. As it is usually only the opening itself that is missing, a veterinarian can usually make an opening surgically. Occasionally, part of the duct itself is missing, a condition known as atresia of the nasolacrimal duct. This can generally be surgically corrected. Although these closely related conditions are congenital, it is not known whether they are inherited.

**Atresia of the nasolac-
rimal duct (dotted lines)
causes tearing.**

Skin

Several equine skin disorders may be at least partially heritable. Some of these conditions are present at birth, while others do not develop until later in life. Also, skin disorders may vary in severity; for example, epitheliogenesis imperfecta (missing skin) is fatal, while the absence of chestnuts on the legs is insignificant. In addition, some skin disorders (even if hereditary in origin) require a certain environment to be revealed. For instance, anhidrosis (dry coat) will not appear unless the horse is moved to a hot, humid climate.

PHOTOSENSITIZATION

Photosensitization refers to the skin's extreme sensitivity to sunlight. Affected skin will become irritated, dried and cracked. Only areas with underlying unpigmented skin are affected because these areas lack protective pigment. Photosensitization is not simply sunburn.

This condition can be caused by the ingestion of a photodynamic substance, as can be found in clover, alfalfa, and buckwheat. In addition, certain drugs (such as phenothiazine) are also photodynamic. When the horse consumes such an agent, his non-pigmented skin becomes hypersensitive to sunlight.

Photosensitization occurs on white markings when the horse is exposed to photodynamic agents (e.g., certain weeds).

Photosensitization may also be due to a congenital malfunctioning of the liver. Bile, a fluid secreted by the liver to aid digestion, contains a photodynamic substance. If the bile duct is blocked, bile will reach the circulatory system and the horse will become photosensitized. A congenital malfunction of the liver which causes this is known to be inherited in sheep.

In any case, only a horse with white markings (areas with underlying pink skin) will be affected; the greater the amount of white, the more severe the condition could be. Therefore, a susceptibility to photosensitization can be considered inherited, as white markings are known to be due to heredity. (Refer to "Coat Color and Texture.")

SQUAMOUS CELL CARCINOMA
Squamous cell carcinoma is a cancer of one of the skin layers and occurs most frequently around body openings, near the mucocutaneous junctions where the skin and mucous membranes meet. It does not occur on the skin itself, but usually on the nasal mucosa, conjunctiva, etc. It most frequently affects horses with white markings, particularly those with white markings around the eyes and nose. Since white markings and white coat color are inherited, the tendency of those horses to develop squamous cell carcinoma is also considered

inherited. In other words, dominant white, Appaloosa, tobiano, overo, and the lightly pigmented cremello or perlino horses are more prone to this disorder.

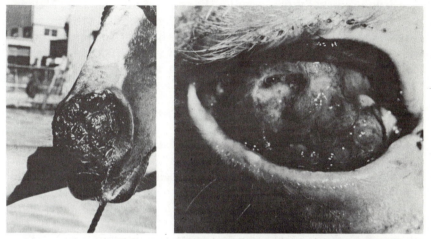

The skin cancer squamous cell carcinoma frequently occurs around the eyes and nostrils, especially in association with white face markings (left). Extensive squamous cell carcinoma of the third eyelid. Note the lack of pigment around the eye (right).

MELANOMA

A melanoma is a tumor caused by the accumulation of the pigment melanin. Although this can occur in horses of any color, it is much more prevalent in grey horses; approximately 80% of all greys over 15 years of age develop melanomas. This high incidence among grey

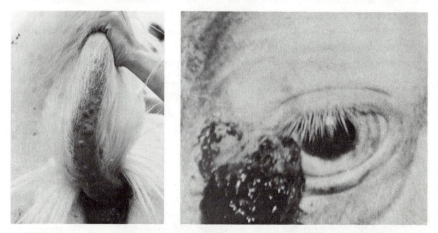

Melanomas frequently occur under the tail and around the anal area (left). Melanomas may also occur around the eye or any other body opening (right).

horses is due to the mechanics of the controlling gene. The greying gene prohibits melanin from entering the hair and, therefore, causes it to accumulate within the skin.

Lesions are usually first seen in the tail and anal regions. Melanomas may be benign for many years, then suddenly metastasize (spread) throughout the body, causing death. For this reason, the life expectancy of a grey horse may be several years less than a horse of another color. Since melanomas are so frequently associated with the greying gene, the tendency to develop them is considered at least partially inherited. (Refer to **"Grey: The G Locus"** under **"Coat Color and Texture**.)

LACK OF CHESTNUTS AND/OR ERGOTS

Chestnuts are the relatively oval-shaped horny growths found just above the knee on the forelegs and just below the hock on the hind legs. Ergots are similar but round-shaped growths located on the back of the fetlock joint. These callosities are probably the remnants of foot pads from the time when ancestors of the horse walked on the entire foot (including what is now the fetlock) instead of on the toes like the modern horse.

Although most horses have both chestnuts and ergots, some Arabian, Thoroughbred, Icelandic, and Connemara horses may lack chestnuts and/or ergots on the forelegs, the hind legs, or both. Horses nearly always have chestnuts on the forelegs. Donkeys have them only on the forelegs, and their presence varies in mules. Both of these growths are more prominent on heavy draft horses.

Chestnuts are sometimes used for identification purposes, since they vary in size and shape between horses. Although no two chestnuts are exactly alike, their size and shape are probably influenced by heredity. For example, the tendency of some chestnuts to be forked

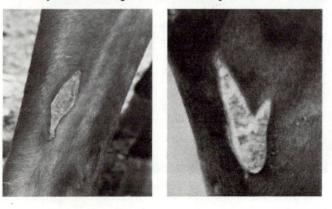

Chestnuts are unique to each horse and are sometimes used as part of an identification system.

Chestnut on the left foreleg of a Damara zebra (Equus burchelli antiquorum).

may be handed down from parents to offspring. Heredity also influences whether chestnuts or ergots are present at all.

Since chestnuts and ergots are non-functional (except for the role they play in an identification system), their presence or absence is only a curiosity.

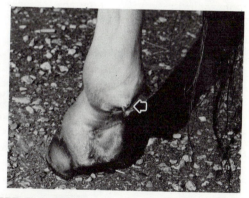

The ergot (arrow) is a soft, horny growth located under the tuft of the fetlock. This tissue is believed to be the remnant of a once functional foot pad.

DRY COAT (ANHIDROSIS)

Anhidrosis is the partial or complete loss of the ability to sweat, resulting in a "dry coat." This condition usually develops in horses that have been moved from a temperate environment to a hot, humid climate. Normally, the adrenal glands, in response to increased temperature, release adrenalin (epinephrine), which causes the horse to sweat. A horse that is susceptible to anhidrosis will sweat excessively when suddenly exposed to constant hot weather. Gradually, he will cease to respond to adrenalin (i.e., he will cease to sweat) and will develop anhidrosis, or dry coat. An affected horse may collapse from exertion, since he is unable to cool his body normally.

A similar condition is seen in cattle, and it is known that the ability of cattle raised in temperate climates to adjust to hot, humid weather is controlled by heredity (e.g., Brahman cattle adjust more readily to these conditions than most other breeds). This suggests that the condition may also be inherited in horses. Anhidrosis in the horse is probably an inherited susceptibility that must be triggered by the stress of a tropical environment.

PINKY SYNDROME

Pinky syndrome is a pigment disorder found primarily in older, grey Arabian horses. The muzzle and eye areas, which were originally black-skinned, will depigment in a blotchy pattern. The pattern often resembles the mottled skin of an Appaloosa horse. This permanent condition should be distinguished from the temporary depigmentation that can occur on the muzzle of a horse of any color, which may be due to exposure to certain chemicals where the horse is fed (e.g., feed troughs or buckets). Because of the restriction of pinky syndrome to grey Arabians, this condition is thought to have a hereditary basis.

Grey Arabian with the pinky syndrome. Note the mottling on the muzzle and around the eye.

VARIEGATED LEUKOTRICHIA

This skin disorder consists of crusty lesions arranged in a variegated or cross-hatched pattern down the back, from the withers to the tail. When the crusts are shed, hair loss on the affected areas occurs. Then, when new hair grows in, it is white. Variegated leukotrichia has been reported only in Quarter Horses, although a similar condition does occur in other breeds. Heredity is considered influential in this disease. Variegated leukotrichia is an uncommon disorder which is only of cosmetic importance.

SUBCUTANEOUS HYPOPLASIA

In this condition, the skin is not firmly attached to the underlying muscle. The affected areas are easily injured. Subcutaneous hypoplasia has been seen only in Quarter Horses, and is believed to be inherited. Whether it is dominant or recessive in nature is unknown.

EPITHELIOGENESIS IMPERFECTA

Epitheliogenesis imperfecta means imperfect development of the skin, and is characterized by missing patches of skin. This rare condition has been reported to occur on the legs, head, and tongue of foals. The missing areas do not heal, and a foal with this disorder usually dies from infection soon after birth. Epitheliogenesis imperfecta is probably due to a rare, recessive gene.

LINEAR KERATOSIS

Linear keratosis is a skin disease characterized by defective formation of the outer skin layer. This condition does not result in missing patches of skin, as in epitheliogenesis imperfecta, but causes permanent spots of defective skin. Linear keratosis has only been reported in Quarter Horses, indicating that the condition is inherited. The mode of inheritance has not yet been discovered, however.

ATHEROMA (SEBACEOUS CYSTS) OF THE FALSE NOSTRIL

An atheroma is a sebaceous cyst that may occur in the posterior portion of one or both false nostrils. (The false nostril is a blind pouch in the lower nasal passage.) They vary in size from a very small cyst to one as large as a tennis ball that may affect respiration and require surgical correction. If drained, the cysts will refill with fluid. Small cysts, however, are only of cosmetic importance. The tendency for these cysts to occur may be hereditary.

MOHAMMED'S THUMBPRINT

Mohammed's Thumbprint is a depression seen in the muscle of the lower neck, either near the midline or between the jugular groove and the midline. This small depression may be quite deep. It has been noted most frequently in Arabian horses, and is thought to be hereditary. Mohammed's Thumbprint in no way hinders the horse, and is of cosmetic importance only.

MALLENDERS AND SALLENDERS

These two disorders, which used to be seen in heavy draft horses, differ only in location. Mallenders affects the back of the knee, while sallenders affects the front of the hocks. Characteristics of these disorders include thickened skin, scabs, and hair loss. These areas sometimes crack, causing pain and subsequent lameness.

Heredity, bacteria, parasites, and friction were all suspected causes of mallenders and sallenders, but this disease, like shivering in draft horses (refer to the discussion on neurological abnormalities) is not often seen today.

Reproductive System

The horse's reproductive fitness is obviously of vital importance to the breeder, since a horse that cannot produce is of no use to a breeding program. The basic reproductive examination is covered in **"Breeding Aspects,"** along with guidelines for selecting stallions and broodmares. Many reproductive abnormalities can be corrected, but those caused by defective chromosomes (e.g., aneuploidy, pseudohermaphroditism, etc.) are permanent. Chromosomal conditions that affect fertility are discussed in **"Sex Chromosomes"** (refer to CY-TOGENETICS AND PROBABILITY).

CRYPTORCHIDISM

Cryptorchidism is an inherited condition characterized by the failure of one or both testes to descend into the scrotum. When one testicle is retained, the condition is referred to as unilateral cryptorchidism. This condition may affect either the right or left testicle, but the left seems to be involved more often. Bilateral cryptorchidism refers to the retention of both testes.

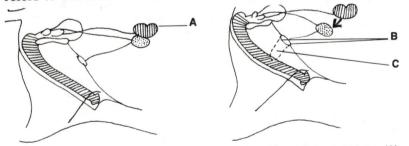

Descent of the testes. Originating in the abdominal cavity near the kidneys (A), the testes drop to the internal inguinal ring (B). The testes normally descend through the inguinal canal (C) and into the scrotal sac (D).

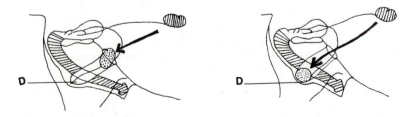

Normally, both testes migrate from the abdominal cavity (where they are formed during embryonic development), through the inguinal canal, and into the scrotum. This process usually takes place prior to the foal's birth. If the testes do not descend, one of three situations may be present:

1) Abdominal Cryptorchidism: The testes are retained within the abdominal cavity. Because the testes increase in size and the inguinal ring (opening to the inguinal canal) constricts shortly after birth, testes which are still located within the abdomen after birth are unable to migrate to the scrotum. They may be located anywhere from adjacent to the kidney to the entrance of the inguinal canal.

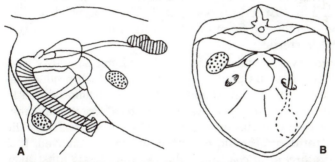

Unilateral abdominal cryptorchidism is the failure of one testis to descend into the scrotum. (A) Side view. (B) Anterior view.

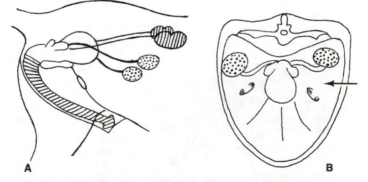

Bilateral abdominal cryptorchidism, side view (A) and anterior view (B). Slow descension of the testes may be completely blocked by the constriction of the internal inguinal ring (arrow).

2) Inguinal Cryptorchidism: The testes are retained somewhere between the scrotum and the internal inguinal ring. Their exact location varies from high in the canal to just below the skin surface. (The subcutaneous types can be palpated.) Occasionally, inguinal testes

will eventually migrate to the scrotum. Cases have been reported in which the foal's testes were in the normal position at birth, but later moved back into the inguinal canal.

3) Monorchidism: When one or both of the testes are missing, the horse is unilateral or bilateral monorchid. The rare condition is usually caused by chromosomal abnormalities, and is not a case of true cryptorchidism. (Refer to the discussion on intersex under **"Sex Chromosomes."**)

Because normal sperm production cannot occur at body temperature, cryptorchid testes are not capable of producing sperm. (The purpose of the scrotum is to protect the testes and regulate their temperature. Refer to **"Breeding Aspects."**) The descended testicle of the unilateral cryptorchid functions normally, but the bilaterally cryptorchid horse is sterile.

Although cryptorchidism may affect the horse's fertility, it does not affect his testosterone production. The cells which produce testosterone (Leydig cells) are functional in both the abdominal and inguinal testes. For this reason, the mature bilateral cryptorchid may show the typical behavior characteristics of the normal stallion. The testes of the bilateral cryptorchid are often surgically removed. If the unilateral cryptorchid is castrated, the retained testis must also be removed. (In most cases, the retained testis is still capable of producing testosterone and contributes to libido, stallion-like behavior and nervousness.)

Because cryptorchidism is limited to the male, it is referred to by geneticists as a sex-limited trait. Even though the female may carry

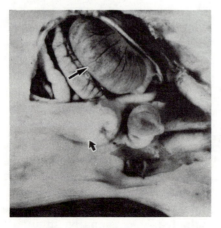

Fetal testes. Location of the testis (bold arrow) before its descent to the scrotum. Note the position of sheath of penis (small arrow).

the cryptorchid gene, she cannot express the defect. She can, however, produce cryptorchid colts. The explanation of how cryptorchidism is transmitted from generation to generation is not clearly understood. Some geneticists believe that cryptorchidism is controlled by two genes. They feel that each gene controls the descent of one testicle. If this is true, breeding unilateral cryptorchids will allow both types of cryptorchidism to continue. The inheritance of cryptorchidism has been studied primarily in the pig and the ox; it has been assumed that the mechanism of inheritance is similar in the horse.

SCROTAL HERNIA

Scrotal hernia is a condition involving the displacement of a piece of intestine through the inguinal canal and into the scrotum. The signs which are often associated with scrotal hernias are swelling on one side of the scrotum, intestinal pain and altered gait. Sometimes the intestinal protrusion is retained within the inguinal canal, and is referred to as an inguinal hernia. The horse with an inguinal hernia will not have the scrotal swelling, but will commonly exhibit distress from intestinal pain.

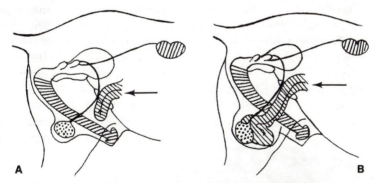

Displacement of a piece of intestine (arrow) through the internal inguinal ring causes either an inguinal hernia (A) or a scrotal hernia (B).

Scrotal (or inguinal) hernias in newborn foals are common, since the opening (inguinal ring) between the abdomen and the inguinal canal does not always constrict immediately after birth. (The time required for constriction of the inguinal ring may be an inherited trait.) Foals with scrotal or inguinal hernias will usually recover without treatment. These causes are not serious, unless the intestine is strangulated by the inguinal ring as it gradually constricts. Gut strangulation cuts off the blood flow to the displaced area, causes extreme intestinal pain and, if left uncorrected, will result in the death of the foal. For this reason, it is important to watch for signs of colic in the foal born with a scrotal hernia.

Scrotal (or inguinal) hernias in stallions can be caused by accidents (e.g., kicked by another horse), by breeding or by an inherited conformation defect of the genital tract. The presence of intestinal tissue within the scrotum raises the testicular temperature and inhibits the stallion's sperm production. Sometimes a veterinarian can correct the condition by rectal manipulation. (Rectal manipulation involves repositioning the intestine by hand, from the inside of the rectal wall.) In some cases, corrective surgery will be required, especially if gut strangulation occurs. If the testes are damaged, they are usually removed during the operation.

Although geneticists believe that some scrotal (or inguinal) hernias are inherited, the genetic action involved is not yet known. Like cryptorchidism, scrotal and inguinal hernias are sex-limited traits. (The female cannot express the trait, although she can pass the controlling gene to her offspring.)

HYPOTHYROIDISM

The thyroid gland produces the hormone thyroxin, which is responsible for the utilization of oxygen within every cell of the body. Many body systems (such as the nervous, the reproductive and the metabolic) depend heavily upon this hormone, so much that a thyroxin deficiency (hypothyroidism) may have widespread effects upon the affected individual's normal physiological functions.

In the mare, hypothyroidism may be associated with "silent" heat and crestiness of the neck. In the stallion, it can cause depressed

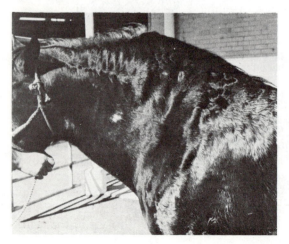

Hypothyroidism in the mare is often characterized by an extremely crested neck. This mare also had fertility problems.

sperm concentration and quality. In both cases, the hypothyroid condition causes an increased tendency to gain weight and, therefore, is
an indirect cause of lowered fertility. (Obesity is an important cause
of lowered fertility in both the stallion and the mare. Refer to
"Breeding Aspects.")

Although rare, hypothyroidism is believed to be inherited in
humans, suggesting that the condition may also be inherited in the
horse. The inheritance pathway is not known at this time.

NYMPHOMANIA

Nymphomania is thought to be an inherited condition involving the
continuous or exaggerated display of estrus in the mare (such as raising the tail, urinating, squealing, etc.). Because the trait can only
occur in the female, it is referred to by geneticists as a sex limited
trait. (The exact mode of inheritance is unknown.) The degree of
nymphomaniac behavior varies, and the condition can be easily differentiated into two types: pseudonymphomania and true nymphomania.

Pseudonymphomaniac mares ovulate throughout the year at regular intervals and have especially long periods of receptivity during
the first part of the breeding season. Maiden mares will often exhibit
this extended estrus in the early spring, but the condition is usually
temporary. In most cases, pseudonymphomania is not considered a
serious condition, and treatment is usually unnecessary. Because
some pseudonymphomaniac mares are sexually receptive when they
are pregnant, a special effort should be made to diagnose pregnancy.
(Breeding or artificially inseminating a pregnant mare could cause
her to abort.)

True nymphomania occurs in many forms and can be divided into
several categories:

1) Mild nymphomania is caused by an excess of reproductive hormones. In this case, the mare behaves normally except that she becomes extremely excited during estrus. During this period, she may
refuse to accept the stallion. If the mare's behavior is considered undesirable, it can usually be modified by the removal of one or both of
her ovaries (ovariectomy). If it is necessary to remove both ovaries,
the mare will obviously be sterile.

2) The mare with another form of true nymphomania will constantly show signs of heat but will not usually conceive. This form is closely associated with cystic ovaries; the cysts inhibit ovulation. If the ab-

normal behavior is caused by a malfunction of one ovary (e.g., cystic ovary), an ovariectomy will modify the condition.

3) Severe nymphomania has been designated as a nervous disorder. Although an endocrine (hormone) problem may influence the initial nymphomaniac reaction, the condition cannot usually be corrected by an ovariectomy. Severe nymphomaniac mares exhibit signs of estrus constantly but, in most cases, will not accept the stallion. These severe cases show extremely violent behavior, especially when handled about the hindquarters or when mounted by a stallion. The severe nymphomaniac mare may be extremely aggressive toward other horses or even toward her handler.

PNEUMOVAGINA

Pneumovagina is a condition involving the presence of air and/or fecal matter within the mare's vagina. This condition increases the mare's susceptibility to reproductive infections, such as cervicitis and endometritis. Pneumovagina is caused by a poorly formed, relaxed or injured vulva (opening to vagina).

Genetically influenced conformation traits, such as flat croup and high tail head, alter the vulva's position so that it is not the desired vertical, but almost horizontal (tipped vulva). In this position, falling fecal matter is likely to enter the mare's vagina, predisposing her entire reproductive tract to infection. Tipped vulva is relatively common in breeds that are characterized by high tail placement (e.g., the Arabian breed). Another inherited conformation trait involves small, underdeveloped vulval lips. Because the underdeveloped lips do not form a tight protective seal over the vagina, air and fecal matter may enter. When the affected mare is in motion, the sound of air rushing past the vuval lips can sometimes be heard.

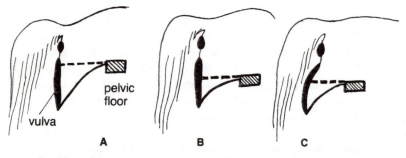

Position of the mare's reproductive tract relative to the pelvic floor. (A) Normal. (B) Low pelvic floor in relation to the vulva. (C) "Tipped" vulva.

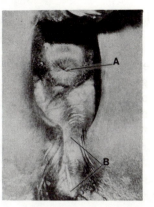

Pneumovagina conformation. A sloping angle from anus (A) to vulva (B), predisposes a mare to pneumovagina, infection and subsequent fertility problems.

Relaxation of the vulva, another cause of pneumovagina, can occur when the mare's pelvic ligaments are relaxed (e.g., during estrus or after foaling). Also, as the mare ages, she naturally loses some of her vulval tone and becomes more susceptible to pneumovagina. Frequently, the poorly conditioned thin mare or the well conditioned, but lean, race mare have relaxed vulvas that allow air to be sucked into the vagina.

Injury to the vulva may also cause pneumovagina. For example, if the mare's vulval muscles are stretched or torn during foaling, her vulva will lose its elasticity. This condition increases the mare's susceptibility to pneumovagina and its related infections.

There are several possible effects of pneumovagina:

1) vaginal, cervical and/or uterine infections
2) lowered fertility
3) urine pooling
4) presence of fecal matter within the vagina
5) soiled hindlimbs (caused by an infection-related discharge)
6) windsucking sound when the mare is in motion
7) ballooning of the vagina caused by the presence of air
8) presence of air and/or fecal matter in the uterus, if ballooning of the vagina occurs

Although pneumovagina is a potentially serious detriment to productivity, it can be remedied by partially suturing the lips of the vulva together. This procedure, known as Caslick's operation, helps prevent fecal matter and air from entering the vagina. The sutured mare can, in most cases, be safely bred (especially by artificial insemination), but of course it is always necessary to remove the sutures before she foals. (If the stitches are not removed, the vulva could be seriously torn and the mare permanently damaged.) Some trainers routinely suture race mares to protect the reproductive tract.

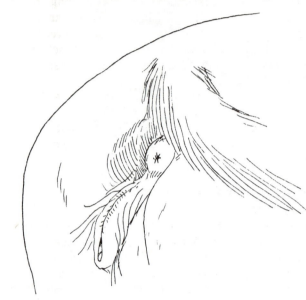

Caslick's operation: The upper portion of the lips of a slanted or poorly formed vulva are often stitched together to aid in the prevention of reproductive infections.

Digestive System

Congenital deformities which may affect the horse's digestive system include cleft palate, atresia coli, atresia recti, atresia ani, and shistosoma reflexum. Although these conditions are thought to be inherited, research has not yet established the inheritance pathways.

CLEFT PALATE

Cleft palate results from a failure of the hard or soft palate (the shelf of tissue separating the mouth from the nasal cavity) to close along its

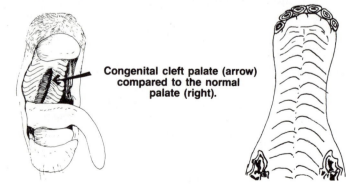

Congenital cleft palate (arrow) compared to the normal palate (right).

midline prior to birth. Nursing results in the expulsion of milk through the nostrils, and sometimes the inhalation of milk into the lungs (with subsequent inhalation pneumonia). Clefts of the soft palate can sometimes be surgically corrected. This is a relatively new procedure, and it is not yet known whether such horses will have a normally functioning soft palate when they reach maturity. If the opening is extensive, the foal will usually die soon after birth.

Although some forms of cleft palate are known to be inherited in humans, research has not yet indicated whether this is also true for horses. Possible causative factors of this defect include prenatal infections, the presence of toxic substances (such as certain drugs during pregnancy), or a complicated mode of inheritance.

ATRESIA COLI, ATRESIA RECTI, AND ATRESIA ANI

These three conditions are grouped together because they are similar defects of the digestive system. Atresia means "absence of," so atresia coli refers to the absence of a piece of the large intestine; atresia recti refers to the absence of the rectum; and atresia ani refers to the absence of the external opening from the rectum. Except for foals with additional birth defects, foals affected with these conditions appear normal at birth, stand, nurse, and seem to be in good health. The foal with atresia coli or atresia recti will not defecate when given an enema. Sometime after birth, usually within 8 to 24 hours, the foal will die after developing a colic that does not respond to treatment.

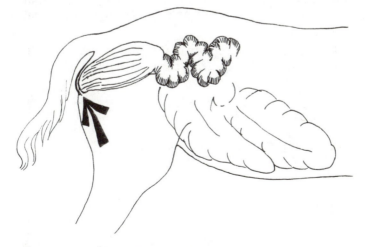

Atresia ani (arrow).

As already indicated, a foal affected with atresia coli will be missing a section of the large intestine, so that the intestine is in two unconnected pieces. Several feet or whole sections of the intestine may be missing. Contents of the intestine cannot pass through the body to be excreted. Since this condition cannot be corrected, it is always fatal.

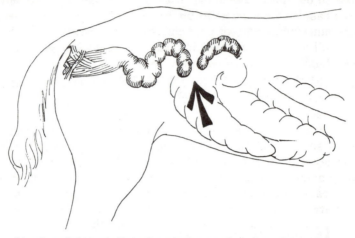

Atresia coli (arrow). Note the missing section of small intestine.

Atresia coli was first observed in Percheron foals (Yamane as quoted in Wreidt), in which it was thought to be due to a single recessive gene (the carrier parents appeared normal). It is also seen in association with the "white foal syndrome," which occurs in about

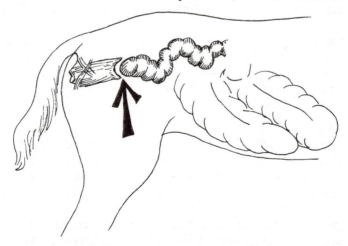

Atresia recti accompanied by atresia ani. The rectum ends in a blind pouch (arrow) and the anal opening is missing.

10% of all overo-overo crosses. The affected foal is born completely white, with atresia coli among other birth defects (such as brain malformation, which also occurs with atresia coli in the Percheron). It has been proposed that the overo parents of a white foal carry a recessive allele, o^e (overo is oo). In the homozygous state, $o^e o^e$, this allele is believed to be fatal.

In the case of atresia recti, part or all of the rectum is missing, so that the anus ends in a blind pouch. This condition is serious because the foal will be unable to defecate. Like atresia coli, atresia recti is not spontaneously or medically correctable.

Foals with atresia ani can frequently be saved, if their muscular sphincter and imperforate anus are adequately developed. An opening from the rectum to the outside can be surgically made.

The exact mode of inheritance of both atresia recti and atresia ani is unknown, since these conditions are infrequent and occur only sporadically. They may be due to either a dominant mutation, or to a rare recessive allele. Although other factors, such as nutrition or drugs, may also cause these defects, there is evidence that heredity is responsible for atresia ani in other species, such as cattle and swine.

ESOPHAGEAL DEFECTS

Esophageal dilatations or strictures are two abnormalities that may affect the ability to swallow, and predispose the horse to choke. Megaesophagus, or esophageal dilatation, is an abnormal widening of an area of the esophagus. This area has no muscle tone, a characteristic which is believed to be caused by a lack of nerve supply to the affected area. This condition is known to be inherited in the German Shepherd dog.

A stricture is an abnormal constriction, or narrowing, of the esophagus and may occur anywhere along its length. This condition is also found in association with persistent right fourth aortic arch (refer to "Circulatory System"). A dilatation may immediately precede the stricture. Whether this condition is inherited in the horse is unknown.

SHISTOSOMA REFLEXUM

Shistosoma reflexum is a condition in which the abdominal contents of the fetus are located outside the body wall; the head is turned back towards the croup. The fetus, as a result, appears to be partially turned inside out. There is also a complete lack of development of the amnion, the fluid-filled membrane that encircles the fetus to protect it from injury. A fetus affected with shistosoma reflexum will be aborted in the early stages of gestation. This defect is believed to be caused by a recessive lethal gene.

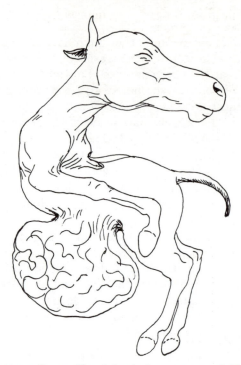

Shistosoma reflexum. The abdominal organs are outside the body wall on the aborted fetus.

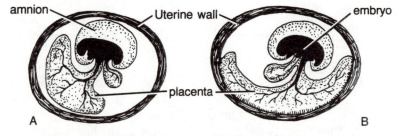

The embryo surrounded by its protective amnion before implantation in the uterus (A) and after implantation (B).

Nervous System

The following discussion examines inherited traits that affect the nervous system. This includes characteristics such as temperament as well as diseases and defects affecting the nervous system. The central nervous system (CNS) includes the brain and the entire spinal cord, which is located in the vertebral column. The cranial nerves are extensions of the brain, and the peripheral nerves are extensions of the spinal cord. Roaring, a neurological disease which affects respiration, in covered in **"Respiratory System."**

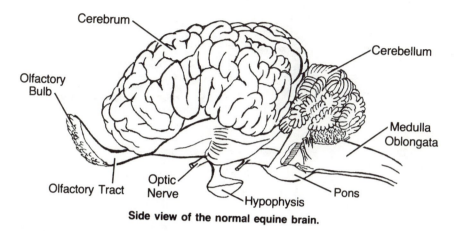

Side view of the normal equine brain.

TEMPERAMENT

Temperament is a general consistency in the performance of specific actions under specific circumstances. It depends on both environmental and genetic factors. As an inherited trait, temperament can be selected for by judicious breeding. The breeder should realize that the use of an untractable broodmare may result in offspring with her undesirable disposition.

The tameness of horses and other domestic animals is partially the result of selection; domestic animals differ genetically from their wild counterparts. For example, zebras are extremely difficult to train by conventional methods, although some species, such as Grevy's zebra, are more trainable than others.

Some authorities believe that inbreeding to a temperamental horse increases the chance of a bad disposition in the offspring. In addition, maternal influence shapes the foal's temperament. If a mare possesses a bad disposition, her foal may mimic the behavior.

Temperament may also be influenced by eye placement. As discussed under **"Evolution,"** a horse's behavior is modified by his acuity of vision. The depth of the eye socket and the width of the forehead can affect vision and thus influence temperament. A deep socket and a wide forehead decrease the degree of binocular vision. This reduction in binocular vision may cause the horse to be nervous and apprehensive, in comparison to the horse with more prominent eyes and a narrow forehead. Since these physical attributes are inherited, it can be inferred that temperament is partly inherited indirectly (a secondary characteristic of conformation).

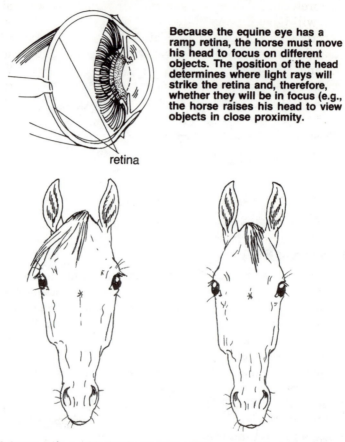

Because the equine eye has a ramp retina, the horse must move his head to focus on different objects. The position of the head determines where light rays will strike the retina and, therefore, whether they will be in focus (e.g., the horse raises his head to view objects in close proximity.

retina

A horse with a wide forehead (left) will have less binocular vision (i.e., will see less with both eyes simultaneously) than a horse with a narrow forehead (right). This may influence the horse's reaction to stimuli and, consequently, his temperament.

HEREDITARY ATAXIA

Ataxia is defined as the loss of muscular coordination. In the horse, spinal ataxia and cerebellar ataxia are suspected as being genetic in origin.

Wobbler Syndrome (Equine Spinal Ataxia)

Ataxia (incoordination) due to a disorder of the cervical spine (neck) is known as the wobbler syndrome, or equine spinal ataxia. The only sign of this condition is an incoordinated gait that worsens when the horse is turned or backed (the hind legs are more severely affected than the forelegs). In all other respects, the horse appears healthy. A "wobbler" will have a proprioceptic defect, meaning that he will not know exactly where his feet are at all times. This will cause him to stumble and hesitate to place his feet. His gait may include sideways lurching and "drunken" movements of the hindquarters. Wobbles may appear either gradually or suddenly; if it appears suddenly, an accident is often a precipitating factor (e.g., the horse fell, became cast in his stall, etc.).

In discussing equine spinal ataxia, reference is made only to those cases with the aforementioned signs and pathology. Wobbler syndrome must be differentiated from fractures of the cervical vertebrae, vertebral abscesses, and the ataxia occasionally associated with rhinopneumonitis, all of which could cause similar signs. Wobbler syndrome has been classified into three types:

Type 1: The horse exhibits permanent flexion of the neck, which functionally compresses the spinal cord. This type is very rare.

Type 2: There is symmetrical overgrowth of the articular vertebral processes. This type is more common in sucklings and weanlings.

Type 3: There is asymmetrical overgrowth of the articular vertebral processes. This type is more common in yearlings, and is the most common of the three types.

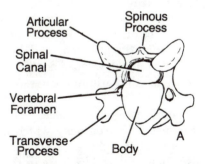

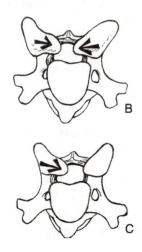

A comparison of a normal vertebra (A) to a wobbler vertebra with symmetrical over-growth (B) and asymmetrical overgrowth (C). The boney overgrowth compresses the spinal cord and causes the wobbler syndrome.

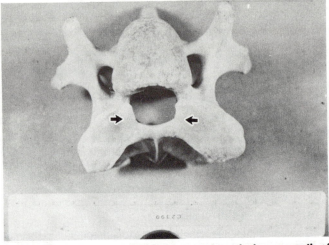

This vertebra from a wobbler shows symmetrical overgrowth of the articular processes (arrows).

Types 2 and 3 may not become apparent until flexion or trauma occurs. It has been theorized that the characteristic malformations of the cervical vertebrae (usually located between C4 and 5, C5 and 6 or C6 and 7) are inherited, and that as many as 10% of all foals may have these malformations (Rooney, quoted by Adams). The foal is often able to compensate until injured, explaining why an animal frequently shows no ataxia until trauma occurs. It seems likely that these foals are more "accident-prone" due to a slight undetected incoordination.

If this condition were due solely to trauma, some horses should recover. This is not the case, however, as afflicted animals usually worsen, sometimes reaching a plateau of disability. Because heavily muscled and well-fed horses are frequently afflicted, and because related horses raised on different farms have developed equine spinal ataxia, the importance of nutrition as the primary cause of this condition is minimized.

In a study (Dimock) of 191 diagnosed cases of wobbles, 164 were Thoroughbreds, (the study was conducted in Kentucky which could explain the high percentage of afflicted Thoroughbreds), and 121 had closely related sires. The study also showed that 43% were in the tail male descent from one of three Thoroughbred sire lines, 14% from the second line and 1% from the third. Wobbler syndrome affects more males than females, with ratios varying from 3:1 to 27:1, depending on the type of wobbler and the study. Some authorities feel that wobbles may be due to a disproportionately long neck, which stresses the vertebral canal and stretches the spinal cord. The sex incidence may be attributable to the fact that males have longer necks and heavier muscles, on the average, than do females.

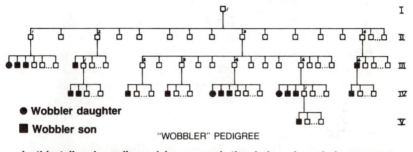

"WOBBLER" PEDIGREE

● Wobbler daughter
■ Wobbler son

In this tail-male pedigree (sires - sons), the darkened symbols represent horses with the wobbler syndrome. Note that a generation III wobbler sired a generation IV wobbler.

Although some mares have produced more than one wobbler foal, there is an absence of dominance in the condition's inheritance, since most wobblers are produced by nonwobbler parents. In one test-mating, a wobbler - normal cross did not produce a wobbler. Since a wobbler - wobbler cross has produced normal offspring, wobbles is not a simple recessive condition either. If the wobbler syndrome was due to a simple recessive gene, wobbler (homozygous recessive) × wobbler (homozygous recessive) should produce only wobbler offspring.

In a recent British study (Falco et. al, 1976) on the inheritance of equine spinal ataxia, a total of 134 Thoroughbreds were examined (67 wobblers and 67 comparable normal horses; 114 males and 20 females). In this study, no evidence of heredity was found. Conclusions were based on the following reasons:

1. The predominance of normal parents producing wobblers rules out an autosomal dominant mode of inheritance.

2. The production of wobbler females from normal sires rules out a recessive X-linked (located on a sex chromosome) mode of inheritance.

3. Since inbreeding increases the presence of recessive defects, the presence of a greater than normal amount of inbreeding in the pedigrees of the affected animals, as compared to the normal population, would have indicated a recessive defect. A relationship between increased incidence of wobbles and inbreeding was not found.

4. Comparing the pedigrees of normal and wobbler horses, there was no significant difference in the incidence of wobbles, suggesting that there was no complex mode of inheritance.

5. There was no genetic basis for a predisposition to vertebral stenosis (narrowing of the vertebral canal).

These results are in almost direct apposition to the earlier study. In regard to a possible breed or sex predisposition to wobbles, the researchers of this 1976 study felt that the high incidence of Thoroughbred wobblers in other studies (e.g., 86%, 98.7% and 87%) was because the reports had been made in areas of high Thoroughbred population. They also felt that many affected females with equine spinal ataxia may be kept for breeding purposes and never reported.

The researcher in the earlier study (Dimock) examined the pedigrees of six families and showed that the wobbler syndrome sometimes occurred within several branches of the same family. In one pedigree, 31% of the wobblers studied were distributed throughout every branch of the family.

Congenital Occipito-Atlanto-Axial Malformation (OOAM)

Congenital occipito-atlanto-axial malformation is a skeletal defect of the joint between the skull and the first cervical vertebra, or between the first and second cervical vertebrae. This defect occurs in Arabian horses. Characteristics include the following:

1. Fusion of the atlas (the first neck vertebra) to the occipitus (back of the skull). The atlas also develops "peg-like" structures on either side.
2. Enlargement of the axis (the second neck vertebra) and its transverse processes, and an abnormally short blunt dens (the dens is a tooth-like projection on the front of the axis, on which the atlas articulates).

An affected foal shows bumps on either side of the neck, just behind the skull, due to the "pegged" atlas. The foal will not be able to flex his neck normally and a clicking sound will be produced when the neck is flexed. Because the joint between the atlas and the axis is unstable, the dens will pith (destroy) the spinal cord. This can occur during parturition, resulting in stillbirth. In this case, the defect should be differentiated from those which also result in stillbirth: Dystocia (malposition of the fetus), infections or other malformations.

The spinal cord may be only partially affected during birth, so that the foal is born alive, but with all four legs spastic or paralyzed. In this situation, the condition should be differentiated from neck trauma, infectious diseases such as meningitis and malformations such as contracted tendons or scoliosis.

If the foal is born without damage to the spinal cord, he will eventually develop progressive ataxia. This resembles the wobbler syndrome, since the affected horse stumbles and falls. These signs may not appear until several months of age. At this stage, the defect should be distinguished from cerebellar hypoplasia (in the latter, the foal also shows head tremors) and cervical vertebral stenosis (narrowing of the spinal canal that causes wobbler signs in large, fast-growing horses). Other conditions that may show similar signs include degenerative myeloencephalopathy, a rare condition that has only been reported in Shetland ponies, and myelitis caused by protozoal infections or rhinopneumonitis.

Because of its restriction to the Arabian breed, congenital occipito-atlanto-axial malformation is believed to be, at least partially, inherited. Of twelve cases that have been reported, two pairs of full siblings and a half sibling to one of the pairs were included. In the nine pedigrees that were available for study, a common ancestral sire was incriminated and inbreeding was higher than the breed average. As a result of this evidence, geneticists believe that the defect is due to a simple autosomal recessive.

Cerebellar Hypoplasia

Cerebellar hypoplasia, the underdevelopment of the cerebellum, is a congenital defect in horses. Because the cerebellum regulates coordination, its underdevelopment is characterized by head tremors, paddling, difficulty in backing and a wide-leg stance. The severe progressive ataxias occurring in Oldenburg horses and in Gotland ponies have signs which are similar to those of cerebellar hypoplasia.

Since affected horses have a close familial relationship, it is possible that a hereditary factor causes this condition. In a study of 21 affected foals (Sponseller), 20 were purebred Arabians and 19 were heavily inbred to two particular stallions. In one case, a mare with cerebellar hypoplasia produced an affected foal, while a normal mare produced two affected foals by the same sire. Another normal mare produced three affected foals by three different sires.

In other species (cattle, sheep, dogs and rabbits), cerebellar hypoplasia has been shown to be inherited. It has also been hypothesized that a viral infection may cause this condition (as it can

388

in cattle and cats) and that the susceptibility to the virus may be inherited.

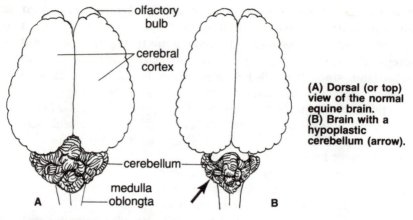

olfactory bulb

cerebral cortex

(A) Dorsal (or top) view of the normal equine brain.
(B) Brain with a hypoplastic cerebellum (arrow).

cerebellum

medulla oblongta

A

B

Cerebellar Ataxia

Cerebellar ataxia refers to incoordination resulting from a disorder of the cerebellum. (As mentioned previously, the cerebellum is the portion of the brain that directs and coordinates movement.) This type of ataxia is uncommon, but it has been well-documented in the Arabian horse, the Gotland pony from Sweden and the Oldenburg breed from Germany. An affected horse has a stiff unsteady gait soon after birth and tends to fall frequently. The foal may also exhibit a fine tremor of the head and usually has poor vision.

Underdevelopment of the cerebellum (cerebellar hypoplasia) is known to be the cause of cerebellar ataxia in the Arabian, and is possibly the cause of the similar ataxia seen in the Gotland pony and Oldenburg horse. The study on cerebellar hypoplasia in the Arabian (Sponseller) showed that the condition is probably genetically controlled. There was also a case in which a Welsh/Arabian foal showed signs of a progressive cerebellar cortical abiotrophy (an abiotrophy is a postnatal, degenerative disorder which is not due to external causes; hereditary but not congenital). Since a similar syndrome (seen in Jersey cattle) is known to be inherited, this type of cerebellar ataxia in horses may also be genetic in origin.

Because of the nature of cerebellar ataxia, affected animals do not usually reach maturity and therefore, cannot perpetuate the condition. The condition is transmitted through apparently normal carriers. Any stallion or mare which have been known to produce even one affected foal should be considered as possible carriers of the disorder.

CONGENITAL HYDROCEPHALUS

Hydrocephalus is the accumulation of excess cerebro-spinal fluid in the cranial cavity. This excess fluid causes pressure on the brain, resulting in an increase in cranial size and a prominent dome shape to the roof of the cranium. Hydrocephalus is usually fatal within 48 hours.

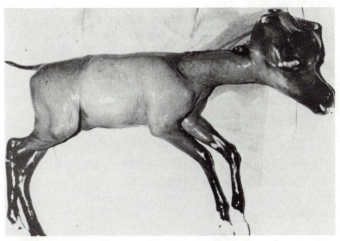

Hydrocephalic fetus; note the increase in size and dome shape of the cranium.

Internal hydrocephalus is the most common form of this condition. In this type, drainage of the cerebro-spinal fluid is blocked. As fluid is produced by the choroid plexus, it accumulates and compresses the cerebellum, causing pressure necrosis or degeneration.

A. Cross section of the normal brain.
B. Cross section of hydrocephalic brain.

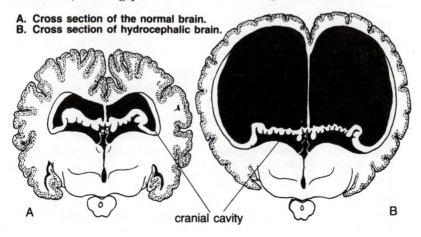

cranial cavity

A   B

Congenital hydrocephalus is rare and may be due to a dominant mutation.

CONVULSIVE SYNDROME

Convulsive syndrome is usually considered to be the result of mismanagement at foaling time; the foal's umbilical cord is sometimes cut too soon after birth, depriving the foal of up to a pint of oxygenated blood. This causes brain hypoxia (oxygen deprivation) and the foal exhibits the signs of convulsive syndrome: jerking the head, aimless movement of the legs, convulsions and "barking."

Convulsions have also occurred in foals that are known to have suffered no blood loss or oxygen deprivation. This has been observed almost exclusively in the Thoroughbred, and is thought to be genetic in origin.

IDIOPATHIC EPILEPSY

Epilepsy is a functional disorder of the brain. Signs of epilepsy include repeated seizures. In a grand mal attack, the horse shows distress and breathes heavily, then falls to the ground and becomes rigid. The affected animal will exhibit muscle jerking, paddling and may foam at the mouth. A grand mal seizure will last anywhere from 1 to 30 minutes. A petit mal attack is not as severe, and may not even be apparent. The horse may either faint, or seem to be in a daze. Epilepsy should be differentiated from seizures due to poisoning, inflammation or neoplasms (tumors).

True epilepsy, as seen in man, is not documented in the horse. Epilepsy in the horse is called "idiopathic" because there is no known cause. This condition may be due to an inherited defect in the development of the brain.

SHIVERING

Shivering is a chronic neuromuscular disorder which most often affects Shires, Clydesdales and other similar draft-type horses. Usually, the muscles of the hind legs and tail are affected, causing the tail, leg or hindquarters of the horse to tremble when the horse moves backwards. Occasionally, one hind leg is held up in a fixed position while the muscles quiver. This spasm lasts a few moments, then the muscles relax and the leg can be lowered. Shivering should be differentiated from stringhalt (which is involuntary overflexion of the hock).

Although the cause of shivering has not been precisely determined, it has been suggested that, because of its occurrence in draft breeds, this condition may be inherited. As a result, the use of afflicted horses as breeding stock is discouraged or even prohibited in some countries.

Respiratory System

A healthy, well-developed respiratory system is necessary for the horse to achieve his full potential. There are several respiratory unsoundnesses which interfere with the horse's performance. Several are thought to be influenced by heredity. These include roaring (laryngeal paralysis), heaves (chronic bronchitis and emphysema), epistaxis (pulmonary hemorrhage) and pharyngitis, each of which is described in the following chapter. Although roaring is by far the most prevalent inherited *nervous* disorder in horses, it is covered here because its most obvious symptom is respiratory distress.

ROARING

Roaring, also known as laryngeal paralysis or laryngeal hemiplegia, is an incurable neurological condition which causes the horse to suffer from a respiratory obstruction. As a result, the horse makes a whistling or roaring noise during exertion. This sound is due to paralysis of the left recurrent laryngeal nerve, which impairs the ability of the horse to fully open the left side of his larynx during exercise. The left recurrent nerve is affected, probably because of the anatomical asymmetry of the two nerves. The left recurrent nerve is longer than the right, and follows a different path. It wraps around the aorta, the major artery leaving the heart, and may be under greater tension than the right nerve.

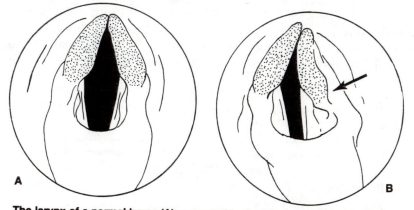

The larynx of a normal horse (A) compared to that of a roarer (B). The roarer's larynx is characterized by sagging laryngeal muscles and cartilage (arrow) which interfere with air flow and cause the whistling or roaring sound.

At one time, laryngeal paralysis was thought to be the aftermath of respiratory diseases, such as strangles and virus infections, but no evidence substantiates this claim. Available evidence points to the probability that the disease is due to an autosomal recessive gene. Taller breeds with long necks are more likely to develop roaring; this may be due to the greater stress a long neck places on the left recurrent nerve. The contributing factor of a long neck may help explain why roaring is seen most frequently in Thoroughbreds and other tall breeds that have proportionately longer necks, and rarely in ponies and mules that have proportionately shorter necks. Roaring is believed to be genetically influenced, since length of neck and body size are inherited conformation characteristics.

Scientific evidence has indicated that roaring may be inherited in Clydesdales, Thoroughbreds, and other tall breeds of horses. In one study (Cook, 1978), 85 test-matings produced the following results:

1) 36 (normal × normal) matings resulted in 10 affected offspring (28%).
2) 47 (normal × affected) matings resulted in 20 affected offspring (43%).
3) 2 (roarer × roarer) matings resulted in 1 affected offspring (50%).

The 28% diseased offspring in the first group is compatible with the theory that roaring is caused by a simple autosomal recessive. The laws of probability state that, for a recessive trait, the mating of two carriers should result in 25% of the offspring being homozygous for the recessive trait. The 25% homozygous recessive offspring exhibit the disease. The 28% diseased offspring produced in the test-matings approximately equals that percentage. However, this result is compatible with the theory only if it is accepted that most of the normal parents were carriers and that a few of the apparently normal horses were, in fact, diseased.

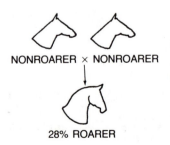

NONROARER × NONROARER

28% ROARER

Mating nonroaring carriers resulted in approximately 28% roaring and 72% nonroaring offspring.

The 43% diseased offspring that resulted from a cross of normal and roarer parents is also compatible with the theory that roaring is due to a simple recessive, if it is accepted that most of the normals were carriers. The laws of probability state that, for a simple recessive trait, the mating of a normal carrier and an affected horse (i.e., homozygous recessive) should result in 50% affected offspring. The 43% approaches that figure, so most of the "normal" parents must have been carriers.

Mating roarers to nonroaring carriers resulted in approximately 43% roarer and 57% nonroarer offspring.

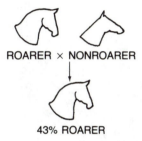

ROARER × NONROARER

43% ROARER

In the third group, since there were only two test-matings, the result is not really significant. However, to fit with the theory that roaring is due to a simple recessive, both offspring should have been roarers. The laws of probability state that, for a simple recessive, affected × affected should result in all diseased offspring. Obviously, the 50% affected offspring in this "experiment" is not close to the theoretical 100%. However, since the two offspring were checked for roaring at three years of age, it is possible that the nonroaring offspring might still develop the condition. More roarer - roarer crosses are necessary to disprove the theory that roaring is due to a simple recessive gene.

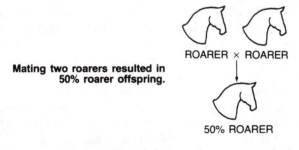

ROARER × ROARER

Mating two roarers resulted in 50% roarer offspring.

50% ROARER

Some well-known Thoroughbred horses have carried the roaring gene: e.g., Pocahontas (the only mare considered in the Vuillier Dosage System for Thoroughbreds) and Herod (an important early Thoroughbred sire).

HEAVES

Heaves, sometimes referred to as "broken wind," is caused by a combination of chronic bronchitis and chronic pulmonary alveolar emphysema. In this condition, the horse has difficulty in breathing and in particular, difficulty in exhaling. Persistent coughing (coughing up sputum) is a common sign, and many horses have a nasal discharge after exercise. Dusty feed, confinement to a dusty stall, and poor stable ventilation are all contributing factors.

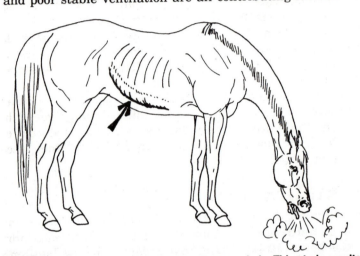

A horse with heaves must make an extra effort to exhale. This strain results in a double expiratory effort and the development of a "heave line" (arrow).

An allergic reaction, perhaps to the dust and mold found in poor quality hay, may cause chronic bronchitis (inflammation of the bronchi of the lungs) and lead to emphysema. In man, it is known that allergies can cause chronic bronchitis. The signs of heaves are similar to those of asthma in man, a disease that often "runs in families." In horses, heaves may also be due to an inherited factor, though there is no scientific evidence to support this hypothesis.

It should be remembered that chronic alveolar emphysema is not inherited. Rather, the affected horse probably had an inherited predisposition to develop this condition when stressed by the presence of allergens (e.g., stable dust, moldy hay, etc.). If environment alone was the cause of heaves, all stabled horses exposed to these conditions should develop the disease. The fact that they do not implies an inherited susceptibility in those that are affected. If heaves is an inherited disease, it is probably caused by both hereditary and environmental factors. (In man, there are many such diseases.)

EPISTAXIS

Epistaxis, or bleeding from the nose, generally observed in racehorses may actually be the result of lung hemorrhage. It may occur during or after a workout or race, and is sometimes seen in conjunction with hypertension (high blood pressure). Of all racehorses suffering from epistaxis, about one-third have hypertension.

Some researchers believe that hypertension, lung disease (such as heaves) or inherited defects in lung capillary strength predispose the horse to epistaxis. Extreme exercise can cause fragile alveolar capillaries within the lungs to rupture. The horse may discharge blood through both nostrils and in some cases, may stagger. Rarely, the horse may fall or even die from suffocation due to blood collecting in the lungs.

In a horse without any genetic predisposition to epistaxis, the condition may be caused by trauma, infection or fatigue. It has been proposed that predisposing factors, such as reduced capillary strength, high blood pressure and an altered clotting mechanism, may be inherited.

PHARYNGEAL CYSTS

Pharyngitis, or inflammation of the pharynx, resulting from bacterial, viral, mechanical or allergic causes is a major problem in the race horse (especially in the Standardbred). The horse may have difficulty in breathing due to the inflamed pharynx, and his performance may be hindered. In more extreme cases, the affected horse may tire easily or stand with the neck extended.

Pharyngeal cysts, which are diagnosed through an endoscopic examination, may have some hereditary basis. These cysts can cause signs similar to those seen from pharyngitis due to mechanical obstruction of the pharynx. However, additional research is required to study the mode of inheritance.

Circulatory System

There are a number of circulatory system abnormalities that can result from defective development during gestation or shortly after birth. These include patent ductus arteriosus, interventricular septal defects, patent foramen ovale and persistent right aortic arch. As with many congenital abnormalities, the role of heredity is unknown. These conditions have been included because they may be directly or indirectly inherited.

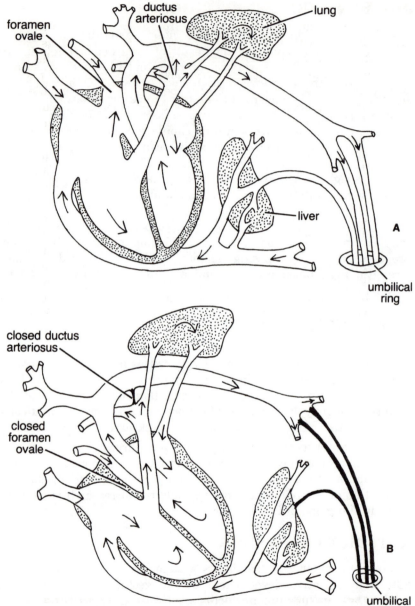

Comparison of the fetal heart (A) and normal mature heart (B). The arrows show the flow of blood through the circulatory system. In the fetus, most of the blood bypasses the lungs through the ductus arteriosus. There is also an opening (foramen ovale) between the chambers of the heart. In the normal, mature heart, the ductus arteriosus and foramen ovale are closed off.

PATENT DUCTUS ARTERIOSUS

In a mature horse, blood is carried from the right side of the heart to the lungs via the pulmonary artery, then from the lungs back to the left side of the heart, which then pumps the blood to the rest of the body via the aorta. In the fetus, however, the lungs are nonfunctional, since the blood has already been oxygenated by the dam. Therefore, a short blood vessel leads directly from the pulmonary artery into the aorta, bypassing the nonoperational fetal lungs. This blood vessel is called the ductus arteriosus.

The opening from the pulmonary artery into the ductus arteriosus normally closes within a few days after birth; the neonatal foal's blood is then oxygenated as it is pumped through the lungs, rather than supplied by the dam via the umbilical artery. Occasionally, the opening into the ductus arteriosus does not seal completely. This patent (open) duct causes nonoxygenated blood (from the pulmonary artery) to be mixed with the oxygenated blood (carried by the aorta). The heart pumps this mixture throughout the body. Since all of the blood is not sufficiently oxygenated, the foal develops cyanosis (becomes blue due to lack of oxygen). Frequently, a heart murmur can be heard as the blood leaves the pulmonary artery and enters the ductus arteriosus.

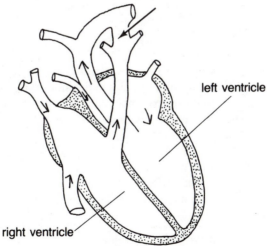

left ventricle

right ventricle

Mature heart with patent ductus arteriosus (arrow). Normally, the ductus arteriosus constricts with maturity.

Patent ductus arteriosus also occurs in man and other species, often in association with other heart defects. The genetic cause of this condition has not been determined.

INTERVENTRICULAR SEPTAL DEFECTS (IVSD)

In the mature horse, the heart has two ventricles (right and left) which are divided by a muscular wall known as the interventricular septum (meaning "the wall between the ventricles"). In the embryo, however, the ventricles form a single chamber. This chamber is eventually divided by the growth of a muscular partition from both the top and bottom of the chamber. Sometimes, the two sections of the wall fail to join in the middle, leaving an opening between the ventricles.

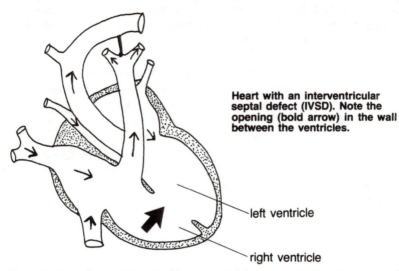

Heart with an interventricular septal defect (IVSD). Note the opening (bold arrow) in the wall between the ventricles.

left ventricle

right ventricle

This defect allows blood to bypass the lungs and flow between the ventricles, so that blood entering the right side of the heart (nonoxygenated) and blood leaving the left side of the heart (oxygenated) become mixed. This decreases the overall amount of oxygen in the blood, and may cause cyanosis. If the defect is not severe, the foal may survive, but may not do well. Occasionally, afflicted horses die of heart failure when older.

As with the other congenital heart defects covered in this discussion, genetic factors involved in defects of the interventricular septum are unknown.

PATENT FORAMEN OVALE

The heart of a mature horse has two atria, or upper chambers. In the embryo, however, the atria are joined in one chamber. A muscular membrane eventually divides the atria, but leaves an opening known as the ostium primum (meaning first opening). Later, a second opening (known as the ostium secundum) appears in the dividing membrane. Finally, another membrane grows alongside the original membrane, closing the ostium primum and partially covering the ostium secundum. The opening that remains is called the foramen ovale (oval opening). In young horses, this opening usually poses no problems, as there is a difference in pressure between the two atria that effectively keeps it closed. The foramen ovale normally closes permanently between two and nine weeks of age. If the foramen ovale does not close completely, the right side of the heart will overload; blood will tend to flow from the left atria into the right atria. Overgrowth of the right side of the heart and high blood pressure may result. If only a small defect is left, it will probably not affect the horse's health significantly.

Patent foramen ovale. This heart defect is characterized by an opening in the wall between the two chambers of the heart (foramen ovale of the fetal heart).

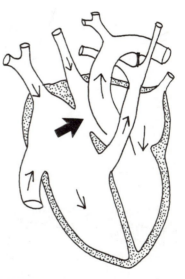

PERSISTENT RIGHT AORTIC ARCH

The blood vessel that carries oxygenated blood from the left side of the heart to the rest of the body is the aorta. The aorta develops from one of the aortic arches (structures formed in the embryo). Some of the aortic arches atrophy during development, while others eventually become part of the circulatory system. The aorta usually forms from one of the left aortic arches, so that it is on the left side of the horse's esophagus and trachea. Therefore, the ductus arteriosus (which leads from the pulmonary artery into the aorta) is on the same side of the body as the aorta.

If the aorta develops from one of the right aortic arches (instead of one of the left), it would be on the right side of the esophagus and tra-

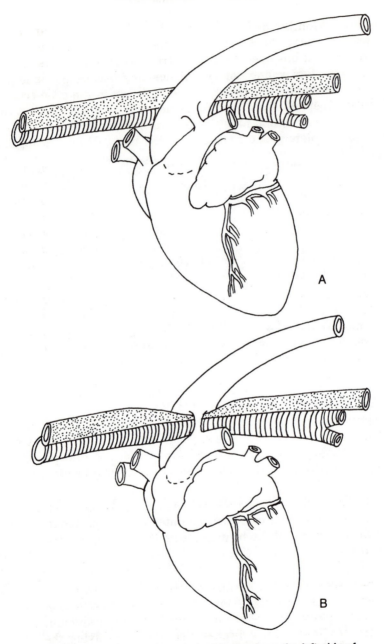

In the normal heart, the aorta originates from the left side of
the body (A). In the defective heart (B) with a persistent right
aortic arch, the aorta rises from the right side of the body,
constricting the esophagus and trachea.

chea, and not on the same side of the body as the ductus arteriosus. Because the ductus arteriosus must connect the pulmonary artery and the aorta, it has to go around the trachea and esophagus, forming a ring. As the foal grows, and the trachea and esophagus enlarge, this ring will constrict and partially close both structures. The constriction interferes with both breathing and swallowing.

Blood

Blood is an essential transport medium for oxygen, nutrients, metabolic byproducts and hormones. It also controls cellular temperature and function, and protects the body from invading organisms (e.g., bacteria and viruses). The discovery that blood group factors are inherited, and that the expression of many of these factors never skips a generation, has led to important practical applications, such as identification, parentage testing and breed differentiation. (Refer to **"Paternity Tests"** within the appendix.)

Genetic studies have also indicated the inheritance of several blood-related defects. Because the blood supports life-sustaining mechanisms, heritable disorders are often lethal. This discussion presents a study of the relationship between genetics and blood-related abnormalities, such as neonatal isoerythrolysis, combined immunodeficiency and hemophilia.

NEONATAL ISOERYTHROLYSIS

Neonatal isoerythrolysis (NI), a disorder caused by inherited blood factors, is manifested by dullness, weakness, acute or severe anemia with or without jaundice, loss of appetite and death if untreated. The reason for this potentially serious defect is outlined below:

1) A stallion and a mare with incompatible blood types (red cell antigen factors) are bred. The resulting embryo inherits, from the sire, a blood factor which is not compatible with the dam's corresponding blood factor.

2) If the placenta (protective membrane around the fetus) is damaged or defective, and if a substantial amount of the foal's blood enters the mare's circulatory system, the mare may become isoimmunized. Isoimmunization is a normal defense mechanism involving the production of antibodies to counteract the presence of any undesirable foreign substance. In this case, the foreign substance is the foal's sire-related blood factor.

3) The mare's isoantibodies are large protein molecules that cannot pass through the placenta to the developing fetus. Unfortunately, they become concentrated in the mare's colostrum along with natural (passive) antibodies. (Passive antibodies protect the newborn foal from infection until he is capable of producing sufficient antibodies on his own.) The structure of the equine placenta also prevents transfer of natural antibodies from mare to offspring during gestation.

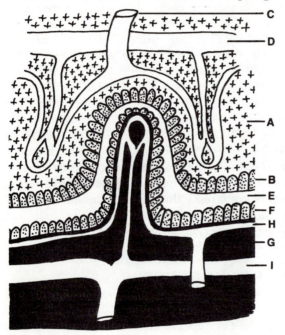

Through the placental attachment between mare and foal, the developing fetus receives nutrition, oxygen and water, and eliminates metabolic waste products. The placenta is also the production site for important hormones. (A) allantoic vesicle, (B) allantochorion (fetal placenta), (C) fetal artery, (D) fetal vein, (E) uterine lumen, (F) uterine epithelium, (G) stroma, (H) maternal artery, and (I) maternal vein.

4) The foal's digestive tract is usually permeable by the large antibody molecules up to 24-36 hours after birth. If the foal receives his dam's colostrum during this period, absorption of the isoantibodies (i.e., those that react against the sire-related blood factors) will cause the hemolysis (destruction) of his red blood cells.

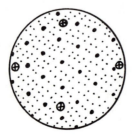

A representation of red cell antigens showing how the different factors might be arranged on the surface of the red blood cell. Each symbol represents a blood factor.

Depending on how concentrated the antibodies are in the mare's colostrum, how much colostrum the foal receives and which antibody is involved, the hemolysis of red blood cells causes either slow or rapid deterioration of the foal's health: progressive weakness, sleepiness and jaundice, among other problems. In many cases, severe anemia and loss of appetite result in the hemolytic foal's death. (Hemolytic foal refers to the foal affected with NI.)

If a mare first receives incompatible blood from a fetus during the foaling process, the newborn foal will not be hemolytic. (The immunization process has not yet concentrated the isoantibodies in the mare's colostrum.) If she later conceives an embryo that carries the same incompatible blood factor (i.e., the sire-related blood factor involved in the original immunization process), the foal will be affected. Some researchers believe that, after primary immunization, the passage of even a small amount of incompatible blood from fetus to dam could cause the potentially lethal hemolytic reaction. Because there are eight different inherited red cell antigen systems, and because each system is controlled by one or more alleles, the chances are remote that an immunized mare will be reimmunized for the exact same factor, unless she is bred back to the same stallion.

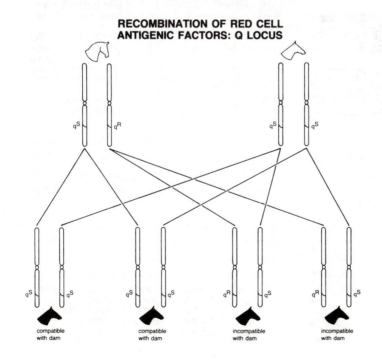

RECOMBINATION OF RED CELL ANTIGENIC FACTORS: Q LOCUS

Although a mare which produces one affected foal will not necessarily produce other hemolytic foals unless rebred to the same stallion, precautionary measures should be taken with each of her pregnancies. For example, if a mare has produced a hemolytic foal, her colostrum can be tested for the presence of undesirable isoantibodies. If the antibodies are present, the foal can be saved by preventing access to his dam's colostrum. This can be done either by stripping the mare's first milk or by muzzling the foal for at least 36 hours and feeding a milk replacer. In both cases, the foal should be given antibiotics or provided with colostrum from another mare. (Some breeding farms keep a supply of frozen colostrum, which can be checked for compatability with the foal's red blood cells.)

COMBINED IMMUNODEFICIENCY DISEASE

Combined immunodeficiency disease (CID) is an inherited defect of the foal's immunity system. This condition, which has been reported extensively in the Arabian breed, involves an inability to produce the infection fighting blood factors (B & T lymphocytes) and, therefore, creates a very high susceptibility to infection. CID always results in the death of the affected foal.

Because they receive temporary immunity from antibodies within the colostrum, CID foals usually appear healthy at birth. Gradually, as the maternal antibodies subside, the foal's condition deteriorates. The disorder may become apparent within about 10 to 50 days, as the affected foal shows an incresed susceptibility to bacterial, viral, fungal and protozoal infections. Commonly, the CID foal contracts pneumonia, dermatitis, or an adenoviral infection. Death is usually due to severe secondary respiratory complications.

Although CID has been observed only in the Arabian, it is possible that undetected cases have occurred within other breeds. Researchers believe that about 2.3% of all Arabian foals are born with CID, indicating that as many as 25% of all Arabian horses are carriers of the CID gene. (Refer to **"Breed Improvement Through Applied Genetics."**) This exaggerated incidence of CID within the Arabian breed indicates that the controlling allele is closely linked to a gene that controls some desirable trait. It is possible that the desirable trait is breed-related and that it has been carefully selected for over the years. CID probably originated as a remote mutation and gradually increased within the Arabian breed through accidental selection.

Geneticists believe that CID is controlled by a recessive allele; therefore, both the sire and the dam each must contribute the allele (designated as CID^R) before their offspring can express the trait.

When both sire and dam are CID carriers (meaning that they carry only one recessive CID allele in their genotype and cannot express the immunity defect), approximately one out of every four of their foals will be born with the CID condition.

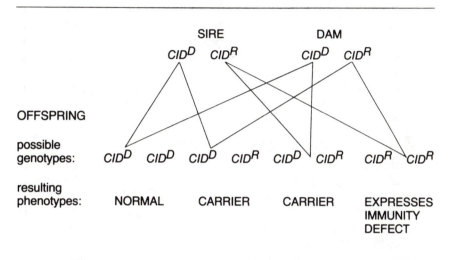

CID^D = the dominant allele that allows normal formation of B & T lymphocytes
CID^R = the recessive allele that causes the immunity defect

Diagnosis of the CID condition can be made by a series of laboratory tests (e.g., determining the level of functional lymphocytes in the foal's blood). A CID diagnosis indicates that the foal will die and that both of his parents are carriers of the undesirable CID^R allele. Researchers are currently working on a test that will determine whether or not a horse is a CID carrier. They believe that CID may be caused by a missing enzyme, a chemical substance necessary for the completion of some biological process. In this case, the process involves changing lymphocytes into B and T lymphocytes. Step II (refer to illustration) is the most probable site of a CID-related enzyme deficiency. If the enzyme (or enzymes) at this point could be identified, and the absence of one or several enzymes could be related to the combined immunodeficiency disease, researchers believe they could formulate a carrier test. (The carrier test would identify horses carrying one recessive allele without expressing the CID condition.) Researchers believe that the test would allow breeders to gradually select against the inherited immunity defect.

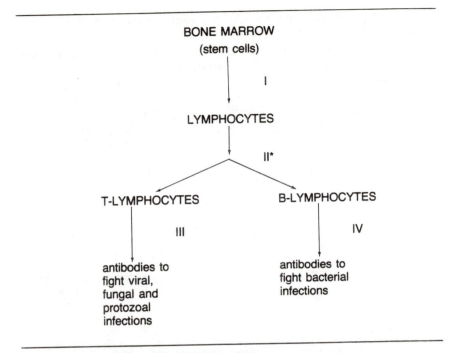

BONE MARROW
(stem cells)

I

LYMPHOCYTES

II*

T-LYMPHOCYTES   B-LYMPHOCYTES

III   IV

antibodies to fight viral, fungal and protozoal infections   antibodies to fight bacterial infections

*The CID condition (lack of B & T lymphocytes) is caused when step II is some-how blocked. An enzyme may be absent at this point.

Large scale genetic studies on the horse are very expensive and un-common. However, extensive research on CID in horses has been undertaken to supplement information on a similar condition in man. Occasionally, children are born without the ability to produce anti-bodies. Unless they are secluded in a sterile environment, they will die from infection.

HEMOPHILIA

Hemophilia is an incurable inherited disorder involving spontane-ous bleeding and a hemorrhagic tendency from birth until death. The hemophiliac often develops large hematocysts (swollen areas caused by internal hemorrhage) and usually dies from severe internal bleed-ing. In the horse, the disease is rare; only a handful of cases have ever been reported. (It should be noted that epistaxis, or nosebleed, is not a symptom of hemophilia. Refer to **"Respiratory System."**)

Clotting (i.e., the mechanism that stops the flow of blood from an injured vessel) is controlled by many complicated biochemical pro-

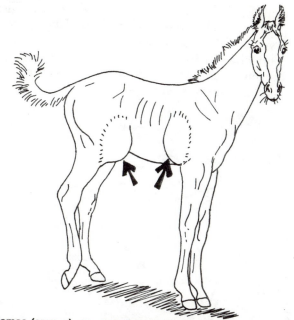

Hematomas (arrows) are caused by internal hemorrhage. The swelling is due to deposits of blood beneath the skin. A foal with hemophilia may develop hematomas from even a mild injury.

cesses and by the presence of various chemical factors (or enzymes). The basic steps involved have been simplified and outlined:

1) The vessel is damaged.
2) Specialized blood cells (platelets) adhere to the vessel at the injured site.
3) As the blood flows through the injury, platelets gather around the platelet-tissue mass to form a "plug" over the injured area.
4) A biochemical reaction causes the platelet mass to become more compact.
5) The surrounding blood plasma coagulates (forms insoluble fibers) through a multi-step biochemical process. The completion of each step depends on the presence of a certain factor, or enzyme. These factors are designated by Roman numerals I through XII.
6) A deficiency in any one of these factors can block the clotting process.

In man, there are many possible types of hemophilia, depending on the missing factor. In horses, only a factor VIII deficiency has been reported. The resulting defect is referred to as "factor VIII deficiency," "hemophilia A," or "true hemophilia." Factor VIII deficiency is described as a delayed lethal (in horses), since it eventually causes the death of nearly all affected individuals. The trait is also X-linked, meaning that its controlling gene is located on the X chromosome (i.e., one of the sex determining chromosomes).

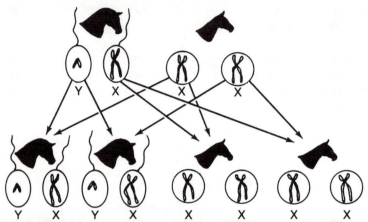

The male sex characteristics are determined by the presence of an X and a Y chromosome, while the female characteristics are caused by the presence of two X chromosomes. X-linked alleles in the male are unpaired (hemizygous). Therefore, recessive X-linked traits (such as hemophilia) in the male are never hidden in the heterozygous state.

As discussed under **"Breeding Aspects,"** the male embryo develops differently than the female because the Y chromosome is present in his genotype. (The male carries an X and a Y chromosome within each cell, while the female carries two X chromosomes within each of her cells.) If the female carries the hemophilia allele on only one of her X chromosomes, she will not be a hemophiliac; the dominant effect of the corresponding normal allele overrides the recessive hemophilia allele. Because she is capable of passing the hemophilia allele to her offspring without expressing any hemophilia characteristics, she is called a hemophilia carrier:

$$X^N X^N \ = \ \text{normal female}$$
$$X^N X^H \ = \ \text{hemophilia carrier female}$$
$$X^H X^H \ = \ \text{hemophilia female (an unlikely occurrence)}$$

X designates the X chromosome; H designates the hemophilia allele found upon the X chromosome. N designates the normal (dominant) allele on the corresponding X chromosome.

The presence of the hemophilia allele upon the male's single X chromosome results in the expression of hemophilia:

$$X^N Y \quad = \quad \text{normal male}$$
$$X^H Y \quad = \quad \text{hemophilia male}$$

Y designates the Y chromosome; it does not carry the sex-linked hemophilia allele.

Because the Y chromosome does not carry the X-linked hemophilia allele, there are no male hemophilia carriers. When a normal stallion is bred to a carrier mare, approximately half of their male foals will be born with hemophilia:

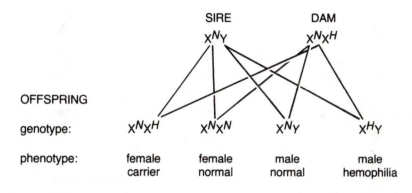

The only possible way a female could express hemophilia is if her dam was a hemophilia carrier and her sire was a hemophiliac that reached reproductive maturity (a very remote possibility that has not been recorded in the horse):

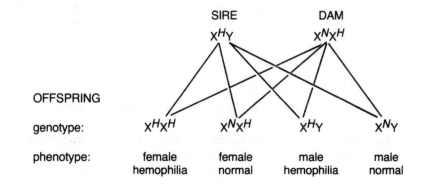

Researchers suspect that the carrier female may show a slight reduction in the amount of factor VIII normally produced. (Remember that a factor VIII deficiency causes the hemophilia condition in horses.) If this proves true, a carrier test may some day be developed. This test would allow detection of a slight factor VIII production defect, or show that an enzyme is deficient. In this way, veterinarians could identify mares that carried a hidden hemophilia allele. ♞

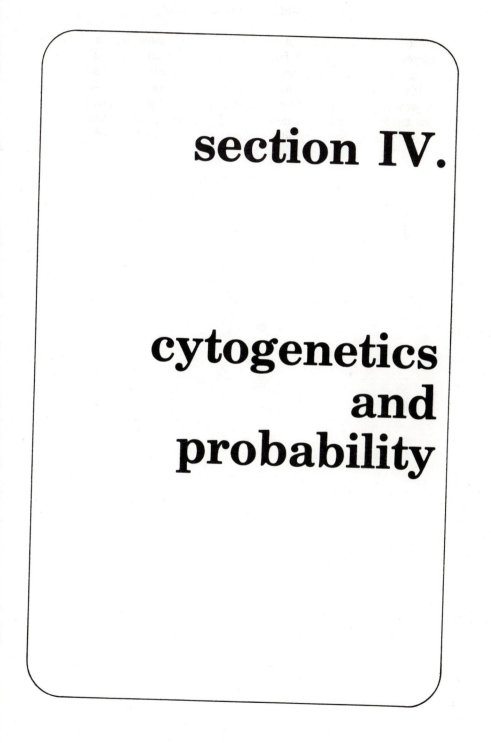

section IV.

cytogenetics
and
probability

14

THE CELL

Due to the nature of the subjects discussed in this section, the necessity for exploring intricate concepts and relying on technical terms was unavoidable. However, a study of genetics is not complete without briefly examining the answers to questions that have puzzled scientists for years:

-What do chromosomes and genes look like?
-What causes cells to divide?
-Why do some of the dividing cells within the developing embryo suddenly change functions?
-Why are hybrid crosses between different species usually sterile?
-What actually causes an individual to have both male and female characteristics?
-What determines whether a developing embryo becomes a male or a female?
-What causes mutations?

This section is presented to answer these and other questions, and to supplement the reader's understanding of basic genetics. Although the serious student may wish to study each concept in detail, a thorough understanding of the following chapters is not essential for practical selection.

Cell Structure

The cell is the basic unit of all plant and animal life. Every organism is composed of one or more cells, from the single-celled ameoba to the multi-billion-celled horse. Even complicated organisms, such as the horse, begin life as a single cell — the fertilized egg, or zygote. The zygotic cell divides and multiplies, developing into an embryo. As the embryo develops, each of the cells divide into two "daughter" cells. Some of the resulting daughter cells differentiate (change their function). Some become brain cells, while others become red blood cells, etc. Cell division and differentiation continue, as the embryo becomes a fetus and finally a foal. Even in the adult horse, new cells (e.g., blood cells, bone cells, skin cells, etc.) are constantly produced as needed.

Growth and maintenance of the entire body is the result of cell activity. There are hundreds of different types of cells in the body, each with a genetically determined function. These functions are coded and determined by the genes that an individual inherits. This genetic inheritance is contained in each and every cell of the body. When a cell divides into daughter cells, the blueprint of inheritance must be duplicated so that the new cell will be able to fulfill its role in the body. The science which studies the genetics of the cell is known as *cytogenetics*.

An understanding of the basic principles of cytogenetics will enable the horseman to relate the horse's phenotype (appearance and performance) to the horse's genotype (genetic inheritance carried in each cell). A change in the inheritance results in cellular change, thus causing the visible or measurable differences in coat color, conformation, performance, etc. All inheritance is basically the passing on of some type of cell characteristic.

The cell consists of a mass of living matter (protoplasm), surrounded by a protective cell membrane. The cell membrane controls which substances enter and leave the cell. The protoplasm has two parts: the cytoplasm and the nucleus. Cytoplasm is a glue-like substance in which the small structures of the cell (organelles) float. The nucleus is a rounded mass of dense matter that contains the chromosomes. It is located in the cytoplasm and surrounded by the nuclear membrane.

The organelles within the cell, such as the mitochondria and the ribosomes, perform various functions. The mitochondria, for example, are the principal sources of energy for the cell, while the ribosomes are responsible for protein synthesis. The cytoplasmic reticulum is a network of tubules (flattened sacs) extending throughout the cytoplasm; ribosomes may be attached to the surface of the cytoplasmic reticulum. The Golgi apparatus is another complex series of tubules, vesicles (blisters) and vacuoles. It is located adjacent to the nucleus and is involved with the production of intracellular secretions.

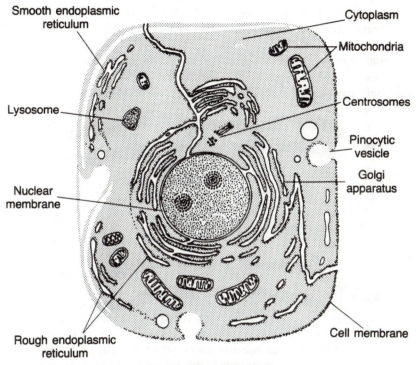

Diagram of a typical cell.

The Structure of Inheritance

The nucleus of every cell contains the individual's genetic inheritance. The structures in the nucleus which carry the units of heredity (genes) are the chromosomes. Each gene codes for a certain cell function resulting in one or more characteristics. The genotype (complete set of genes in each cell) is duplicated within the nucleus of every cell as the cell divides. (Refer to **"Mitosis"** and **"Meiosis."**)

The genes are arranged linearly on the chromosomes. The specific location of a gene on a chromosome is known as a locus; (plural is loci). Chromosomes are paired within the cell, meaning that there are two chromosomes with corresponding genetic information found together. These pairs are known as homologous chromosomes. The total number of chromosome pairs varies between species; the horse has 32 pairs, the equivalent of 64 individual chromosomes.

Although the traits affected by the genes of a particular locus are related (e.g., grey-nongrey, short-tall, etc.), the genes may not be. A gene may have many forms resulting from various mutations. These forms are referred to as alleles. For example, one chromosome of a pair could carry the allele for greying, while the other carries the allele for nongreying. In this case, the loci are identical, but the alleles are not.

A homozygous locus has identical alleles on corresponding points of homologous chromosomes. For example, the following chromosome pair has identical chromosomes for greying at identical loci.

A heterozygous locus has different alleles on the corresponding points of homologous chromosomes. The following chromosome pair has an allele for nongreying and an allele for greying, but both are at the same locus.

Each chromosome has a centromere, or constriction, that divides it into two arms. The centomere may be acrocentric (near one end of the chromosome), metacentric (near the middle of the chromosome), or telocentric (very close to the end of the chromosome).

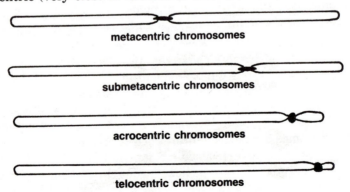

metacentric chromosomes

submetacentric chromosomes

acrocentric chromosomes

telocentric chromosomes

When a chromosome divides into two identical chromosomes during cell division ("replication"), the sister chromosomes (chromatids) remain attached at their centromere.

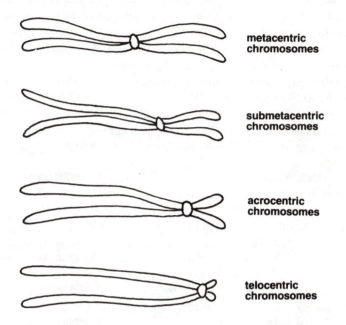

metacentric
chromosomes

submetacentric
chromosomes

acrocentric
chromosomes

telocentric
chromosomes

In the horse, 36 of the autosomes (nonsex chromosomes) are acrocentric, and are V-shaped when paired (18 pairs). There are 26 metacentric autosomes which are shaped like X's when paired (13 pairs). The male sex chromosomes (commonly referred to as the "Y" chromosome) is acrocentric. The female sex chromosome (known as the "X" chromosome) is a submetacentric chromosome, so that a *pair* of "X" chromosomes in the mare would look as follows:

Homologous
X chromosomes

sister
chromatids

It should be noted that there is no such thing as a pair of "Y" chromosomes; the male's pair of sex chromsomes consists of one "X" and one "Y" chromosome.

Each chromosome is made up of deoxyribonucleic acid (referred to as DNA). DNA is one of the two types of nucleic acid, a chemical substance found in the cell nucleus. (The other type, RNA, will be examined later.) It is the DNA that codes hereditary characteristics by directing the production of specific proteins. Proteins form many of the body's vital products including enzymes, hormones and cell structures. Proteins also direct cellular activity: cause chemical reactions, build new cells, etc. A change in the DNA message results in a change in the protein produced. This, in turn, changes the function of the cell. Eventually, these cellular changes affect the phenotype of the horse.

The DNA molecule is divided into small units (genes), each of which codes for a certain protein. Therefore, a gene affects a specific trait by determining the type of protein produced in the cell. Protein makes up about 75% of the dry weight of most body cells; it is composed of long chains of hundreds of chemical compounds, known as amino acids. To make a particular protein, the gene codes for a specific sequence of amino acids.

Each amino acid is coded for by a *codon*. A codon is a sequence of three bases, or a "triplet" (a base is a chemical compound which forms a salt when mixed with an acid). A gene consists of many amino acids and, therefore, hundreds of codons. The difference between alleles (forms) of the same gene is due to differences in one or a few codons. The a^t allele (which causes the seal brown color) probably differs from the A allele (which causes the bay pattern) by one or two codons. In other words, the a^t allele of the A gene produces a different protein than the A allele, due to a change in one or two amino acids.

Genes vary in length, according to the protein involved. In addition, genes may overlap, with the end of one gene functioning as the beginning of the next. For example, assume that the gene for coat texture (curly vs. straight) overlaps the gene for eye color:

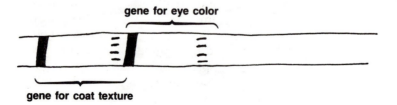

gene for eye color

gene for coat texture

To summarize, the chromosome is made up of a long strand of DNA, which is subdivided into a great number of genes. The gene causes the production of a specific protein by determining the arrangement of amino acids. Each amino acid is coded for by a codon (group of three bases) so that a gene is composed of many codons.

DNA

Deoxyribonucleic acid (DNA) is a chemical compound of a base which contains nitrogen, a sugar and phosphoric acid. The sugar is deoxyribose, from which DNA gets its name. The phosphate group and the sugar group alternate in a long chain:

$$S - P - S - P - S \ldots$$

The bases are attached to the sugars:

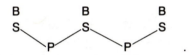

The sugars and the phosphate groups are all identical (S=S, P=P), but the bases differ. There are four bases that may be bound to the sugar, either adenine (A), guanine (G), thymine (T) or cytosine (C). These four bases form the codons, or triplets, mentioned earlier. Each combination of three bases codes for a specific amino acid. For example, GAG and TTT are codons. Since genes can overlap, codons may also overlap.

overlapping codons

There are four bases, so there are 64 possible combinations (4×4×4) of three of the bases. There are only 20 amino acids, so some amino acids can be coded for by more than one codon.

Two of these long chains bond together to form the DNA molecule.

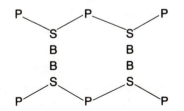

The bases can bond together only in certain combinations, which are A with T and C with G.

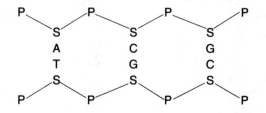

Actually, the DNA molecule is not a long, flat chain as the diagram above suggests. Instead, it is a spiral of the two long chains of sugars, phosphate groups and bases, known as a double helix.

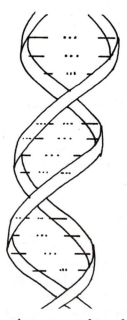

The DNA molecule can be compared to a ladder, with the sugar and phosphate groups the sides of the ladder and the bases the rungs. The ladder is twisted to make the double helix.

DNA is formed into chromosomes only during cell division when chromosomes are duplicated (see **"Mitosis"** and **"Meiosis"**). During interphase (resting period between cell divisions), the DNA is in the form of long strands.

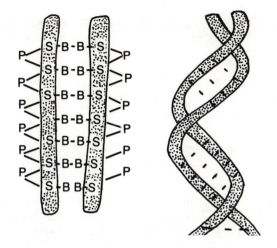

During meiosis, the chromosomes are duplicated (a process known as replication) allowing each daughter cell to receive a complete set of chromosomes:

1. The DNA molecule splits up the middle in the same way that a zipper opens.

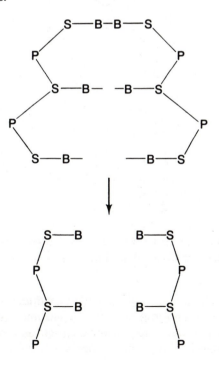

2. Bases, sugars and phosphate groups form the second half of each molecule, creating two complete DNA molecules — one for each daughter cell.

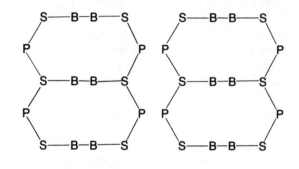

RNA

RNA is ribonucleic acid, a chemical compound similar in structure to DNA. The sugar is RNA is ribose and the four bases are adenine (A), uracil (U), cytosine (C) and guanine (G).

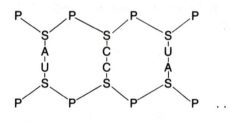

The structure of RNA.

Although DNA, located in the nucleus, contains genes which code for protein production, proteins are actually synthesized by the ribosomes, which are located in the cytoplasm. The genetic material must somehow be carried into the cytoplasm and to the ribosomes. This is accomplished by RNA.

RNA duplicates the genetic information of the DNA molecule, then leaves the nucleus and travels to the ribosomes. The ribosomes then follow the code carried by RNA and build proteins out of amino acids.

KARYOTYPES

A karyotype is a photograph of an individual's chromosomes, as seen through an electron microscope. The chromosomes from a single cell nucleus are photographed, then each chromosome is cut apart and arranged in descending order of size and according to the location of the centromere (all the acrocentric chromosomes are grouped together, etc.).

Regardless of breed, the karyotype of every horse consists of 32 pairs of chromosomes. There are 31 pairs of homologous autosomes (nonsex chromosomes) and one pair of sex chromosomes ("XX" or "XY"). Karyotypes are helpful in diagnosing certain physical defects that are caused by chromosomal abnormalities (intersex, small hard ovaries, etc.). For example, a karyotype might reveal that a sterile mare has only one X chromosome, a condition known as aneuploidy 63 XO. This would explain the mare's sterility, small ovaries and show that the condition is permanent.

Karyotypes differ between species. A species is usually defined as a group of organisms that bear a close resemblance to each other in their more essential features. When mated, the members of a species normally produce fertile offspring. Although there are usually greater differences between species than between breeds, this is not always the case. For example, there appear to be fewer differences between various zebra species than between the Shire and the Arabian horse. Yet a karyotype reveals that the Grant's and the mountain zebras are separate species, while the Shire and the Arabian are different breeds of the same species.

The accompanying table lists the various species of the equine family with their respective number of chromosome pairs.

Scientific Name	Common Name	Chromosome Pairs
Equus przewalski	Przewalski's horse (Mongolian wild horse)	33
Equus caballus	Domestic horse	32
Equus asinus	Domestic Ass	31
Equus hemionus	Kulan (Mongolian wild ass)	28
Equus kiang	Kiang (Tibetan wild ass)	28
Equus onager	Persian Onager (Persian Wild Ass)	28
Equus grevyi	Grevy's zebra (Somaliland zebra)	23
Equus burchelli	African zebra (Burchell's, Damara, Chapman's, Grant's, Selous')	22
Equus zebra	Mountain zebra (Cape, Hartmann's)	16

Notice that the horse has 32 pairs of chromosomes, while the donkey has only 31 pairs. When the horse and donkey are crossed to produce a hybrid, the resulting offspring (a mule or a hinny) has an uneven number of chromosomes (63, or 31½ pairs). This is because the gametes of the parents (sperm and egg) are haploid, meaning that they contain only one chromosome from each pair. So the horse's gamete has 32 chromosomes (not 32 pairs) and the donkey's gamete has only 31 chromosomes (not 31 pairs). This results in 63 chromosomes in the hybrid offspring.

The shape of the chromosomes also varies between species. This is important because acrocentric, metacentric and submetacentric chromosomes will pair only with others of their same shape. The Przewalski horse, for example, has 26 submetacentric and metacentric autosomes and 38 acrocentric autosomes, as compared to the 30 submetacentric and metacentric autosomes and 32 acrocentric autosomes of the domestic horse. The mule has 13 metacentric chromosomes from the mare and 19 from the jack. Because these shapes do not correspond, the chromosomes cannot pair properly. This is believed to be one of the reasons why mules are sterile.

Another factor that affects successful hybridization is the amount of homologous DNA between the two species. To pair correctly, chromosomes must be homologous (have identical loci) and must be the same shape. In any case, species must be closely related, as are the horse and donkey, to produce viable offspring. Usually, differences in number or shape, and an inadequate amount of homologous DNA, mean that a hybrid will be sterile; it will be unable to produce viable gametes. (Refer to **"Mitosis"** and **"Meiosis."**) Although male mules and male hinnies are sometimes able to produce a few spermatozoa, they are usually small and deformed. There have been a few reports of fertile female mules, but none of the females were karyotyped. Therefore, scientific proof that they were mules is lacking. It is theoretically possible for a mule or hinny to be fertile, when bred to a horse, if all the hybrid's horse chromosomes end up in one gamete. The chances of a hybrid's horse and donkey chromosomes segregating during gametogenesis (sperm and egg production) are negligible.

Hybrids have also been produced by mating zebras to horses, ponies and donkeys, resulting in offspring that have been described as zebroids, zeonies and zeonkies. Like mules, these hybrids are sterile. Thus far, only one hybrid from the mating of two equine subspecies, the domestic and Przewalski's horses, has been fertile. In this case, a Przewalski stallion was crossed with a Norwegian Fjord mare. A fertile hybrid stallion was produced from this mating. Karyotypes of the hybrid stallion's parents revealed that the chromosome differences between the two subspecies were due to a variation in the number of acrocentric chromosomes. It is possible for two acrocentrics to join (a process known as centric fusion, which will be discussed later under **"Mutations"**) and form a submetacentric or metacentric chromosome. When certain lines of Przewalski's horse evolved into the domestic horse, the Przewalski's extra acrocentric chromosomes may have fused, reducing the number of chromosomes in the modern horse. The amount of similarity between the karyotypes of these two species allows the production of fertile hybrids.

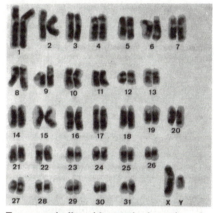

Equus caballus (domestic horse) male

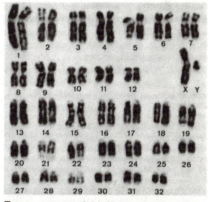

Equus prezewalski (Przewalski's horse) male

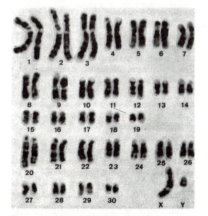

Equus asinus (donkey) male

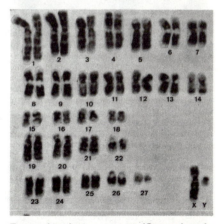

Equus hemionus onager (Onager) male

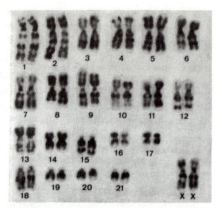

Equus grevyi (Grevy's zebra) female

Equus burchelli boehmi (Grant's zebra) female

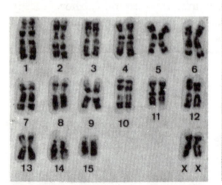

Equus zebra hartmannae (Hartmann's zebra) female

E. grevyi x E. burchelli antiquorum hybrid male

Mitosis

As discussed earlier, chromosomes are protein frames within the nucleus, which carry important messages (genes) for the surrounding cell. The messages may instruct the cell to produce a needed chemical, to support growth by careful duplication, to form the egg or sperm or to perform some other vital duty. Although some cells function throughout the life of the individual (e.g., nerve cells), many are short-lived (e.g., skin cells). For this reason, many cells must be constantly duplicated without losing their genetic instructions on "how to perform." The chromosome's ability to duplicate its entire structure, and the cell's division into two identical sister cells, are referred to as mitosis. Intricate cellular changes that occur during mitosis (e.g., movement of chromosomes) can be observed through the electron microscope. Cytologists (scientists who study cells) still have many unanswered questions, however. (What triggers the division? What causes cells to divide haphazardly in cancerous tissues?) Five distinct steps of mitosis have been identified: interphase, prophase, metaphase, anaphase and telophase.

STAGES OF MITOSIS
Interphase

Interphase is the resting, or dormant, stage between cell divisions. During interphase, the chromosomes are not visible through the electron microscope (except as tiny granules located throughout the nucleus). During this period, the chromosomes are believed to be in their normal diploid state. (Diploid refers to the normal chromosome number, such as 64 chromosomes in the horse. In the diploid state, all chromosomes are present and none are duplicated.) Although it is generally accepted that interphase is a resting period, some sources think that, near the end of this phase, the undetected chromosomes begin to replicate (duplicate their entire structure).

Interphase - Chromosomes are not visible during this dormant period.

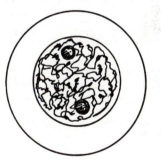

Prophase

During prophase, the cell undergoes several changes. Small granular particles within the nucleus begin to concentrate into long strands, forming the visible chromosomes. A small cytoplasmic organelle, called the centriole, divides into two similar structures. Each centriole appears to be surrounded by several ray-like extensions. Together, the centriole and its rays are called the astral body. Each astral body migrates to opposite ends of the cell, and long protein molecules form a spindle (three-dimensional arch) between the two bodies.

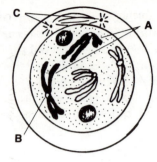

Prophase - Appearance of homologous chromosome pairs (A). A precise replication of each chromosome is connected to the original by the centromere (B). Division of the centriole to form two migrating astral bodies (C), which later become poles for the protein spindle.

As the chromosome strands enlarge, each completes a complicated replication process to form another identical chromosome strand, or *chromatid*. (Refer to **"DNA".**) Gradually, the identical chromosome strands partially separate, remaining attached at a point of constriction called the *centromere*. The chromosome number is now temporarily doubled (e.g., 128 chromosomes in the horse) and is referred to as the tetraploid chromosome number. As the chromatids steadily become thicker, they can be identified by their relative lengths and by the position of their respective centromeres. (Refer to **"Karyotyping."**) Next, the nuclear membrane disappears and the chromatids enter the cytoplasm, migrating toward the spindle fibers. The centromere of each chromatid pair moves to the spindle's equator.

Metaphase

Metaphase is a relatively short stage involving, simply, the presence of the chromatids at the spindle's equator.

Metaphase - Chromatids are located at the spindle equator (A).

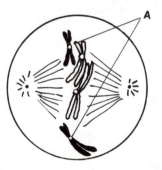

Anaphase

During this period, the sister chromatids should separate (the centromere divides) and move along the spindle fibers to opposite poles (astral bodies) of the cell. Occasionally, two chromatids will not part, and the segregation of chromosomes is unbalanced. This nondisjunction of chromatids results in the formation of daughter cells with missing or duplicated chromosomes.

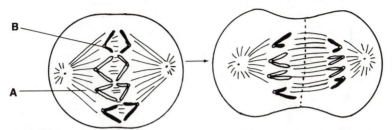

Anaphase - Attached to the spindle fibers (A) by their centromeres (B), the chromatids separate and migrate to opposite ends of the cell.

Telophase

The last stage in mitosis involves the collection of a complete set of chromosomes (one chromatid from each pair of sister chromatids) at opposite ends of the cell and the division of the cytoplasm into two identical cells. The nuclear membrane reappers in both cells, separating the chromosomes from the cytoplasm. Telophase completes the cell division and results in the transition of one cell into two genetically identical cells.

Telophase - Separation of the cytoplasm to form cells with identical chromosomes.

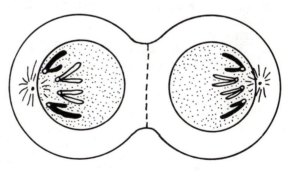

Meiosis

Transmitting copies of genetic material from parent to offspring is the basis of inheritance. Conveyance of genetic material is made possible by the formation of gametes (the ova and sperm), which unite to form the zygote. (The zygote is the immediate combination of the egg and sperm before the embryo begins to develop.) Each gamete provides the zygote, and therefore future offspring, with a sample of genetic material from the respective parents. To insure that the characteristic chromosome number (i.e., 64 chromosomes in every body cell of the horse) remains the same from generation to generation, the sperm and the egg carry only half the normal chromosome number. This reduced chromosome number is referred to as the *haploid* number (e.g., 32 chromosomes in the horse).

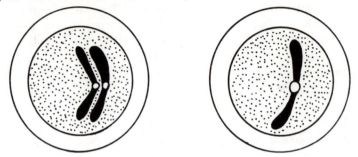

The diploid chromosome number (left) occurs when each chromosome has one corresponding homologue. The haploid chromosome number (right) is found in the ovum and sperm; each chromosome is without its homologue.

The process by which the egg and sperm are formed, with only half the normal chromosome number, involves mitosis and two specialized cell divisions called meiosis. This reduction-division process can only occur within special cells of the ovary and testes. Because the overall results of meiosis are slightly different in the male and female, gamete formation is divided into two categories: Spermatogenesis and Oogenesis.

SPERMATOGENESIS — PART I

Spermatogenesis originates within the small twisting seminiferous tubules of the testes. Special cells (spermatogonium) along the inside walls of the tubules divide by mitosis to produce a second layer of similar cells. Some of these "new" spermatogonium (primary spermatocytes) begin the sperm production process: 1) each cell divides into two secondary spermatocytes and 2) each secondary spermatocyte divides into two spermatids. The spermatids are premature sperm cells.

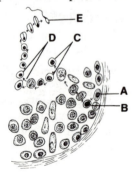

Spermatogenesis. Within the testicular tubules, spermatogonium (A) may divide into two primary spermatocytes (B). Each primary spermatocyte performs the first meiotic division to form two secondary spermatocytes (C). The second meiotic division forms two spermatids (D) from each secondary spermatocyte. These divisions are followed by maturation of the spermatid to form motile sperm (E).

Their development involves concentration of nuclear material (e.g., chromosomes), loss of much of the cytoplasm and the production of a long propelling tail or flagellum.

The transition from testicular tissue to independent haploid sperm involves both mitosis and meiosis. The development of the second layer of spermatogonium (primary spermatocytes) depends on mitosis, the simple division of a cell to form two identical sister cells. When the primary spermatocytes divide to form secondary spermatocytes, a more complicated process (the first half of meiosis) is involved.

Prophase I

During the first meiotic prophase, the chromosomes become apparent (under an electron microscope) and begin to duplicate their long structures (replicate), slowly forming sister chromatids. This period is characterized by its relatively long duration and by the *synapse* (attraction) between homologous chromosomes. These corresponding chromosomes (chromosomes with similar genes, but not necessarily carrying the same alleles) migrate toward each other as they gradually reproduce. Because two pairs of chromatids are closely associated, the resulting structure appears to have four strands. For this reason, the structure is referred to as a tetrad.

Prophase I. Attraction between homologous chromosomes.

During the synapse of homologous chromosomes, an exchange of genetic material between the paired chromosomes may occur.

Synapse of double stranded homologous chromosomes to form the tetrad during prophase I. Crossing over (arrow) may occur at this time.

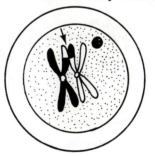

This event, which is referred to as crossing over, increases genetic variation by suddenly rearranging each chromosome's structure (and, hence, changing genetic messages on both chromosomes). This mixing of genetic material allows offspring to inherit new gene combinations from their parents.

As prophase progresses, the astral bodies form and migrate to opposite ends of the cell. (Refer to **"Mitosis."**) The spindle also develops, and the nuclear membrane disappears. Gradually, the homologous chromosomes separate and move into the cytoplasm, steadily migrating toward the spindle equator.

Metaphase I

During ths period, the homologous chromosomes are completely separated. The sister chromatids are still attached by their centromeres which, in turn, are attached to the spindle's equator.

Metaphase I. Tetrads are attached to the spindles equator by their centromeres.

Anaphase I

Anaphase I involves the movement of homologous chromosomes along the spindle fibers to opposite astral bodies, without separation of chromatids. Failure of the homologous chromosomes to separate during late prophase I, metaphase I or anaphase I could result in the formation of gametes with missing or duplicated chromosomes.

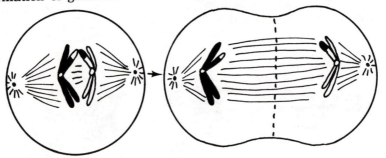

Anaphase I. Homologous chromosomes migrate to opposite ends of the cell.

For example, if the male's X and Y chromosomes do not separate, spermatogenesis results in formation of sperm with both X and Y, or with no sex chromosomes at all. Affected gametes sometimes live to produce individuals with unusual karyotypes that are closely related to certain phenotypic abnormalities. (Refer to **"Sex Chromosomes."**)

Telophase I

The separation of the cytoplasm to form two sister cells completes the first meiotic division. Nuclear membranes form within each cell to enclose the respective set of chromosomes. Note, however, that each nucleus contains only half the original chromosome *types* (one chromosome from each homologous pair) and that each type is doubled, so that the chromosome number is not altered. The migration of homologous chromosomes to opposite daughter cells allows gamete cells (in this case, sperm) from the same individual to carry different chromosome combinations — an important cause of genetic variation.

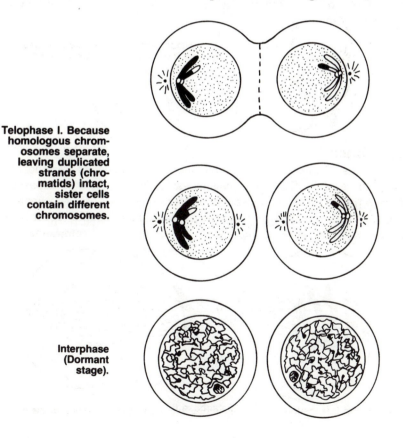

Telophase I. Because homologous chromosomes separate, leaving duplicated strands (chromatids) intact, sister cells contain different chromosomes.

Interphase (Dormant stage).

SPERMATOGENESIS — PART II

The secondary spermatocyte is the end-product of the first meiotic division. These new spermatocytes lie within the testicle's tubules just above the primary spermatocytes. Throughout spermatogenesis, the developing cells are pushed closer to the center of the tubule.

After telophase I, each secondary spermatocyte enters a period of dormancy (interphase) before the second meiotic division begins. The second division, which results in the formation of spermatids, resembles mitosis.

Prophase II

During prophase II the dividing cells carry one chromosome from each homologous pair along with its connected chromatid. During this stage, the chromosomes do not replicate. They replicated previously during prophase I. Prophase II is characterized by the appearance of astral bodies, the formation of a spindle, and the disappearance of the nuclear membrane. This period ends as the chromatid pairs seek the spindle equator.

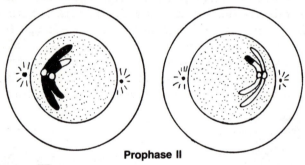

Prophase II

Metaphase II

During this period, the chromatid pairs are attached to the spindle equator by their centromeres.

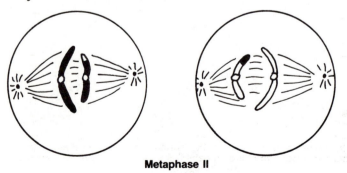

Metaphase II

Anaphase II

At this point, each centromere splits and the sister chromatids move along the spindle toward opposite ends of the cell.

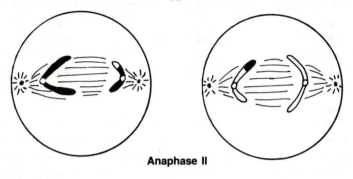

Anaphase II

Telophase II

This final step in meiosis involves the division of cytoplasm and the appearance of nuclear membranes to form, once again, two sister cells. These cells are the spermatids which mature to form spermatozoa (motile sperm). Each cell now carries only one sample from each of the homologous chromosome pairs. This segregation of chromosomes is necessary for the recombination of homologous chromosome pairs upon fertilization. In the horse, 32 chromosomes within the dam's egg pair with 32 corresponding chromosomes within the sire's sperm.

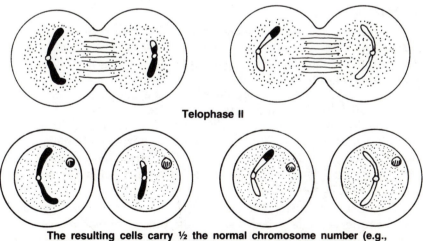

Telophase II

The resulting cells carry ½ the normal chromosome number (e.g., there are no homologous pairs). Some of the cells will become mature sperm.

Through the process of meiotic division, one primary spermatocyte forms four sperm cells, each of approximately equal size. This rapid propagation of viable sperm (four-fold increase) is the major difference between spermatogenesis and oogenesis.

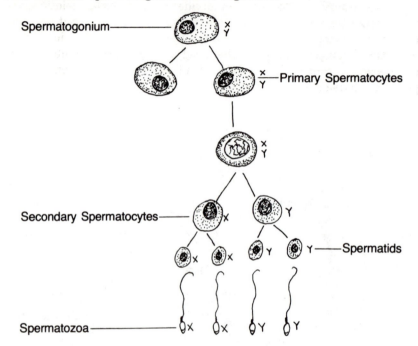

OOGENESIS

The purpose of oogenesis is to produce gametes — ova containing the haploid number of chromosomes and carrying unusually large amounts of cytoplasm. This cytoplasm provides the early embryo with nutrients and protection. To form this specialized structure, oogenesis is somewhat modified, but organized into the same basic steps: mitosis, first meiotic division and second meiotic division.

In the female, special cells within the ovary divide by mitosis, producing primitive egg cells (primary oocytes) during embryonic development. Because this process occurs before birth, the female is born with a specific number of primary oocytes (only a few of these will ever develop, however).

After sexual maturity, the primary oocytes start to divide by meiosis, forming secondary oocytes. During this growing process, each oocyte is protected and nourished by a surrounding follicle cell. As the primary oocyte completes the first meiotic division, the follicle cell bursts and expells the partially developed egg (secondary oocyte) into the reproductive tract (oviduct). The broken follicle forms a scarred area (corpus luteum) on the ovary's surface which produces reproductive hormones (e.g., progesterone).

The secondary oocyte does not divide again (second meiotic division) until after fertilization. If a sperm penetrates the egg, it rests its head in the cytoplasm while the egg's second meiotic division forms the ootid. Then, the nuclei of the ootid and the sperm combine, the chromosomes from each are paired and embryonic development begins (i.e., division by mitosis).

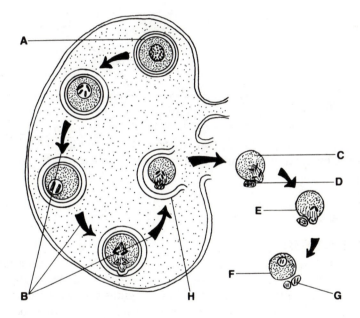

Oogenesis: (A) primary oocyte, (B) first meiotic division, (C) secondary oocyte, (D) first polar body, (E) second meiotic division — occurs after fertilization, (F) ootid, (G) second polar body, (H) follicle cell.

The most important difference between oogenesis and spermatogenesis is the resulting gamete. One primitive sperm cell can produce four mature sperm, while the primary oocyte develops into just one mature egg. The reason for this is that, as each meiotic division occurs during oogenesis, one daughter cell receives most of the cytoplasm, while the other is left with its share of chromosomes and very little cytoplasm. The small cells are called *polar bodies* and are usually retained within the larger cell just below its outer membrane. The first polar body (product of the first meiotic division) usually disappears, but the second polar body (product of the second meiotic division) persists and may, in most unusual cases, be fertilized by a sperm.

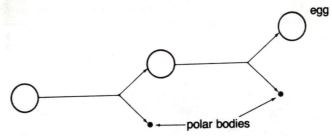

The end result of these modified cell divisions is the production of a large egg with an increased amount of cytoplasm. If the egg is fertilized, the cytoplasm reinforces the embryo's development by nourishing and protecting it, until transmission of nutrients and the formation of a protective placenta occurs within the uterus.

15

SEX CHROMOSOMES

Sex Determination

The moment the egg is penetrated by a sperm, the future embryo's sex is decided. Several factors lead to the sex determination: 1) the movement of two X chromosomes to opposite ends of a dividing cell during oogenesis or the movement of the X and Y chromosomes to opposite ends of a dividing cell during spermatogenesis; 2) the formation of haploid gametes; and 3) the combination of the sire's X or Y chromosome to the dam's X chromosome during fertilization.

When the first meiotic cell division occurs during oogenesis or spermatogenesis (collectively called gametogenesis), the sex related chro-

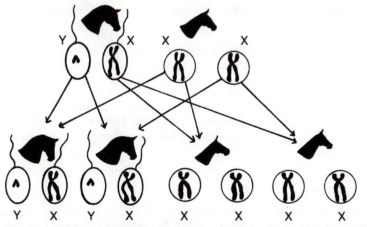

Fertilization involving normal sperm and ova results in one of two possible recombinations of the X and Y chromosomes.

mosomes are separated as homologous chromosomes migrate to opposite ends of the cell. (Although X and Y are not true homologous chromosomes, they do carry a small portion of homologous or similar genes.) This separation of sex chromosomes allows each resulting gamete to carry only one sex chromosome. (Female gametes carry only X, while the male gametes carry either X or Y.) When the male and female gametes unite and their chromosomes combine to form the normal diploid chromosome number, there are two possible combinations for the sex chromosomes. The XX combination results in the female phenotype, while XY instructs the formation of male characteristics.

Although, statistically, the odds should be 50% chance female and 50% chance male, the sex ratio at conception is somewhat altered. The number of males *conceived* is actually greater than the number of females. (Note: More male fetuses die from sex-linked lethals and incompatible uterine environment than do females.) Several interesting theories have been suggested to explain this altered conception ratio:

1) The X sperm may be heavier than the Y sperm. Because of the weight difference, the Y sperm moves to the oviduct (where the ovum is held) much faster than its X-bearing counterpart. If the ovum is present and conditions favor conception, the Y sperm should have a better chance at fertilization.

2) Another theory states that the Y sperm has a positive electrical charge, while both the ovum and X sperm are negatively charged. Physics states that opposites attract and, according to this, the ovum and Y sperm would be drawn together, while the X sperm would meet some resistance.

3) A third theory involves the acidity or alkalinity of the mare's reproductive tract. The slightly acidic reproductive tract of the mare might favor the Y sperm.

4) Because the X sperm have slightly greater longevity than the Y sperm, breeding time relative to ovulation time may counteract the altered conception ratio (i.e., early breeding increases the chances of producing a female).

X-Linked Traits

When a certain trait is controlled by a gene (or genes) located on the X chromosome, its pattern of inheritance is quite distinctive. For this reason, X-linked traits (with high expressivity) are easily identified, and their appearance in future generations can sometimes be predicted. A basic understanding of sex determination can be applied to the appearance of sex linked traits.

The male can only receive the X chromosome and, hence, his X-linked traits from his dam:

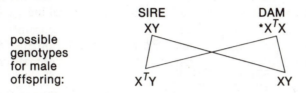

	SIRE	DAM
	XY	$^*X^TX$
possible genotypes for male offspring:	X^TY	XY

(*X^T indicates a gene located on the X chromosome)

When a female is conceived, she inherits an X chromosome both from her sire and dam. Therefore, she can receive X-linked traits from her sire and/or her dam:

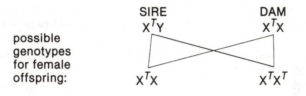

	SIRE	DAM
	X^TY	X^TX
possible genotypes for female offspring:	X^TX	X^TX^T

Although the transmission pattern for the X chromosome can easily be illustrated, the expression of X-linked traits depends on many other factors: gene interactions, penetrance, lethal capacity, environment, and so on. As a case in point, an X-linked dominant allele (designated as X^D) is expressed more frequently in the female:

POSSIBLE MALE
GENOTYPES

ratio		
1	X^DY	50% express the dominant allele
1	X^RY	

POSSIBLE FEMALE
GENOTYPES

ratio		
1	X^DX^R	75% express the dominant allele
2	X^DX^D	
1	X^RX^R	

An X-linked recessive allele is expressed more frequently in the male than in the female, simply because the male has only one X chromosome (a situation which eliminates the possibility of a corresponding dominant allele overriding the effects of the recessive):

POSSIBLE MALE
GENOTYPES

ratio
1          $X^D Y$
1          $X^R Y$ ———— 50% express the recessive allele

POSSIBLE FEMALE
GENOTYPES

ratio
1          $X^D X^D$
2          $X^D X^R$ ———— 25% express the recessive allele
1          $X^R X^R$

This point is illustrated, in the horse, by the transmission of the X-linked recessive allele for hemophilia. (Refer to **"Blood"** under **IN-HERITED ABNORMALITIES.**) In the case of hemophilia, however, other complicating factors are involved. Because hemophilia is a partial lethal condition, most hemophilic males (and theoretically females) do not reach sexual maturity. Hence, most matings which result in hemophilic offspring involve normal stallions crossed with hemophilic carrier mares. Matings of this type result only in hemophilic males, normal males, carrier females, and normal females:

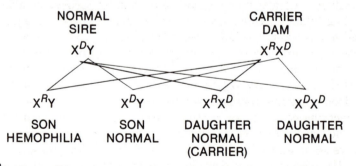

(${X^R}$ indicates the recessive hemophilia allele on the X chromosome; ${X^D}$ indicates the normal dominant allele at a corresponding point on the X chromosome.)

Environment may also affect the expressivity of the X-linked hemophilia allele. Theoretically, the hemophilia condition may never be expressed if the affected horse is not bruised or injured. From a practical standpoint, however, incidental environmental factors may determine how long a horse with hemophilia will live. But the chances of a horse completely escaping injury are remote.

X-linked inheritance is characterized by the fact that X-linked traits are never passed from sire to son. These traits have been studied extensively in human genetics (e.g., some forms of color blindness, deafness, diabetes, anemia, etc.), but similar knowledge of X-linked traits in horses is limited.

Y-Linked Traits

When genes are located on the Y chromosome, they are referred to as *holandric* ("holos" = entire + "aner" = man) genes. Y-linked traits can only be expressed in the male, but specific characteristics have not yet been identified in horses. (Very few have been identified in humans.) Complicating factors include confusion with sex limited and sex influenced traits, low expressivity and limited research. Some researchers believe that the Y chromosome (whose limited size indicates the presence of relatively few genes) carries genes which depress the expression of female sex characteristics and trigger the expression of masculinizing autosomal genes. There is no conclusive research to support this theory, however.

Sex Influenced Traits

Sex influenced traits include any trait whose expression is affected by the individual's gender (sex). In many cases, these traits are influenced by reproductive hormones. For example, the stallion's characteristic behavior is caused by the presence of testosterone. The normal (or the inconsistent) behavior of the cycling mare is controlled by levels of the reproductive hormones, estrogen and progesterone.

Appaloosa coloring may be a sex influenced trait (i.e., sex hormones may affect the color and/or pattern). Male Appaloosas usually have more varied patterns than the females, suggesting that testosterone may have some effect on the expression of Appaloosa genes. In other species, the effects of sex on coloration are readily seen (e.g., male birds often have brighter plumage than females).

Sex Limited Traits

Sex limited traits are those which appear only in one sex. For example, sperm production and lactation are limited to the male and female respectively. Except in unusual cases of intersex, reproductive anatomy is characteristic for each sex. Although the genes which control these sex limited traits may be located either on the sex chromosomes or the autosomes, the presence or absence of the Y chromosome somehow influences their expression.

Sex Chromosome Abnormalities

During cell division, the loss of a chromosome and/or nondisjunction can cause genetic changes that result in severe physical abnormalities. With advanced techniques in studying an individual's chromosomes (refer to "**Karyotyping**"), scientists have related several reproductive syndromes to chromosome abnormalities. Aneuploidy and intersex are believed to be caused by abnormal combinations of the sex chromosomes.

SMALL HARD OVARIES (ANEUPLOIDY, 63 X0)

In the mare, the condition known as "small hard ovaries" (a form of aneuploidy) is characterized by normal external reproductive anatomy, slightly reduced body size, inconsistent sexual behavior and sterility. Sometimes the affected mare displays the typical estrus behavior and accepts the stallion. During other estrus periods, the same mare may refuse the stallion, or she may react passively to both teasing and mounting.

When veterinarians discovered that these physical and behavioral characteristics were closely related to the presence of small firm ovaries, further investigation was warranted. Researchers discovered that the undersized ovaries contained no germ cells (cells which produce the eggs). Hence, the mares were sterile. Lack of follicles (cells which surround each developing egg) caused an estrogen deficiency and resulted in small flaccid uteri. Photographs of chromosome sets (karyotypes) from various body tissues showed that aneuploid mares had a missing X chromosome in their karyotype. The abnormality is represented as 63 XO, with the O indicating the missing X chromosome.

The condition 63 XO might be caused by nondisjunction during gametogenesis. If, for example, the X and Y chromosomes within the primary spermatocyte do not separate during anaphase I, one of the resulting secondary spermatocytes receives both an X and a Y chromosome, while the sister spermatocyte receives no sex chromosomes. As a result, two spermatozoa develop — one that contains an extra sex chromosome and one that contains only autosomal chromosomes.

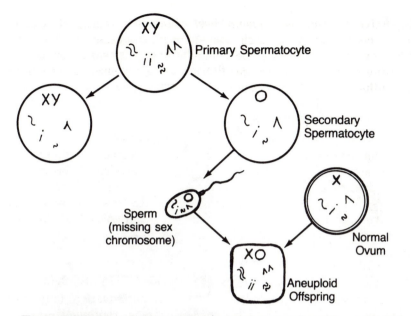

The aneuploid zygote is formed when a normal ovum unites with a sperm that lacks either sex chromosome. O indicates that the X chromosome has been lost.

If a secondary spermatocyte that contains no sex chromosomes forms a sperm and unites with a normal ovum, the resulting zygote would lack one chromosome. This XO embryo would develop aneuploidy characteristics.

If nondisjunction occurs during the early stages of embryonic development, the resulting embryo could carry different combinations for the same homologous chromosome pair (e.g., XX, XO, XY, XXY, etc.). Affected individuals are referred to as mosaics. Karyotyping several of the individual's cells reveals a variety of chromosome combinations. Several laboratories are now able to karyotype such animals, enabling breeders to identify and remove the non-productive individuals from their breeding herds.

A new technique in karyotyping, referred to as chromosome banding, has shown that many mares thought to be XY females were actually 63 XO mares. The Y chromosome was a small piece of the lost X chromosome.

INTERSEX

Male Pseudohermaphroditism

When a genetic and gonadal (i.e., physical) male appears with incomplete masculinization, he is referred to as a male pseudohermaphrodite. For years, this condition has been confused with cryptorchidism, since the testes are frequently retained. A close look at these unusual cases shows that the affected males have both partially developed male and female genitalia (e.g., clitoris/small glans penis, both under-developed testes and ovaries, partially developed uterus/character-istic stallion behavior, etc.). Because the pseudohermaphrodite's testes contain no germ cells, he is sterile.

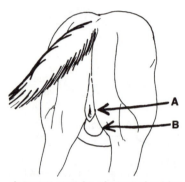

Intersex may be characterized by the presence of both an exaggerated clitoris (A) and underdeveloped testes (B).

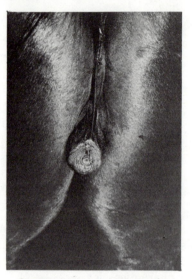

Genitalia of an intersex horse.

As with aneuploidy, unusual sex chromosome combinations have been associated with male pseudohermaphroditism. Karyotypes have shown that a wide variety of sex chromosome abnormalities can exist between similarly affected individuals. (Perhaps several closely re-lated syndromes are involved.)

Identified karyotypes:

65XXY
66XXXY
64XX/64XY
64XX/65XXY

The most probable explanation for these chromosome combinations is nondisjunction or, perhaps, complete loss of a chromosome during cell division. For example, the XXY combination could be caused by nondisjunction during oogenesis to form an egg with two X chromosomes. If the XX egg is fertilized by a Y sperm, the resulting zygote would contain three sex chromosomes, XXY. Other causative factors could include: 1) the transmission of blood between twin fetuses (chimerism), 2) the combination of two zygotes (fertilized ova) to form one developing embryo with contradictory karyotypes or 3) the combination of a fertilized egg and a fertilized polar body (refer to **"Oogenesis"**) to form one individual with unusual sex chromosome combinations.

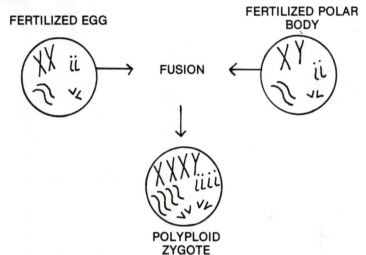

FERTILIZED EGG

FERTILIZED POLAR
BODY

FUSION

POLYPLOID
ZYGOTE

Although it is possible for one horse to produce several intersex foals (due to a malfunction in the parent's gametogenesis), male pseudohermaphroditism has not been established as an inherited condition.

Testicular Feminization

Although the intersex condition known as testicular feminization is not caused by sex chromosome omissions or duplications, its signs are similar to male pseudohermaphroditism. In this case, the female phenotype is predominant, and male behavior characteristics are displayed. As with male pseudohermaphroditism, the affected individuals possess underdeveloped genitalia and are always sterile.

Scientists believe that this intersex condition originates when the male embryo fails to respond to testosterone. This lack of response results in feminization (development of the female sex characteristics) to the extent that many affected horses are registered as females. Although both the stallion's physical and behavioral characteristics are controlled by testosterone levels, testicular feminization does not diminish stallion-like behavior.

Testicular feminization cannot be transmitted from sire to son, and is never expressed in the female. This suggests that the disorder is controlled by a rare allele located on the X chromosome (i.e., X-linked trait). There is also a possibility that the trait is controlled by a recessive sex limited autosomal allele (i.e., the sex limited gene could be present but could not be expressed in the female). ♞

16

COMPLICATING FACTORS OF INHERITANCE

Multiple Factors

Variation, or the extent to which a trait is expressed between different individuals, depends largely on the number of genes controlling the trait. Continuous variation is displayed in quantitative traits such as fertility, speed and size. Many genes (polygenes) affect fertility and, depending on the type of allele at each controlling locus, the horse may be anywhere from extremely fertile to permanently sterile. Because several loci combine efforts to produce one trait, the exact genotype related to that trait is difficult to define. Genetic interactions (i.e., dominant, recessive, etc.) are also complicated by polygenetic control. The environment seems to play a very important part in determining the final expression of traits with continuous variation.

Multiple Alleles

In the chapter "**Evolution**," the origin of genetic variation was attributed to the mutation of genes to other forms, or alleles. These mutations may change the expression of a trait slightly or drastically, and may be detrimental or helpful to the overall phenotype (i.e., physical appearance). In any case, the existence of several forms of the same gene allows various loci to influence the same trait differently, depending on the type of alleles present. These differences are referred to as genetic variation.

Hence, genetic variation depends on a variety of gene types between individuals. As the number of alleles at one locus increases, the expression of the trait becomes more complicated. The identification of the various alleles and their specific actions involves time-consuming studies and extensive comparative analysis.

Modifying Genes

Although the presence of a trait may depend primarily on one or several genes ("main" genes), the action of these controlling genes may be altered by the presence of modifying genes. Modifiers have been discussed in "Coat Color and Texture" with respect to certain color patterns, such as Appaloosa spotting. These genes can cause a relatively simple pattern of inheritance to appear extremely complicated. Differentiation of modifying genes from so called "main" genes simplifies the identification of inheritance pathways.

Gene Interactions

Gene interactions, such as dominance and recessiveness, complicate the study of genetics by changing the overall effects of different allelic combinations at the same loci. Because each combination interacts independently of the others, and because there may be several alleles for one gene, one allele may show several types of gene action. Therefore, genetic action is defined with respect to only a certain allele or group of alleles. An allele may be completely dominant, with respect to a limited number of alleles, and recessive to other alleles at the same locus. This possibility of dual roles for each allele complicates inheritance pathways, making it difficult to explain the presence or absence of a certain trait. For example, a geneticist may have to explain the presence or absence of a trait and its relationship to other traits by saying, ". . . the controlling allele, B, is completely dominant to alleles B_1, B_2 and B_3, incompletely dominant to allele B_6, and recessive to alleles B_4 and B_5 . . .".

DOMINANCE

The term "dominance" has been used throughout the text. Its definition has been presented, simply, as an allele whose influence overrides the influence of another allele at a corresponding point on the homologous chromosome. The definition is accurate, but four distinct types of dominance should also be defined.

Complete Dominance

Complete dominance involves the ability of one allele to completely override the effects of a corresponding recessive allele. For example, an allele, represented as D, is completely dominant to the recessive allele, d. Therefore, when an individual's genotype includes two dominant alleles, designated as DD, the resulting trait is expressed the same as if it was controlled by Dd.

Codominance

Codominance refers to a situation where two alleles are equally dominant. The resulting trait is, therefore, a compromise between the two allelic influences. An example is the dilution of coat color in horses. As explained in "Coat Color and Texture," the presence of the C allele at the C locus results in the expression of color, while the c^{cr} allele causes dilution. If the C allele is homozygous (CC), color is completely expressed. (The expression of color may be limited by the presence of other modifying genes, however.) If the c^{cr} allele is homozygous, the individual is almost completely diluted (e.g., cremello and perlino). When C and c^{cr} are both present (heterozygous), the resulting color (palomino, buckskin, etc.) is caused by an equal mixture of both effects.

Incomplete Dominance

Incomplete dominance is similar to codominance except that the two alleles do not contribute equally to the phenotype. While both alleles influence the expression of a trait, one has more control than the other.

Overdominance

The chapter "Inbreeding and Outbreeding" included an explanation of heterosis, the sudden improvement in performance or appearance caused by an increase in heterozygosity. Overdominance is partially responsible for this sudden improvement in traits (e.g., fertility and vigor). Alleles which show overdominance somehow compliment each other so that, in the heterozygous state, the two alleles produce an unusually outstanding characteristic (in comparison to their influence in the homozygous states). As explained earlier, overdominance is helpful to the improvement of performance, but the resulting high quality cannot usually be passed consistently from parent to offspring. (Refer to "Inbreeding and Outbreeding.")

EPISTASIS

When an allele completely overrides the effects of an allele at some other locus, either on the same or separate chromosomes, it is referred to as epistatic. The allele whose influence is diminished is referred to as the hypostatic allele. These interactions between separate genes may be detrimental to breed improvement, since they sometimes inhibit the expression of desirable traits, and therefore limit selection. For example, the coat color extension-restriction (black) allele, E^D, is epistatic to the bay and brown alleles, A and a^t, so that a black horse could carry a hidden gene for bay or brown coat color patterns. Another example is the dominant white allele, which is epistatic to all color genes. The roan and grey alleles are epistatic to all color genes except dominant white.

PENETRANCE AND EXPRESSIVITY

Occasionally, an individual may inherit genes for a particular trait, live in an environment that favors that trait, but fail to exhibit the characteristic. The trait's absence is due to a "lack of *penetrance*." By definition, penetrance is the frequency of a trait's appearance among a large group of individuals who carry the necessary genetic material. A trait may be penetrant, incompletely penetrant, or may have a complete lack of penetrance.

The terms penetrance and *expressivity* are often used interchangeably, although degrees of expressivity for a certain trait are only possible when the trait is penetrant. Expressivity is the extent to which a trait appears in just one individual (in comparison to other individuals with similar genetic potential). This variation is affected greatly by environmental factors. Epiphysitis, wobbler syndrome, and early maturity are examples of traits with variable expressivity.

Linkage and Independent Assortment

In the discussion "**Sex Chromosomes,**" the concept of linked genes was explained, and the inheritance of traits whose controlling genes are located on the X or Y chromosomes was examined. These traits follow a pattern of inheritance directly related to the sex of the individual (e.g., X-linked traits are never passed from father to son). Linkage, in general, refers to genes positioned on the same chromosome. If these linked genes are in close proximity, the chances of separation by crossing over (refer to "**Mutations**") or loss of a chromosome section, are remote. Therefore, closely linked genes are usually inherited together.

On the other hand, genes which are located on separate chromosomes are inherited independently of the other (independent assortment). A horse inherits half his genetic material from his sire and half from his dam. When sexually mature, he will begin gametogenesis to produce gametes (egg or sperm) with a random selection of one chromosome from each homologous (corresponding) pair. (Refer to "**Mitosis**" and "**Meiosis.**") Although it is possible for these gametes to carry only the sire-related or only the dam-related chromosomes, they usually carry a mixture of both. The following diagram illustrates this concept of independent assortment.

Independent Assortment of Two Allelic Pairs

SAMPLE OF
THE SIRE'S
GENE PAIRS

SAMPLE OF
THE DAM'S
GENE PAIRS

| R r |
| J J' |

| r r |
| J J' |

GAMETOGENESIS

HAPLOID GAMETES

| R |
| J |
| R |
| J' |
| r |
| J |
| r |
| J' |

| r |
| J |
| r |
| J' |

POSSIBLE COMBINATIONS

SPERM	+	EGG	=	OFFSPRING
R J	+	r J	=	Rr JJ
R J	+	r J'	=	Rr JJ'
R J'	+	r J	=	Rr J'J'
R J'	+	r J'	=	Rr J'J'
r J	+	r J	=	rr JJ
r J	+	r J'	=	rr JJ'
r J'	+	r J	=	rr J'J
r J'	+	r J'	=	rr J'J'

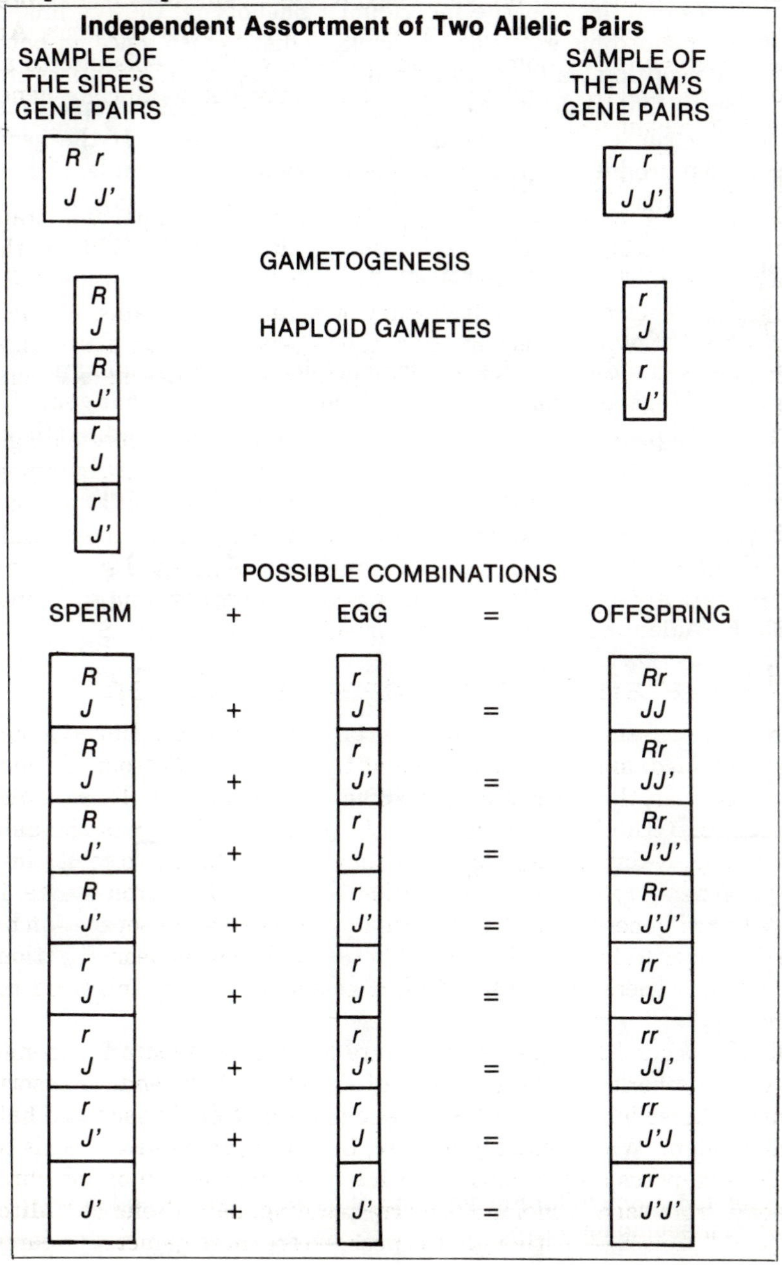

Pleiotropic Effects

A pleiotropic ("pleio" = many; "tropic" = affinity for) gene causes more than one effect on the individual's phenotype. Closely linked genes and pleiotropic genes are difficult to distinguish since, in both cases, the resulting traits are usually inherited together. Scientists are beginning to find evidence that the multi-effects of a pleiotropic gene is probably caused by one simple action of that gene. If the gene controls the production of an important enzyme, for example, it may indirectly influence the function of several body systems initiated by that enzyme. An example might be the simultaneous occurrence of "white foal syndrome" and "atresia coli" in overo crosses.

Prepotency

When an individual carries the same allele on corresponding points of two homologous chromosomes, he is homozygous for that gene. In random matings (i.e., no intentional inbreeding or outbreeding), it is

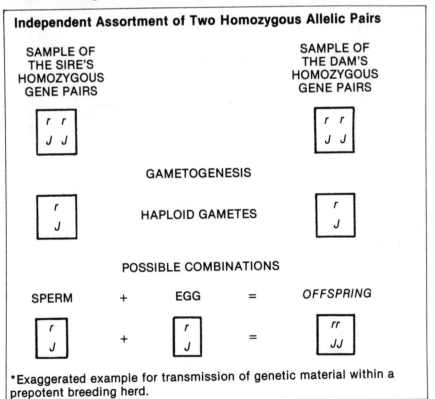

Independent Assortment of Two Homozygous Allelic Pairs

SAMPLE OF
THE SIRE'S
HOMOZYGOUS
GENE PAIRS

SAMPLE OF
THE DAM'S
HOMOZYGOUS
GENE PAIRS

r r
J J

r r
J J

GAMETOGENESIS

r
J

HAPLOID GAMETES

r
J

POSSIBLE COMBINATIONS

SPERM   +   EGG   =   *OFFSPRING*

r
J   +   r
J   =   rr
JJ

*Exaggerated example for transmission of genetic material within a prepotent breeding herd.

normal for the resulting offspring to receive some homozygous gene pairs. When an individual carries a higher than average number of homozygous dominant gene pairs, he is prepotent. Prepotency increases the ability to breed true (produce identical offspring) by providing only one possible allele from each homozygous locus during gametogenesis. For example, all foals of a homozygous grey stallion (*GG*) will be grey since all of the stallion's gametes will contain *G*. Because the gametes of a prepotent horse are quite similar, its offspring tend to resemble one another.

On the other hand, a decrease in homozygosity causes decreased prepotency, since the genetic content of the gametes varies greatly. The effects of decreased homozygosity (increased heterozygosity) are discussed in "**Inbreeding and Outbreeding.**"

Combination of Traits

In cases where a trait is actually the combination of several interacting traits, genetics becomes even more complicated. An example is speed, a trait which depends on lung capacity, soundness, attitude, strength, blood pressure, etc. Variation in each of these traits is probably controlled by several loci (and by the effects of the environment). The heritability of the overall result, speed, has been well documented. Because of the complex nature of this trait (effects of genetic interactions, modifiers, multiple alleles, polygenes and the environment upon several traits), its exact mode of inheritance is not understood. Similarly, the inheritance pathways of other complex traits are difficult to define.

17

MUTATIONS

A mutation can be defined as an alteration in gene or chromosome structure. These changes can be perpetuated through cell division, and may eventually affect physical appearance or function.

Since all mammals, including the horse, seem to have had a common ancestor (refer to the discussion on Theriodant under "**Evolution**"), the genotype of the prehistoric ancestor must have undergone many changes, or mutations, throughout evolution. These changes resulted in the vast number of mammalian genotypes now in existence. Mutations increase genetic variability, then natural and artificial selection determine which genotypes are most favorable. Each species evolved through the years to the form that best suited its particular environment. It is important to remember that throughout evolution, natural variation originated from some form of mutation; obviously, some of these changes must have been beneficial. Today, however, changes in genotypes which alter physical form or function, are usually detrimental to the animal. These mutations are harmful if the affected individual cannot adapt to the environment. White markings and parti-colored coats are two mutation-caused traits that might be considered desirable, while combined immunodeficiency disease is harmful.

Mutations are usually characterized by the following factors:

1. They permanently change the gene to some degree.
2. They are sudden changes.
3. They are relatively rare.

Mutations can be grouped into two basic categories: natural and induced. Natural mutations occur spontaneously and are caused by factors such as background radiation (the normal amount of radiation present in the environment to which all animals are exposed daily), high temperatures and the waste by-products of metabolism. Induced mutations, on the other hand, are caused by an external stimulus, such as x-ray radiation or ultraviolet light. Induced mutations are occasionally referred to as iatrogenic which, in this context, means "man-made."

For example, some drugs which man administers to the horse may cause mutations. It is for this reason that many drug products are provided with inserts which caution against the administration of a particular drug to a pregnant mare. These warnings should be strictly heeded. For instance, corticosteroids are not recommended for pregnant mares, particularly during the early stages of gestation. They may cause malformation of the fetus or abortion. (Mice treated with steroids during pregnancy have produced offspring with cleft palates.)

It is possible to purposefully induce a mutation. This, however, is impractical in animal breeding because the mutation process cannot be controlled. A specific gene cannot be purposefully mutated in a particular *desirable* way. In other words, a breeder cannot cause only beneficial mutations. In addition, the fertility of animals exposed to mutagens (the substances that cause mutations) will be lowered.

Mutations may occur either in the body cells (known as the somatic cells) or in the gametes (sperm and egg cells). Although somatic mutations may affect the horse's phenotype, they are not passed to his offspring. For example, a chestnut horse may have a somatic mutation which causes a small black spot on his coat. Since this mutation is not carried in the chromosomes of the gametes, the horse's offspring will not have the black spot.

Somatic mutations which occur in a single cell may eventually be found in many cells, if the mutated cell divides into many daughter cells. The original mutant cell may divide into enough daughter cells (which also carry the mutation), so that a physical change occurs. If only one cell carried a mutation, phenotypic change would probably never be noticed.

Structural Mutations

Mutations are responsible for the changes in the structure and composition of the genetic material. Most alterations in gene structure cause changes in gene action and eventually lead to changes in phenotype. Therefore, most new phenotypes result from new genes (alleles) produced by structural changes of existing genes.

Mutations act on the DNA molecule. A mutation may alter only one base-sugar-phosphate group (known as a nucleotide) or several of these structures. If only one nucleotide is affected, the mutation is known as a point mutation; if several nucleotides are changed, it is referred to as a gross mutation.

The larger the number of nucleotides affected by the mutation, the greater the change in the protein produced by that gene. The mutation's location upon the gene also affects its severity. For example, a mutation located near the end of a gene will not have as much effect as one situated near the beginning. If the mutation is near the end of a gene, the resulting protein will be almost identical to that produced by the original gene. This is because the protein produced by the mutated gene will be identical to that produced by the original gene, up to the point at which the mutation occurred. If the mutation is located near the beginning of the gene, the resulting protein will be changed drastically.

Genes may mutate from the original "wild-type" to a new allele (represented by A to a). The mutant allele may also revert back to the original form (a to A), a process which is referred to as reverse or back mutation. The allele produced by a back mutation is identical to the original allele.

There is a constant, unique mutation rate for every allele, meaning that A mutates to a at a set rate, as does a to A. Alleles at the same locus do not always mutate at the same rate. For example, A may tend to mutate to a much more frequently than a mutates to A. Each gene varies in its tendency to mutate, with some genes mutating much more easily than others. The mutation rate is usually stated in terms of the number of mutations per locus per generation. The rate is low for all genes, probably between 10 to 80 mutations per million gametes per generation. Fortunately, the stable double-helix structure of the DNA molecule tends to limit mutations. Only dominant gametic mutations are revealed in the next generation, since recessive mutations are hidden by the corresponding normal allele. A recessive mutation cannot be manifested unless two individuals carrying the mutation produce a homozygous recessive offspring. Even then, the chances of producing a mutant offspring are only one in four.

For example, suppose that M = normal dominant (nonmutant), and m = mutant recessive: MM individuals are normal noncarriers, Mm individuals look normal, but carry the mutation, and mm individuals express the mutation. The typical 1:2:1 ratio results from breeding two heterozygotes, so only one offspring out of four will be physical mutants (two others out of the four will carry the mutation, however).

$$Mm \quad \times \quad Mm$$

$$1 \;\; MM \;\; : \;\; 2 \;\; Mm \;\; : \;\; 1 \;\; mm$$

Since most mutations are recessive, they are usually not expressed for several generations. This is because the mutation must spread throughout the breed to increase the chances that two carriers will be mated.

The most harmful structural mutations are those which affect the enzyme systems of the cell. Enzymes are proteins which are coded for by genes. The production of other vital proteins can only occur in their presence, so a mutation-caused defect in an enzyme results in the production of many defective genes and proteins. Harmful mutations cause the following conditions:

1. They change the ability of the DNA molecule to initiate, continue, or terminate the synthesis of nucleic acid (remember that DNA and RNA are the two nucleic acids). Therefore, the DNA molecule will be unable to replicate properly.

2. They increase the number of errors that occur during the replication of DNA, and during transcription. Transcription occurs when RNA forms the code to carry genetic information out of the nucleus and to the ribosomes for protein synthesis.

3. They change the amino acid content of the protein because of the altered information carried by the codon.

These changes cause not only intracellular (within the cell) effects (such as changes in codons), but also intercellular (between cells) effects. For example, mutations adversely affect the ability of the cell to differentiate into different types, and frequently slow general growth of the affected animal. Although mutations have a wide variety of effects at the molecular level, they are usually detected only by their effects on the individual's external characteristics.

Point Mutations

A point mutation causes the substitution of one base in a nucleotide for another. When G and A are substituted for each other (G for A or A for G), or when C and T are substituted for each other (C for T or T for C), it is known as a transition point mutation.

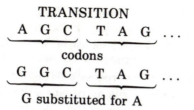

TRANSITION

A G C ‿ T A G ...

codons

G G C ‿ T A G ...

G substituted for A

When A and T, or C and G, are substituted for each other, the change is a transversion point mutation.

TRANSVERSION

A G C ‿ T A G ...

codons

A G C ‿ T T G ...

T substituted for A

Substituting one base for another results in a change in the codon. A mutation may cause a "sense" codon, which codes for the same amino acid as the original codon, or it may result in a "mis-sense" codon, which codes for a different amino acid. Eventually, the "mis-sense" codon causes the production of a different protein. A mutation could also change the triplet to a "non-sense" codon, which does not correspond to a code for any amino acid. Protein synthesis stops when it reaches a non-sense codon.

If an entire nucleotide is either deleted or added to the DNA molecule, every codon after the insertion or deletion is shifted.

For example, if the gene reads as follows:

A C C ‿ G T A ‿ C T T ...

codons

and the second C is deleted (since there is one base to a nucleotide, each base in the gene represents a nucleotide, so the deletion of the C represents the deletion of an entire nucleotide):

A C _ G T A C T T ...

the gene would now read as:

A C G ‿ T A C ‿ T T ...

codons

Each base following the deletion has been shifted to the left, thereby changing all the codons following the deletion. Since ACG, TAC, etc. probably code for different amino acids than do ACC, GTA, CTT, etc., a different protein will probably be produced or, if a "non-sense" codon is encountered, protein synthesis may stop.

If the original is: A   C   C   G   T   A   C   T   T   ...,

and the base T is added: A   C   C   T   G   T   A   C   T   T   ...,

the codons subsequent to the addition are changed. This results in the production of a different protein than that coded for by the original gene, or in the termination of protein synthesis.

Gross Mutations

Gross mutations are similar to point mutations except that, since they affect more nucleotides, their effects are more severe. Instead of simply substituting one base for another, a gross mutation may change several bases at one time.

Chromosomal Aberrations

Mutations can also affect the horse's genotype by changing the physical structure of the chromosome. Normal structure and function of the horse's phenotype depend on a balance between the genetic material on homologous chromosomes. Chromosomal aberrations upset this balance. Since the process by which the DNA molecule forms the chromosome is still a mystery, the cause of chromosomal aberrations is unknown. Fortunately, very few chromosomal abnormalities have been observed in the horse.

Chromosomal aberrations can be classified as one of the following:

1) Changes in the amount of genetic material (usually a change in the number of chromosomes), due to an addition or deletion. A technique known as chromosome banding of karyotypes has shown that mares with what appeared to be "XY" genotypes were actually "XO" females. Geneticists discovered that the small chromosome was a piece of the lost X chromosome, rather than a true "Y" chromosome.

2) Changes in the arrangement of genetic material on the chromosome. (This type usually does not cause an abnormality unless it results in an actual change in the amount of genetic material.)

Some of these mutations, such as crossing over, are relatively harmless and may occur frequently. Others, such as ring chromosomes, are infrequent but can be very detrimental.

NONDISJUNCTION

A change in the number of chromosomes most often results from nondisjunction, a failure of the chromosome pairs to separate during cell division. This results in one of the daughter cells having both homologous chromosomes, while the other has none. Nondisjunction during mitosis may cause mosaicism, a condition in which an individual's cells have different chromosome numbers. (Refer to "Sex Chromosomes.")

CENTRIC FUSION

Centric fusion (also known as Robertsonian translocations) results when acrocentric chromosomes break near the centromere and exchange unequal amounts of genetic material. This process causes the formation of a new, V-shaped chromosome and subsequent loss of the very short chromosome formed by the fusion of the two short arms.

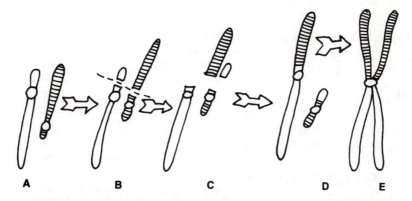

A   B   C   D   E

(A) Two acrocentric chromosomes in close proximity. **(B)** Breakage of the acrocentric chromosomes near the centromeres to form four arms. **(C)** Exchange of genetic material. **(D)** Two metacentric chromosomes are formed by the transition. The small chromosome will be inert if it contains no genes. **(E)** The large metacenric chromosome replicates normally during cell division. The individual will have one less chromosome in each resulting daughter cell.

The fragment produced by the fusion of the two short arms is composed almost entirely of heterochromatin from the chromosome area surrounding the centromere. The heterochromatin is believed to be genetically inert (contains no genes), which is why the loss of the short-arm chromosome does not harm the organism.

Centric fusion occurred frequently during evolution and caused the basic differences in chromosome numbers and shapes between species. In many instances, this process reduced the total number of chromosomes without decreasing the number of genes.

Chromosomal Rearrangement

The genetic material may be rearranged due to breaking and rejoining of the chromosomes. These breakages are usually caused by one of the following:

1) radiation
2) chemicals
3) extreme changes in cell environment
4) viral infections

Since broken chromosomes frequently reunite in their original pattern, many breaks probably go undetected. Types of chromosomal rearrangements include crossing over, translocations, deletions, inversions and ring chromosomes.

CROSSING OVER

Crossing over is the reciprocal exchange of genetic material between two paired homologous chromosomes during meiosis. This process causes the transfer of a block of genes from each chromosome to its homologue (i.e., its homologous chromosome). This exchange occurs frequently. Although crossing over does not change the amount of genetic material on the chromosome, it may cause the genes to segregate and recombine differently, resulting in new combinations of traits. (Refer to "**Mitosis**" and "**Meiosis**.")

TRANSLOCATION

A translocation is the transposition of two segments between non-homologous chromosomes. Abnormal breakage occurs in both chromosomes and, somehow, the pieces get switched and reattached to the wrong chromosomes.

The chances that two breaks will occur on non-homologous chromosomes (translocation) are greater than the chances that two breaks will occur on homologous chromosomes. This is because every chromosome normally has 62 non-homologous chromosomes, but only has one that is homologous.

The quantity of genetic material is not changed by a translocation, only the position is changed. A mere change in position is sometimes

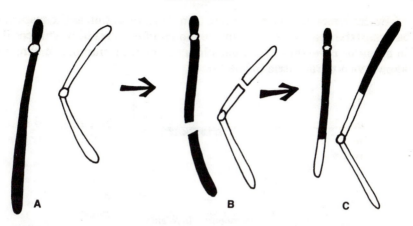

(A) Two non-homologous chromosomes. (B) Breakage occurs, forming four arms of genetic material. (C) Rearrangement of the genetic material results in two new chromosomes (i.e., results in two new combinations of genetic material).

enough to produce abnormalities; this is known as the position effect. If the genetic material at the actual point of breakage is damaged, physical abnormalities may result. Even if animals with translocations appear normal, they frequently have lowered fertility.

DELETIONS, INVERSIONS AND RING CHROMOSOMES

When two breaks occur on a single chromosome and they do not reheal in the original pattern, the following four alternatives may result.

1) Both breaks may occur on one arm, and the piece between the breaks may drop out and become lost. This is known as a deletion. The consequences of a deletion depend on the length of the material deleted, and on the role of the genes involved. Obviously, a long deletion, or one containing especially important genetic information, is most serious.

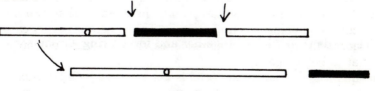

Deletion.

2) Both breaks may occur on one arm, then reheal after the fragment between the breaks has turned around end for end (180°). This results in an order reversal of the genes in a segment of chromosome, and is known as a paracentric inversion.

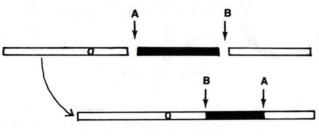

Paracentric inversion.

3) The breaks may occur on different arms, thus including the centromere. The fragment between the breaks may rotate 180° and then re-fuse. This is a pericentric inversion. This type of inversion causes the animal to be less fertile than does a paracentric inversion because the centromere is included and meiosis cannot proceed normally.

Pericentric inversion.

4) The breaks may occur on different arms, causing the terminal pieces to drop off, leaving the central portion with the centromere. The central portion then fuses at the broken ends, causing the formation of a ring chromosome (the terminal ends are lost).

For example, the base at each end of the following chromosome may be deleted.

```
...C  A  T  G  T  A  A...
......A  T  G  T  A......
```

The ends then "stick" together and form a ring chromosome.

```
        A  A
      A        T
        T  G
```

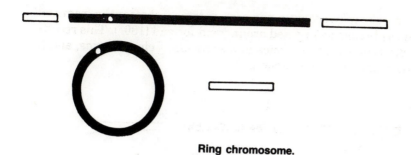

Ring chromosome.

Because of its shape, a ring chromosome cannot properly pair with its homologue during meiosis. The presence of ring chromosomes causes greater fertility problems than any other chromosomal aberration.

18

POPULATION GENETICS

Elements of Progress

There are five factors which influence the amount of progress a breeder can make in his herd: 1) genetic variability within the herd, 2) selection intensity employed by the breeder, 3) environmental influences, 4) breeding values of the breeding stock and 5) heritability values for the traits under consideration.

Genetic variability is a necessary element of progress, since animals that carry similar genetic information produce offspring with similar characteristics. If there are no above average or below average members in the herd, selection will not improve the quality of future generations.

Selection intensity is also important; the number of horses selected for breeding influences the overall quality of the new herd. The selection intensity is usually higher for stallions than for broodmares, meaning that the requirements for broodmares are usually less stringent than the requirements for stallions. The selection differential is determined by the selection intensity. It is the difference between the selected horse, or horses, and the average of the populaion from which they were selected. If a stallion with a time of 2:00 for a mile at the trot is selected from a herd whose trotting average is 2:05, the selection differential is five seconds. If this stallion is used as the herd sire, the herd average should improve. As the stallion's offspring improve the herd average, they also cause a decrease in the selection differential — the average speed of the herd approaches the sire's speed. For this reason, the progress that can be made decreases from generation to generation. To improve the rate of progress, the breeder might decide to select a stallion with a faster trotting speed, thereby increasing the selection differential.

Any visible trait is the result of the horse's genetic capacity, coupled with temporary and permanent environmental influences:

$$\text{Trait} = \text{Breeding Value} + \text{environment}^P + \text{environment}^T$$

Any temporary influence upon a trait can be determined by measuring or observing the considered trait over a period of time; as the environmental influences subside, the trait may change. Good management and favorable environment aid progress by allowing the full expression of the horses' potential. Once the breeder has provided a superior environment, he realizes that most of the variation between horses is due to differences in genotypes.

The breeding value of each horse in a herd also affects the rate of progress. Horses with low breeding values contribute very little to the improvement of a herd, while horses with high breeding values are desirable. If the genetic variability and selection differential within a herd decrease to a point where progress is limited substantially, the breeder can make some progress by improving his accuracy in estimating breeding values for his stock. To accomplish this, the breeder must eliminate those environmental influences that disguise the genetic potential of herd members. He then studies performance and production records of each horse's close relatives. Accurate and unbiased records are essential for maximum progress. The more records a breeder consults, the more accurate his breeding value estimates will be. (Records of close relatives are most significant because the proportion of common genes between individuals increases as the degree of relationship increases.)

As discussed earlier, heritability is the percent of a trait's expression that is due to inheritance. A trait with high heritability improves with selection, while a trait with low heritability is especially responsive to a favorable environment. When selecting a lowly heritable trait, pedigrees and performance records of collateral relatives should be consulted to predict the chances of a particular horse carrying desirable genes. (The genes which control lowly heritable traits are often disguised by environmental influences.)

Observing progeny for quality with respect to particular traits (e.g., racing, conformation, typiness, etc.) can increase accuracy in predicting breeding values of the parents. The greater the number of offspring tested, the closer the estimate will be to the true breeding value of the sire or dam. For a trait high in heritability, only a few offspring are needed. For a trait that is low in heritability, a greater number of offspring should be tested to accurately estimate the horse's breeding value. If the heritability of the considered trait is high, an extensive study of progeny is not essential, because the horse can be evaluated from his own phenotype and from the phenotype of a small number of offspring. If heritability is low, greater herd improvement will be made through management, rather than through selection.

A thorough analysis of a horse's offspring may be limited by the fairly lengthy equine generation time (birth to reproductive age) and by the economic costs involved. However, a progeny study is of great value in identifying stallions or mares that are carriers of harmful recessive traits (e.g., CID). If an inherited disorder that is known to be controlled by a recessive allele occurs in even one offspring, both the sire and dam are carriers of the recessive. (Refer to the discussion on predicting breeding values under **"Breed Improvement Through Applied Genetics."**)

Laws of Probability

If the genotypes of a mare and stallion are known, it is possible to determine what genotypes can occur in their offspring. The probability that a particular gene combination will occur can also be calculated. During gametogenesis, the selection of alleles for each gamete is random. If one sperm of a heterozygous Bb stallion contains a B allele, there is a 50% chance that the next gamete will contain a B allele. The same is true for the third sperm, for the fourth, etc. If a horse is palomino, he possesses the gene pair Cc^{cr}. During meiosis, approximately half of his gametes will receive a C allele, and half will receive a c^{cr} allele. Therefore, the probability of receiving a C allele is ½ or .5, the same as the probability of receiving a c^{cr} allele. If this palomino is mated to another palomino (also Cc^{cr}), the probability that the resulting foal will be CC (nondiluted chestnut) can be determined by multiplying the probability that one gamete will contain C by the probability that the other gamete will contain C:

Probability of a gamete containing C		Probability of a gamete containing C		Probability of a CC offspring
	×		=	
.5	×	.5	=	.25 = 25%

There is a 25% chance that the offspring will be CC when two heterozygous Cc^{cr} parents are mated.

The chances of a $c^{cr}c^{cr}$ (cremello) being produced by these same parents can also be determined:

Probability of a gamete containing c^{cr}		Probability of a gamete containing c^{cr}		Probability of a $c^{cr}c^{cr}$ offspring
	$\times$		$=$	
.5	$\times$	.5	$=$	.25 = 25%

Any palomino offspring (Cc^{cr}) will have received a C allele from the sire and a c^{cr} allele from the dam, or vice versa. In this case, the probability of both events must be added:

1.	Probability of C entering the sperm		Probability of c^{cr} entering the ovum		Probability of a Cc^{cr} offspring
	.5	$\times$	.5	$=$	.25 = 25%

2.	Probability of C entering the ovum		Probability of c^{cr} entering the sperm		Probability of a Cc^{cr} offspring
	.5	$\times$	.5	$=$	.25 = 25%

3.	Probability of Cc^{cr} (step 1)		Probability of Cc^{cr} (step 2)		.25 + .25 .50 = 50%
		$+$		$=$ $=$	

There is a 50% chance that the offspring will be palomino (heterozygous Cc^{cr}) like the parents.

For completely dominant traits (such as greying vs. nongreying), the heterozygous horse which carries only one dominant allele appears the same as the horse which carries both dominant alleles. The probability of the dominant allele occurring in the offspring equals the probability of the homozygous dominant genotype plus the probability of the heterozygous genotype occurring in the offspring. The chances of two heterozygous grey horses (Gg) producing grey offspring can be calculated by adding the chances of the offspring being homozygous dominant (probability of *GG*) to those of the offspring being heterozygous (probability of *Gg*):

$$P\ (GG) + P\ (Gg) = P\ (G\_)$$

The following diagram shows the probability of two heterozygous horses producing grey offspring:

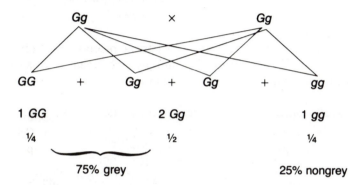

1 *GG*   2 *Gg*   1 *gg*

¼   ½   ¼

75% grey   25% nongrey

The same method can be used to determine the inheritance of two or more traits. In the following example, the probabilities of a grey/nonroan-roan/nongrey cross producing grey, roan, or grey/roan offspring is examined.

Heterozygous grey   ×   Heterozygous roan
Gg/rr (grey/nonroan)   ×   *gg/Rr* (roan/nongrey)

possible combinations:
Gg/Rr (grey/roan)
gg/rr (nongrey/nonroan)
gg/Rr (nongrey/roan)
Gg/rr (grey/nonroan)

The probability of each of these genotypes occurring is ¼. This factor is determined by multiplying the probability of a grey or nongrey gene occurring (½) by the probability of a roan or nonroan gene occurring (also ½): ½ × ½ = ¼.

$$\tfrac{1}{2}\,G \times \tfrac{1}{2}\,R = \tfrac{1}{4}\ \text{grey/roan}$$
$$\tfrac{1}{2}\,G \times \tfrac{1}{2}\,r = \tfrac{1}{4}\ \text{grey/nonroan}$$
$$\tfrac{1}{2}\,g \times \tfrac{1}{2}\,r = \tfrac{1}{4}\ \text{nongrey/nonroan}$$
$$\tfrac{1}{2}\,g \times \tfrac{1}{2}\,R = \tfrac{1}{4}\ \text{nongrey/roan}$$

In this example, the probability of each phenotype is equal. This is not always the case, as the following examples show.

$$Gg/rr \times Gg/Rr$$

$Gg \times Gg$ = ¾ chance of grey
and a ¼ chance
of nongrey

$rr \times Rr$ = ½ chance of roan
and a ½ chance
of nonroan

All possible combinations and their probabilities are listed:

¾ G	×	½ R	=	⅜ grey/roan
¾ G	×	½ r	=	⅜ grey/nonroan
¼ g	×	½ R	=	⅛ nongrey/roan
¼ g	×	½ r	=	⅛ nongrey/nonroan

This means that there is a 12.5% (⅛) chance of producing a nongrey/nonroan foal by crossing a grey/nonroan and a grey/roan. There is a 12.5% chance of producing a nongrey/roan, a 37.5% (⅜) chance of producing a grey/roan and a 37.5% chance of producing a grey/nonroan.

Another example of this is a mating between a grey/roan and a nongrey/roan:

$$Gg/Rr \times gg/Rr$$

$Gg \times gg$ = ½ chance of producing a grey and
a ½ chance of producing a nongrey

$Rr \times Rr$ = ¼ chance of producing a nonroan,
½ chance of producing a roan, and
a ¼ chance of producing the lethal
allelic combination (RR)

Because no RR individuals are
born, there are 2 roans (⅔) for
every 1 nonroan (⅓).

$$\frac{1}{2} \, Gg \, \times \, \frac{1}{4} \, RR \, = \, \frac{1}{8} \text{ lethal}$$
$$\frac{1}{2} \, Gg \, \times \, \frac{1}{2} \, Rr \, = \, \frac{1}{4} \text{ grey/roan}$$
$$\frac{1}{2} \, Gg \, \times \, \frac{1}{4} \, rr \, = \, \frac{1}{8} \text{ grey/nonroan}$$
$$\frac{1}{2} \, gg \, \times \, \frac{1}{4} \, RR \, = \, \frac{1}{8} \text{ lethal}$$
$$\frac{1}{2} \, gg \, \times \, \frac{1}{2} \, Rr \, = \, \frac{1}{4} \text{ nongrey/roan}$$
$$\frac{1}{2} \, gg \, \times \, \frac{1}{4} \, rr \, = \, \frac{1}{8} \text{ nongrey/nonroan}$$

Note that in this case, only ¾ of the matings result in viable foals, since the ⅛ GgRR and the ⅛ ggRR die in utero. The above calculations suggest that approximately 2 out of 8 conceptions may not survive (RR), 2 may be grey/roan (GgRr), 2 may be nongrey/roan (ggRr), 1 may be grey/nonroan (Ggrr), and 1 may be nongrey/nonroan (ggrr). There will probably be 6 viable offspring from every 8 conceptions, of which 2/6 (33% chance) may be nongrey/roan, 2/6 (33% chance) grey/roan, 1/6 (16.7% chance) grey/nonroan, and a 1/6 (16.7% chance) nongrey/nonroan.

Gene Frequency

Gene frequency refers to the occurrence of a certain allele within a population. If only one allele occurs in a population, the gene frequency = 1. For example, a herd of overo horses has two recessive overo alleles in their genotypes. Therefore, the frequency of the O allele at the O locus is 1.0 for that herd. In a herd of 100 tobiano horses, of which 50 are homozygous (TT) and 50 are heterozygous (Tt), the gene frequency of T is .75, while that of t is .25 (.75 + .25 = 1).

Selecting for a gene increases its frequency within a population. If a desirable characteristic is dominant, selection for that characteristic tends to reduce the frequency of the undesirable recessive:

D = desirable dominant
dd = undesirable recessive

$Dd \times Dd$	= 1 DD : 2 Dd : 1 dd
(.5 D and .5 d frequency)	= (.5 D and .5 d frequency)

1. $Dd \times Dd = $ D D   There are 8 alleles within the
                    D d   population; 4 are D and 4 are d.
                    D d   Therefore, the gene frequency =
                    d d   .5 D and .5 d.

2.   D D   The undesirable dd offspring
                    D d   are culled. Out of the 6 remaining
                    D d   alleles, 4 are D and 2 are d.

$\downarrow$ *d d*		
culled		

Therefore, the new gene
frequencies are: 2/3 *D* = .67 *D*
and 1/3 *d* = .33 *d*.
(.67 + .33 = 1.0)

3.

	2/3 *D*	1/3 *d*
2/3 *D*	4/9 *DD*	2/9 *Dd*
1/3 *d*	2/9 *Dd*	1/9 *dd*

The remaining offspring can be
mated, and the genotypes of their
offspring can be determined.

4.

D D	*D d*
D D	*D d*
D D	*D d*
D D	*D d*
d	*d*

There are 18 alleles, 12 *D* and 6 *d*.
Therefore, the allelic frequencies
are 12/18 *D* = .67 *D* and 6/18 *d*
= .33 *d*.

5.

D D	*D d*
D D	*D d*
D D	*D d*
D D	*D d*

The 1/9 *dd* individuals can be
culled, leaving 16 alleles: The new
allelic frequencies are 12/16 *D* =
.75 *D* and 4/16 *d* = .25 *d*.

$\downarrow$ *d d*	
culled	

75% of the alleles are *D* and 25%
are *d*: 75% + 25% = 100%

The Hardy-Weinberg law states that, in a random-mated population the gene frequencies remain constant. If the gene frequency of a
certain allele is 0.5, the frequency will remain at 0.5 from generation
to generation.

The formula used to calculate the frequency of an allele within a population is the Hardy-Weinberg Law: $p^2 + 2\,pq + q^2 = 1$.

p^2 = the frequency of the recessive genotype
q^2 = the frequency of the dominant genotype
$2\,pq$ = the frequency of the heterozygous genotype

Therefore, p = the frequency of the recessive allele, and
q = the frequency of the dominant allele.

The number of recessive alleles can be found because it is known that
every homozygous recessive has 2 and every heterozygote has 1.

$$p = rr + \tfrac{1}{2}\,(Rr)$$

The number of dominant alleles can be found because it is known that every homozygous dominant has 2 and every heterozygote has 1.

$$q = RR + \tfrac{1}{2}(Rr)$$

It is also known that $2pq = Rr$, so $\tfrac{1}{2}Rr = \dfrac{2pq}{2} = pq$.

Because $p + q = 1$, $p = 1 - q$, and $q = 1 - p$.

Therefore, $pq = (p)(1-p)$

With this information, the equation $p = rr + \tfrac{1}{2}(Rr)$ can be solved:

$$p \quad rr_{freq} + \tfrac{1}{2}(Rr_{freq})$$

$$= rr_{freq} + pq$$

$$= rr_{freq} + p(1-p)$$

$$= p^2 + p(1-p)$$

$$= p^2 + p - p^2$$

$$= p$$

Since $p^2 = rr_{freq}$, $p = \sqrt{rr_{freq}}$. If the number of rr genotypes in the population is represented by rr_{freq}, the frequency of the recessive gene is the square root of rr_{freq}, or $\sqrt{rr_{freq}}$.

The preceding calculations prove that the frequency of the recessive allele in a population is equal to the square root of the number of recessive individuals in a population ($p = \sqrt{rr}$). This concept can be used to prove that the Hardy-Weinberg law is correct and that the gene frequencies of a randomly mated population will remain constant from generation to generation.

For example, suppose that 4% of a foal crop of 100 Arabians are born with combined immunodeficiency disese (CID). The frequency of the lethal condition in the foal crop is .04, and the frequency of the lethal CID allele is the square root of .04 = $\sqrt{.04}$ = .2. The frequency of the corresponding nonlethal dominant allele = .8. Substitute $p = .2$ and $q = .8$ into the Hardy-Weinberg equation:

$$p^2 + 2\,pq + q^2 = 1$$
$$(.2)^2 + 2(.2 \times .8) + (.8)^2 = 1$$
$$.04 + 2(.16) + .64 = 1$$
$$.04 + .32 + .64 = 1$$
$$1 = 1$$

The above equation implies that 4% (i.e., $p^2 = .04$) of the foal crop die of the lethal CID condition. It follows that 32% (i.e., $2\,pq = .32$) of the foals are carriers, and that 64% (i.e., $q^2 = .64$) of the foals are non-carriers.

If the herd of Arabians is mated at random, the allelic frequency for CID would remain approximately at 2% if the condition were not lethal. Purebred horses are not mated at random with respect to most traits; the Hardy-Weinberg law is only applicable in studying a trait that has not been altered through selection. The Hardy-Weinberg law can be applied to the frequencies of various blood types, since horses are usually mated at random with respect to blood types.

Selection for a recessive allele, and against its corresponding dominant allele, can result in a frequency of 1.0 for the recessive (the dominant is eliminated). Selection for a dominant, on the other hand, causes an increase in the frequency of the dominant allele. In a large population, the allelic frequency will probably never reach 1.0. As a case in point, the production of an occasional nongrey Lippizaner indicates that there are a few heterozygous grey (*Gg*) individuals within the breed. Selection against, and elimination of, the recessive allele becomes progressively less effective as the frequency of the recessive allele decreases. This is because the recessive alleles are carried by a progressively smaller number of apparently normal heterozygotes. These heterozygous individuals are identified as carriers only when they produce homozygous recessive offspring. When mated to another heterozygote, the heterozygous individual has a one in four chance of producing a homozygous recessive offspring.

Selection is most effective (i.e., early rapid progress can be made) when gene frequencies are intermediate (approximately 0.5). When selecting for dominant alleles (i.e., against recessives), 0.33 is the most effecient gene frequency at which to begin selection. If the dominant allele is common in the population, most of the phenotypes are dominant, leaving little room for improvement. If the frequency of the dominant allele is high, the frequency of the recessive allele will be low. In addition, many of the recessive alleles will be hidden in the heterozygous state. When selecting for a recessive, the allelic frequency should be at least 0.67. This is because the number of recessive phenotpes in the population is very small if the recessive allelic frequency is low. As a result, there would be very few individuals to select.

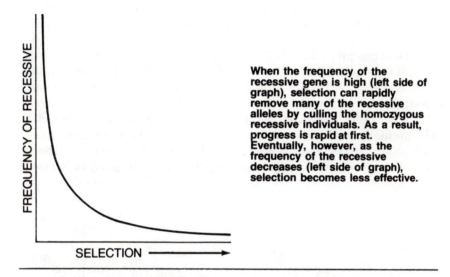

When the frequency of the recessive gene is high (left side of graph), selection can rapidly remove many of the recessive alleles by culling the homozygous recessive individuals. As a result, progress is rapid at first. Eventually, however, as the frequency of the recessive decreases (left side of graph), selection becomes less effective.

The number of generations required to change a gene frequency by a specified amount can be calculated. Assume that the incidence of congenital night blindness (an eye defect described in "Inherited Abnormalities") is 1%, and that the frequency of the recessive allele responsible for this condition is 0.1 (i.e., $p = 0.1$). This means that 10% of the population carries the recessive allele (1% incidence in the population $= p^2$; if $p^2 = .01$, $p = \sqrt{.01} = 0.1$).

The number of generations necessary to reduce the frequency of the condition by 90%, to one×tenth of 1%, can be calculated as follows:

If the new genotypic frequency desired is 0.1%, the new allelic frequency must be 0.032 (0.1% = 0.001; $\sqrt{0.001} = .032$). The number of generations necessary to attain this frequency can be found by using this formula:

$$\frac{1}{\text{desired gene freq.}} - \frac{1}{\text{original gene freq.}} = \text{number of generations}$$

$$\frac{1}{0.032} \qquad \frac{1}{0.1} \qquad = \quad 31 - 10$$

$$= \quad 21$$

Twenty-one generations are necessary to reduce the incidence of congenital night blindness from 1% to 0.1%.

As the example shows, it is often impractical to attempt absolute eradication of an inherited condition when the frequency of the undesirable allele is already small. If the frequency of a gene is intermediate, however, the same change in frequency will require much less time. Suppose that a breeder has a herd of overo mares that he wishes to mate to a solid-colored Thoroughbred stallion. The mares are all *oo*, while the stallion is *OO*. All of their offspring will be *Oo* (overo carriers). The frequency of *o* is .5, as is the frequency of *O*. In subsequent generations (i.e., produced by the heterozygous offspring), the breeder may decide to cull the overo pattern, but to keep other traits of the herd intact. If the breeder wanted to reduce the frequency of the *o* allele from .5 to .1, the generations required are calculated as follows:

$$\frac{1}{0.1} - \frac{1}{0.5} \ = \ 10 - 2 = 8 \text{ generations}$$

Within 8 generations, the breeder could reduce the frequency of the *o* allele to .1. Since the frequency of *o* is .1 (i.e., $p = .1$), the frequency of the *O* allele must be .9 (i.e., $q = .9$). The frequency of the *oo* genotype in the herd is .1 × .1, or 1%. The frequency of the *OO* genotype is .9 × .9, or .81%. Heterozygous overo carriers *Oo* now comprise only .18, or 18%, of the herd.

Inbreeding and Genotypic Frequency

Inbreeding does not change the frequency of the genes in a population, but it does change the frequency of genotypes (gene combinations).

Consider two traits (located on separate chromosomes) that are represented by *Mm* and *Nn* respectively. The entire herd is *MmNn*, so that there is a .5 frequency of each allele in the population. (Half of the alleles at the *M* locus are *M* and half are *m*; half of the alleles at the *N* locus are *N* and half are *n*.) Mating two members of this herd will produce the following genotypes in the stated ratios:

MmNn × *MmNn*

1 *MMNN*	2 *MmNN*	2 *mmNn*
2 *MMNn*	4 *MmNn*	1 *mmnn*
1 *MMnn*	2 *Mmnn*	1 *mmNN*

Notice that a minimum of 16 offspring will reveal all 9 genotypes in the proper ratios. Between these 16 offspring, there are 32 alleles for the *M* locus (either *M* or *m*) and 32 alleles for the *N* locus (either *N* or *n*). Of the 32 alleles at the *M* locus, 16 are *M* and 16 are *m*, so that the frequency of each allele is still .5. The same is true of the *N* locus. Therefore, the allelic frequency within the population has not been changed by mating two individuals of identical genotype. The *genotypic* frequencies have been changed, however. Originally, there were no *MM*, *mm*, *NN*, or *nn* horses. Instead, both parents (which represent the entire population) were *MmNn*. Therefore, the original genotypic frequencies were *MM* = 0, *mm* = 0, *Mm* = 1, *NN* = 0, *nn* = 0 and *Nn* = 1. With only one generation of inbreeding, the frequencies are *MM* = .25, *Mm* = .5, *mm* = .25, *NN* = .25, *Nn* = .5 and *nn* = .25. This means that 25% of the offspring are *MM*, 50% are *Mm*, 25% are *mm*, 25% are *NN*, 50% are *Nn* and 25% are *nn*.

Remember that, although inbreeding does not change the allelic frequencies of a population, selection does. Once inbreeding has been used to reveal undesirable recessives, horses carrying these alleles can be culled. This action will alter the allelic frequencies. For example, the breeder could cull all *mm* and *nn* individuals, leaving only *MM*, *NN*, *Mm* and *Nn* animals. This selection process would reduce the frequency of *m* and *n* in the population, and increase the frequency of *M* and *N*.

Additive Genes

Many traits are affected by several pairs of genes, each of which has a small cumulative effect. This type of trait is described as *quantitative*; each pair of genes has an *additive effect*. Suppose that there are four pairs of genes that contribute to the horse's height. The alleles for tall are represented by H_1, H_2, H_3, H_4; the alleles for short are represented by h_1, h_2, h_3 and h_4:

$$
\begin{array}{llll}
H_1H_1 & H_2H_2 & H_3H_3 & H_4H_4 = \text{ very tall} \\
h_1h_1 & h_2h_2 & h_3h_3 & h_4h_4 = \text{ very short} \\
H_1H_1 & H_2H_2 & h_3h_3 & h_4h_4 = \text{ intermediate}
\end{array}
$$

The *additive effects* of these alleles can be shown also. Suppose that each allele for short is worth 2, while each allele for tall is worth 4. The very tall horse would have a value of 32, while the miniature horse would have a value of 16. The intermediate horse's value would be 24. Each subsequent substitution of a higher value allele for a lower value allele has less effect on the horse's total value. For example, if $a = 2$, $b = 2$ and $c = 2$, and if $A = 6$, $B = 6$ and $C = 6$:

genotype *aabbcc*   = 12
genotype *AABBCC* = 36

The substitution of one uppercase allele for a lowercase allele causes an increase in value of 4:

genotype *Aabbcc*   = 16

This is a 33⅓% increase. An additional substitution, which also raises the value by 4, is only a 25% increase:

genotype *AAbbcc*   = 20

The third substitution is a 20% increase:

genotype *AABbcc*   = 24

Eventually, a limit is reached. In the preceding example, the limit is *AABBCC* = 36. After the limits of inheritance have been obtained, the breeder can make improvement only through superior environment and management. (Refer to **"Quantitative Inheritance"** under **"Breed Improvement Through Applied Genetics."**)

Population Mean

A population is a group of individuals that are observed or considered for a specific purpose. The broodmare band on a breeding farm or all horses within a certain state are examples of designated populations.

The mean of a population is the average value for a trait under consideration. This value is obtained by adding each individual value and dividing the total value by the number of horses in the population. For example, if the trait under consideration is weight, the average weight might be 1000 pounds:

1. Each individual weight is designated as an x value:

$$x_1 = 1000$$
$$x_2 = 1100$$
$$x_3 = 850$$
$$x_4 = 1050$$

2. The sum of these values is 4000. This total is divided by the number of values considered:

$$\frac{4000}{4} = 1000 = \text{the population mean.}$$

3. The population mean is referred to as $\bar{x}$. In mathematical terms, $\bar{x} = \frac{\Sigma x}{n}$, where $\Sigma x =$ the sum of all the x values, and $n =$ the number of horses in the population.

The population mean describes the population's average phenotype, a reflection of the average genotype. The genotypes of most horses within the population will be similar to the average. The average phenotype of the offspring is a good measure of the parents' breeding value, and usually tends to be near the population mean. In fact, the average phenotype of the offspring tends to be nearer the population mean phenotype than to the parents' phenotype. Consequently, superior parents often produce relatively average progeny. In other words, the offspring tend to revert from the parents' superior level to the average of the population. "Regression" is the term used to describe this unfortunate tendency.

Variance and Standard Deviation

Most of the measurements and observations used to determine a population mean do not agree exactly with the average; some values are a little higher; others are somewhat lower. The difference between each value (x) and the population mean ($\bar{x}$) can be measured. The measurement is known as *variance:*

$$\text{Variance} = \frac{\Sigma (x - \bar{x})^2}{n - 1}$$

The difference between each value (x) and the population mean is found by subtracting the mean ($\bar{x}$) from each individual value (x). Each resulting deviation is squared $(x - \bar{x})^2$, and the products are added $\Sigma (x - \bar{x})^2$. This sum is divided by the number of values in the population minus one (n − 1). The value (n − 1) is known as the degrees of freedom; 1 is subtracted from the number of values to compensate for error that may result from measuring the deviations.

The above formula can be applied to a practical example. Suppose that a population of four horses is measured for weight. Each horse has an x value, which is equal to that horse's weight.

1. $x_1 = 1000$ The sum of these values is
    $x_2 = 1100$ 4000, and the mean ($\bar{x}$) is 1000.
    $x_3 = 850$ $\bar{x} = \dfrac{\Sigma x}{n}$
    $x_4 = 1050$

2. $x_1 - \bar{x} = 1000 - 1000 = 0$
    $x_2 - \bar{x} = 1100 - 1000 = 100$
    $x_3 - \bar{x} = 850 - 1000 = -150$
    $x_4 - \bar{x} = 1050 - 1000 = 50$

3. $(x_1 - \bar{x})^2 = 0^2 = 0$ pounds squared
    $(x_2 - \bar{x})^2 = 100^2 = 10,000$ pounds squared
    $(x_3 - \bar{x})^2 = -150^2 = 22,500$ pounds squared
    $(x_4 - \bar{x})^2 = 50^2 = 2500$ pounds squared

4. $\Sigma(x - \bar{x})^2 = 0 + 10,000 + 22,500 + 2500 =$
    35,000 pounds squared

5. degrees of freedom $= n - 1 = 4 - 1 = 3$

6. $\dfrac{\Sigma (x - \bar{x})^2}{n - 1} = \dfrac{35,000}{3} = 11,666$

Therefore, the variance for this population equals
11,666 pounds squared.

The variance can be used to find the standard deviation — the approximate difference between all values and the population mean. It is found by taking the square root of the variance:

$$\text{Standard deviation} = \sqrt{\text{variance}}$$
$$= \sqrt{11,666}$$
$$= 108$$

Since 11,666 is in pounds squared, the standard deviation ($\sqrt{\text{variance}}$) is in pounds. The standard deviation for this population of horses is approximately 108 pounds. In other words, most horses in the population weigh between 892 and 1108 pounds (a deviation from the average by about 108 pounds).

Variance may be due to both inheritance and to environmental influences. The breeder must be able to differentiate between these variations. Obviously, if all variation were caused by the environment, the breeder would not need to select for a trait, but would concentrate on providing optimum conditions and superior management. The genetic component of variance is known as the breeding value — breeders should select horses with superior breeding values for highly heritable traits.

Errors in Measurement

Any measurement or observation is subect to error; even when several measurements of the same animal are made, some degree of variation is likely. A horse that stands approximately 15-2 hands may have been measured as 15-1 hands, 15-2.5, etc. Frequently, the effects of one error will cancel the effects of another, since there are often nearly equal errors in both directions. Because of this balancing of errors, the determination of a value is most accurate when a large number of measurements or observations have been made. When a large number of measurements are graphed, the results usually resemble a bell-shaped or "normal" curve.

For example, if only two values have been taken, they will form a straight line:

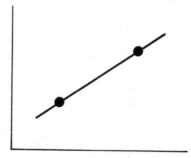

The greater the number of values, the closer the distribution will resemble a normal curve:

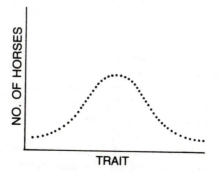

The bell-shaped curve is caused by many small deviations, or errors in measurement, and by relatively few large ones. A measurement of a horse's height, for example, is more likely to err by ¼ or ½ an inch, than by several inches. In addition, when measuring several horses (rather than taking multiple measurements of one horse), most of the values will be near the population mean. A tall, narrow curve implies that the deviation of each measurement from the mean is small; thus, most of the measurements are in close agreement:

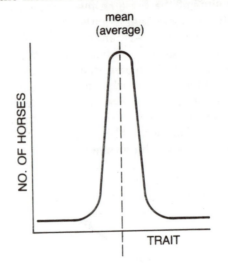

A low broad curve implies that there are large deviations from the population average:

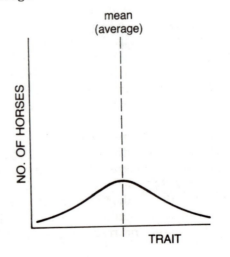

The area under the curve is designated as 1, since it includes all values of the population. The mean of the populaton is found at the center of the bell-shaped curve, with equal numbers of measurements above and below it. Each measurement can be compared to the population average to see how much it deviates; these measurements are used to determine the standard deviation for the population (i.e., standard deviation $= \sqrt{\text{variance}}$).

For any bell-shaped curve, it is known that .67 of one standard deviation will include 50% of the population, while 1.96 standard deviations include 95% and 2.58 deviations include 99%. This means that the population average, plus or minus one standard deviation, includes a certain percentage of the population. If a herd averaged 1000 pounds with a standard deviation of 100 pounds, the statement "1000 pounds ± 67 pounds" means that 50% of the horses in the population weighed between 933 and 1067 pounds. (The value 67 pounds = .67 × one standard deviation = .67 × 100 pounds.) The statement "1000 pounds ± 196 pounds (i.e., 1.96 standard deviations)" means that 95% of the horses weighed between 804 and 1196 pounds. Also, "1000 pounds ± 258 pounds (i.e., 2.58 standard deviations)" means that 99% of the horses weighed between 742 and 1258 pounds.

A sample is used to estimate the frequency of a trait in a herd when the population is too large for every animal to be measured. Serious errors result if the sample used to judge a trait is too small, or if the sample is not representative of the considered population. A breeder with 500 horses on a range may not want to measure each horse to determine their average height. Instead he could select and measure a random 10% of his herd. If representative, the sample should incorporate measurements in proportions similar to those found in the whole population. This results in a bell-shaped curve identical to one that would result from measuring the total population. Should the breeder measure only 10 horses, the resulting average could be far removed from the true mean of the herd. In addition, the breeder may unknowingly choose 50 horses that are disproportionately large or small, or that otherwise fail to represent the herd accurately. This is known as a biased sample.

When selecting a stallion, a breeder should look at an unbiased sample of his offspring. Frequently, only the outstanding offspring (which may be out of superior mares) are brought to the buyer's attention. The breeder needs to consider a representative sample before he can accurately judge the stallion's breeding value. The breeder should see a cross-section of the stallion's progeny, not just those that are out of high-quality mares and reared under optimum conditions.

Association of Traits

Frequently, inherited traits are closely associated. Height, for example, may influence stride length. It is possible to correlate a measurable trait with one that cannot be measured. A correlation can be made between height and stride, so that an increase in height would indicate a proportionate increase in stride. This is a positive correlation. (If stride length decreased as the height increased, it would be a negative correlation.) The known, measurable value is referred to as "x," while the value which is determined by x is designated as "y." In the preceding example, x = height, and y = stride length. The letter y is referred to as the dependent variable, since its value depends on x; x is known as the independent variable. A definite relationship can be established between these two variables. This relationship can be shown on a graph:

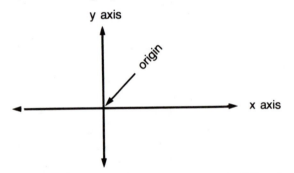

The horizontal line is called the x-axis, while the vertical line is the y-axis. The point of intersection is known as the origin. The axes are labled, so that units increase to the right of the origin on the x-axis and above the origin on the y-axis.

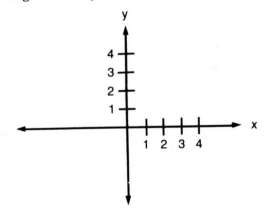

For every value x, there is a value y. If all x and corresponding y values are graphed, a straight line is formed. Consider, for example, stride length and height. Assume that a 2-inch increase in height results in a 6-inch increase in stride length:

$$
\begin{array}{ll}
2 = x & 6 = y \\
4 = x & 12 = y \\
6 = x & 18 = y \\
8 = x & 24 = y
\end{array}
$$

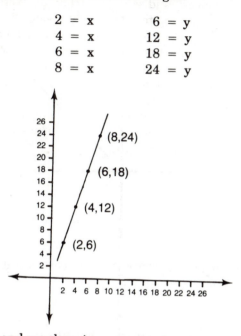

Each pair of x and y values is represented as a point on the graph; a straight line can be drawn between these points. The mathmatical description of the line is y = a + bx, where b measures the slope of the line (i.e., the angle). For instance graph A shows a greater slope than graph B:

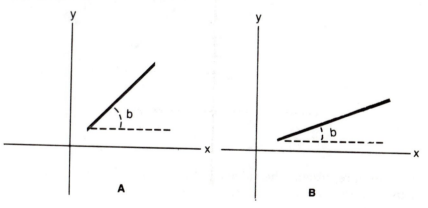

Consequently, the equation for graph A has a higher b (slope) value than the equation for graph B. If b is a positive value, an increase in one measurement causes an increase in the other:

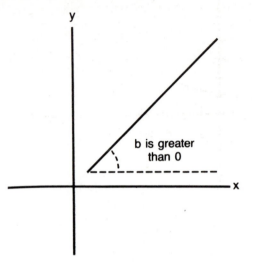

If b is negative, an increase in one value causes a decrease in the second characteristic:

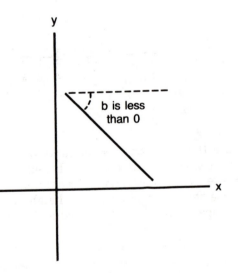

The letter a represents the value at which the staight line crosses the y-axis (i.e., the value of y when x = 0).

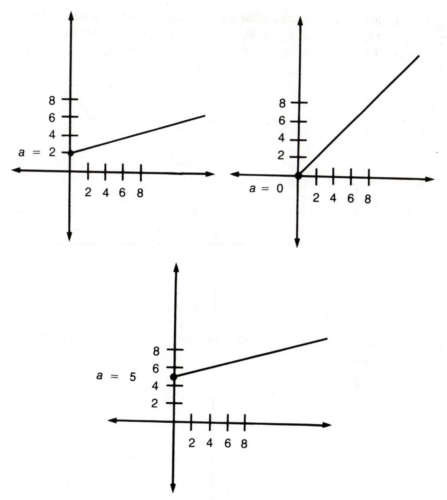

Once the equation a + bx is known, y values can be determined for all corresponding x values. If height increase (x) = 2 and stride increase (y) = 6, the value a (value for y when x equals 0) is 0 because there is no increase in height without an increase in stride. The slope (b) can be determined as follows:

$$y = a + bx$$
$$6 = 0 + 2b$$
$$6 = 2b$$
$$6/2 = b$$
$$3 = b$$

If all the x values are known, the y values can be calculated with the equation $y = a + bx$. Substituting $a = 0$ and $b = 3$, the equation becomes $y = 0 + 3x$, or $y = 3x$. If $x = 2$, $y = 3(2) = 6$. This calculation can be repeated for all x values:

$$x = 4 \qquad y = 3(4) = 12$$
$$x = 6 \qquad y = 3(6) = 18$$
$$x = 8 \qquad y = 3(8) = 24$$

Association between traits in the horse is also illustrated by height and the incidence of laryngeal paralysis (roaring). This is a positive association, since laryngeal paralysis increases in incidence with an increase in height. The condition is virtually unknown in ponies (horses under 14-2 hands), rare in horses between 14-2 and 15-2, more common in horses between 15-2 and 16-2 hands and most common in horses over 16-2 hands. By selecting for increased height, a breeder may unknowingly select for a greater predisposition to laryngeal paralysis. The incidence of this disease is so much greater in the "16-2 hands and taller" group that a breeder could effectively select against this condition by culling horses over that height. In this situation, a breeder should make a choice between the benefits and risks of increased height within his herd.

BIBLIOGRAPHY

Acta Agriculturae Scandinavica. Supplement, Vol. 19 1967

Persson, S. "Blood Volume and Working Capacity."

Acta Agriculturae Scandinavica. Vol. 18 1968

Bengtsson, S., B. Gane and J. Rendel. "Genetic Studies on Transferrins, Albumins, Prealbumins, and Esterases in Swedish Horses," pp. 60-64.

Adams, O.R. **Lameness in Horses.** 2nd ed. Philadelphia: Lea and Febiger, 1972.

Adams, O.R. **Lameness in Horses.** 3rd ed. Philadelphia: Lea and Febiger, 1974.

American Journal of Physiology. Vol. 209 No. 4 1964

Marcilese, N.A. "Blood Values in Light and Heavy Horses," pp. 727-730.

Animal Blood Groups and Biochemical Genetics. Vol. 3 1972

Sandberg, K. "A Third Allele in the Horse Albumin System," pp. 207-210.

Animal Blood Groups and Biochemical Genetics. Vol. 4 1973

Sandberg, K. "The D Blood Group System in the Horse," pp. 193-205.

Schleger, W. and G. Mayrhofer. "Genetic Relationships between Lipizzan Horses, Haflinger, Noriker and Austrian Trotters." pp. 3-10.

Animal Blood Groups and Biochemical Genetics. Vol. 5 1974

Sandberg, K. "Linkage Between the K Blood Group Locus and the 6-PGD Locus in Horses," pp. 137-141.

Australian Veterinary Journal. Vol. 43 March 1967

Hutchins, D.R., E.E. Lepherd and I.G. Crook. "A Case of Equine Haemophilia," pp. 83-87.

Australian Veterinary Journal. Vol. 50 January 1974

Baird, J.B. and C.D. Mackenzie. "Cerebellar Hypoplasia and Degeneration in Part-Arab Horses," p. 25.

Australian Veterinary Journal. Vol. 51 March 1975

Thompson, D.B., M.J. Studdert, R.G. Beilharz and I.R. Littlejohns. "Inheritance of a Lethal Immunodeficiency Disease of Arabian Foals," pp. 109-113.

Australian Veterinary Journal. Vol. 52 October 1976

Thompson, D.B., P.B. Spradbrow and M.J. Studdert. "Isolation of an Adenovirus from an Arab Foal with Combined Immunodeficiency Disease," pp. 435-437.

Australian Veterinary Journal. Vol. 53 1978

Steel, J.D. "The Inheritance of Heart Score in Race Horses," pp. 306-309.

Belschner, H.G. **Horse Diseases.** London: Angus and Robertson, LTD., 1969

Belschner, H.G. **Horse Diseases.** 1976 ed. Hollywood: Wilshie Book Company, 1976

Blood, D.C. and J.A. Henderson. **Veterinary Medicine.** 4th Ed. Baltimore: William and Wilkins Co., 1974.

Blood-Horse. Vol. 63 No. 22 May 31, 1952

Pirri, John Jr. and D.J. Steele. "The Heritability of Racing Capacity," pp. 976, 977 and 990.

Blood-Horse. Vol. 85 No. 13 1963

Churchill, E.A. "The Causes of Lameness," pp. 602-606.

Blood-Horse. Vol. 85 No. 15 1963

Solomon, J.A. "The Turning Out Syndrome — a Hypothesis," p. 714.

Blood-Horse. Oct. 6, 1975.

Cunningham, E.P. "Equine Genetics," excerpts taken from remarks at Royal Dublin Society Genetics Symposium, pp. 4210-4213.

Blood-Horse. Vol. 102 No. 12 March 21, 1977

"Blood Typing," pp. 1341-1342.

Blood-Horse. March 28, 1977

"Report on Furosemide," p. 1416.

Blood-Horse. January 16, 1978

Hollingsworth, K. "Average Yearling Price Up 25.5%."

Bogart, R. **Improvement of Livestock.** New York: The MacMillan Co., 1959.

British Veterinary Journal. Vol. 126 1970

Genhouschi, N., M. Bistriceanu, A. Sugiu and M. Bratu. "A Case of Intersexuality in the Horse With Type 2A + XXXY Chromosome Formula," p. 552.

Burns, S.J. and W.C. McMullan. **Junior Clinics: Equine Section.** College Station: College of Veterinary Medicine, Texas A&M University.

Canadian Journal of Comparative Medicine. Vol. 33 October 1969

Basrur, P.K., H. Kanagawa and J.P.W. Gilman. "An Equine Intersex with Unilateral Gonadal Agenesis," pp. 297-306.

Canadian Journal of Comparative Medicine. Vol. 34

Vandeplassche, M. "Some Aspects of Twin Gestation in the Mare," pp. 218-226.

Chromosoma. Vol. 13 1962

Trujillo, J.M., C. Stenius, L.C. Christian and S. Ohno. "Chromosomes of the Horse, the Donkey and the Mule," pp. 243-248.

The Chronicle of the Horse.
Huggins, F.J. "A Look at the Wobbler Syndrome," p.8

Clinical Orthopaedics and Related Research. Vol. 62 1969
Rooney, J.R. "Congenital Scoliosis and Lordosis," pp. 25-30.

Colin, E.C. **Elements of Genetics.** New York: McGraw-Hill, 1956.

Cornell Veterinarian. Vol. 35 No. 4 1945
Britton, J.W. "An Equine Hermaphrodite," p. 373.

Cornell Veterinarian. Vol. 53 No. 4 1963
Rooney, J.R. "Epiphyseal Compression in Young Horses," pp. 567-574.

Cornell Veterinarian. Vol. 55 1965
Stormont, C., and Y. Suzuki. "Paternity Tests in Horses," pp. 365-377.

Cornell Veterinarian. Vol. 56 No. 2 1966
Rooney, J.R. "Congenital Contracture of Limbs, pp. 172-187.

Cornell Veterinarian. Vol. 57 No. 3 1967
Rooney, J.R. and M.E. Prickett. "Congenital Lordosis of the Horse," pp. 417-428.

Cornell Veterinarian. Vol. 61 No. 4 October 1971
Rooney, J.R., C.W. Raker and K.J. Harmany. "Congenital Lateral Luxation of the Patella in the Horse," pp. 670-673.

Cornell Veterinarian. Vol. 63 No. 1 1973
Vitums, A., B.D. Grant, E.C. Stone and G.R. Spencer. "Transposition of the Aorta and Atresia of the Pulmonary Trunk in a Horse," pp. 41-57.

Craig, Dennis. **Horse-racing.** London: J.A. Allen & Co. LTD., 1963

Crowell, A. **Dawn Horse to Derby Winner.** New York: Praeger Publishers, 1973.

Cytogenetics. Vol. 6 1967
McFeely, R.A., W.C. Hare and J.D. Biggers. "Chromosome Studies in 14 Cases of Intersex in Domestic Mammals," pp. 242-253.

Cytologia 13. Vol. 1 1943
Makino, S. "The Chromosomes of the Horse (Equus caballus) (Chromosome Studies in Domestic Mammals, I)," pp. 26-38.

Davidson, J.B. **Horseman's Veterinary Advisor.** New York: ARCO, 1975.

Edwards, E.H. and C. Geddes. **The Complete Book of the Horse.** New York: ARCO Publishing Co., 1973.

Edwards. G.B. **Anatomy and Conformation of the Horse.** Croton-on-Hudson, New York: Dreeman Press LTD., 1973.

Emery, A.E.H. **Elements of Medical Genetics.** 4th ed. Berkeley: University of California Press, 1976.

Emery, L., J. Miller and N. Van Hoosen. **Horseshoeing Theory and Hoof Care.** Philadelphia: Lea and Febiger, 1977.

Ensminger, M.E. **Horses and Horsemanship.** Danville, Ill.: The Interstate Printers and Publishers, 1969. 4th ed.

Ensminger, M.E. **Horses and Tack.** Boston: Houghton Mifflin Co., 1977.

Equine Conference Notes. University of Illinois, College of Veterinary Medicine, 1978.
Cook, W. R. "Hereditary Diseases in the Horse."

Equine Medicine and Surgery. 2nd ed. E.J. Catcott and J.F. Smithcors, ed. Wheaton, Ill.: American Veterinary Publications Inc., 1972.

Equine Professional Topics. No. 2 1978
Scoggins, R.V. "Stallion Evaluation and Management," pp. 1-2.

Equine Veterinary Journal. Vol. 7 No. 2 1975
Hughes, J.P., P.C. Kennedy and K. Benirschke. "XO Gonadal Dysgenesis in the Mare (A Report of Two Cases)," pp. 109-112.

Equine Veterinary Journal. Vol. 8 1976
Kieffer, N.M., S.J. Burns and N.G. Judge. "Male Pseudohermaphroditism of Testicular Feminizing Type," pp. 38-41.

Equine Veterinary Journal. Vol. 8 No. 3 1976
Fretz, P.B: and W.C.D. Hare. "A Male Pseudohermaphrodite Horse with 63 XO?/64 XX/65 XXY Mixoploidy," pp. 130-132.

Equine Veterinary Journal. Vol. 8 No. 4 1976
Falco, M.J., K. Whitwell and A.C. Palmer. "An Investigation into the Genetics of Wobbler Disease in Thoroughbred horses in Great Britain," pp. 165-169.
McIlwraith, C.W., R. APR. Owen and P.K. Basrur. "An Equine Cryptorchid with Testicular and Ovarian Tissues," pp. 156-160.

Equine Veterinary Journal. Vol. 10 No. 2 1978
Mayhew, I.G. et al. "Congenital Occipitoatlantoaxial Malformation (in the Horse)," pp. 103-113.

Equine Veterinary Journal. Vol. 1 1969.
Jeffcott, L.B. "Haemolytic Disease of the Newborn Foal," pp. 165-170.

Evans, J.W., A. Borton, H.F. Hintz and L.D. Van Vleck. **The Horse.** San Francisco: W.H. Freeman and Co., 1977.

Frandson, R.D. **Anatomy and Physiology of Farm Animals.** Philadelphia: Lea and Febiger, 1972.

Genetics. No. 33, January 1948

   Castle, W.E. "The ABC's of Color Inheritance in Horses," pp. 22-35.

Genetics. No. 39 January 1954

   Castle, W.E. "Coat Color Inheritance in Horses and Other Mammals," pp. 35-44.

Genetics. No. 4 September 1961

   Castle, W.E. and W.R. Singleton. "The Palomino Horse," pp. 1143-1150.

Genetics. Vol. 50 November 1964

   Stormont, C. and Y. Suzuki. "Genetic Systems of Blood Groups in Horses," pp. 915-929.

Genetics. Vol. 53 April 1966

   Gahne, G. "Studies on the Inheritance of Electrophoretic Forms of Transferrins, Albumins, Prealbumins, and Acid Plasma Esterases of Horses," pp. 681-694.

Genetics. Vol. 65 1970

   Braend, M. "Genetics of Horse Acidic Prealbumins," pp. 495-503.

Goldstein, P. **Genetics Is Easy.** 4th ed. New York: Viking Press, 1972.

Hafez, S.E. and I.A. Dyer. **Animal Growth and Nutrition.** Philadelphia: Lea and Febiger, 1969.

Hanauer, E. **Disorders of the Horse and What to Do About Them.** Hollywood: Wilshire Book Company, 1976.

Harrison, J.C. ed. **The Care and Training of the Trotter and Pacer.** Columbus, Ohio: The United States Trotting Association, 1968.

Hayes, Capt. M.H. **Points of the Horse.** New York: ARCO Publishing Co., 1977. 7th ed.

Hayes, Capt. M.H. **Veterinary Notes for Horse Owners.** New York: ARCO Publishing Co., 1974. 16th ed.

Herskowitz, I.H. **Principles of Genetics.** New York: Macmillan & Co., 1973.

Hewitt, A.S. **Sire Lines.** The Thoroughbred Owners and Breeders Ass., 1968.

House, C.A. **Inbreeding, What It Is and What It Does.** London: Fur and Feathers Publishers (Watmoughs LTD.) 12th ed.

I.S.U. Veterinarian. Vol 29 No. 1 1961

   Lundvall, R.L. "Lamenesses of the Upper Hind Leg," pp. 7-11.

Johansson, I. and I. Rendel. **Genetics and Animal Breeding.** San Francisco: Freeman.

Jones, W.E. and R. Bogart. **Genetics of the Horse.** East Lansing: Caballus Publishers, 1971.

Journal of Animal Science. Vol 1 No. 2

   Blakeslee, L.H. and R.S. Hudson. "Twinning in Horses," pp. 155-58.

Journal of Animal Science. Vol. 10

   Rollins, W.C. and C.E. Howell. "Genetic Sources of Variation in the Gestation Length of the Horse," pp. 797-806.

Journal of Animal Science. Vol. 27 No. 3

   Hutton, C.A. and T.N. Meacham. "Reproductive Efficiency on Fourteen Horse Farms," pp. 241-62.

Journal of the American Veterinary Medical Association (AVMA). Vo. 95 No. 750 September 1939

   "Arthritis Deformans in Colts," p. 267.

   Dimock, W.W. and B.J. Arrington. "Incoordination of Equidae: Wobblers," pp. 261-67.

Journal of the AVMA. Vol. 101 No. 787 October 1942

   Jones, T.C. and F.D. Maurer. "Hereditary and Periodic Ophthalmia," pp. 248-49.

Journal of the AVMA. Vol. 119 No. 894

   Stocking, G.G. "Observations Concerning Conception in the Mare," pp. 354-55.

Journal of the AVMA. Vol. 137 No. 6 1960

   Baker, R.H. "Osteochondrosis," pp. 354-55.

Journal of the AVMA. Vol. 138 No. 3

   Bardwell, R.E. "Osteomalacia in Horses," pp. 158-62.

Journal of the AVMA. Vol. 140 No. 12 June 1962

   Morgan, J.P., W.D. Carlson and O.R. Adams. "Hereditary Multiple Exostosis in the Horse," pp. 1320-22.

Journal of the AVMA. Vol. 144 No.3

   Sangar, V.L., R.E. Mairs and A.L. Trapp. "Hemophilia in a Foal," pp. 259-64.

Journal of the AVMA. Vol. 146 No. 12 June 1965

   Evans, L.H., J. Jenny and C.W. Raker. "Surgical Correction of Polydactylism," pp. 1405-08.

   Fowler, M.E. "Congenital Atresia of the Parotid Duct in a Horse."

Journal of the AVMA. Vol. 149 No. 2 July 15, 1966

   Rubin, L.F. "Cysts of the Equine Iris," pp. 151-54.

Journal of the AVMA. Vol. 153 No. 12 1966

   McFeely, R.A. "Chromosomes and Infertility," p. 1672.

Journal of the AVMA. Vol. 155 No. 4 August 15, 1969

   Gelatt, K.N., H.W. Leipold and J.R. Coffman. "Bilateral Optic Nerve Hypoplasia in a Colt," pp. 627-31.

Journal of the AVMA. Vol 156 No. 2 January 1970

   Finocchio, E.J. "Congenital Patellar Ectopia in a Foal," pp. 222-23.

Journal of the AVMA. Vol 164 No. 1

   McGuire, T.C., M.J. Poppie and K.L. Banks. "Combined (B- & T-Lymphocytes) Immunodeficiency: A Fatal Genetic Disease in Arabian Foals," pp. 70-76.

Journal of the AVMA. Vol. 165 No. 7 October 1, 1974
   Gelatt, K.N., V.S. Myers and J.R. McClure. "Aspiration of Congenital and Soft Cataracts in Foals and Young Horses," pp. 611-16.
Journal of the AVMA. April 1973
   McChesney, A.E. "Adenoviral Infection in Foals," pp. 545-49.
Journal of the AVMA. Vol. 168 No. 1 January 1976
   Crowell, W.A., C. Stephenson and H.S. Gosser. "Epitheliogenesis in a Foal," pp. 56-58.
Journal of the AVMA. Vo. 172 No. 3
   Kenney, R.M. "Cyclic and Pathological Changes of the Mare's Endometrium as Detected by Biopsy, With a Note on Early Embryonic Mortality," pp. 241-62.
   Stickle, R.L. and J.F. Fessler. "Retrospective Study of 350 Cases of Equine Cryptorchidism," pp. 343-46.
Journal of Equine Medicine and Surgery. Vol 1 1977
   Moore, J.N., J.H. Johnson, H.E. Garner and D.S. Traver. "A Case Report of Inguinal Herniorrhaphy in a Stallion," p. 391.
Journal of Equine Medicine and Surgery. Vol. 1 November 1977
   Witzel, D.A., R.C. Riis, W.C. Rebhun and R.B. Hillman. "Night Blindness in the Appaloosa: Sibling Occurrence," pp. 383-86.
Journal of Equine Medicine and Surgery. Vol. 2 January 1978
   Jeffcott, L.B. "Disorders of the Equine Thoracolumbar Spine — A Review," pp. 9-19.
Journal of Heredity. Vol. 8 December 1917
   Wright, S. "Color Inheritance in Mammals," pp. 561-64.
Journal of Heredity. Vol. 19 No. 4 April 1928
   Gonzalez, B.M. and V. Villegas. "'Big Head' of Horses a Heritable Disease," pp. 159-67.
Journal of Heredity. Vol. 26 No. 6 June 1935.
   Amschler, W. "The Oldest Pedigree Chart."
Journal of Heredity. Vol. 27 No. 10 October 1936
   Prawochenski, R. "A New Lethal Factor in the Horse," pp. 411-14.
Journal of Heredity. Vol. 30 October 1939
   Gremmel, F. "Coat Color in Horses," pp. 437-45.
Journal of Heredity. Vol 31 March 1940
   Castle, W.E. "The Genetics of Coat Color in Horses," pp. 127-28.
Journal of Heredity. Vol. 32 July 1941
   Salisbury, G.W. "The Inheritance of Equine Coat Color," pp. 235-39.
Journal of Heredity. Vol. 32 August 1941
   Salisbury, G.W. and J.W. Britton. "The Inheritance of Equine Coat Color," pp. 255-60.
Journal of Heredity. Vol. 33 No. 11 November 1942
   Hazel, L.N. and J.L. Lush. "The Efficiency of Three Methods of Selection." pp. 393-99.
Journal of Heredity. Vol. 34 No. 4 April 1943
   Blakeslee, L.H., R.S. Hudson and H.R. Hunt. "Curly Coat of Horses," pp. 115-18.
Journal of Heredity. Vol. 35 No. 7 July 1944
   Lerner, I.M. "Lethal and Sublethal Characters in Farm Animals," pp. 219-24.
Journal of Heredity. Vol 36 No. 3 March 1945
   "Temperament Gene in the Thoroughbred," R.C. p. 82.
Journal of Heredity. Vol. 37 February 1946
   Castle, W.E. "Genetics of the Palomino Horse," pp. 35-38.
Journal of Heredity. Vol. 39 December 1948
   Domanski, A.J. and R.T. Prawochenski. "Dun Coat Color in Horses," pp. 367-70.
Journal of Heredity. Vol. 41 July-August 1950
   Castle, W.E. and F.H. Smith. "Silver Dapple, A Unique Coat Color Variety Among Shetland Ponies," pp. 139-45.
Journal of Heredity. Vol. 41 No. 12 December 1950
   Dimock, W.W. "Wobbles — An Heredity Disease in Horses," pp. 319-23.
Journal of Herdity. Vol. 42 January-February 1951
   Castle, W.E. "Dominant and Recessive Black in Mammals," pp. 48-49.
   Castle, W.E. and F.L. King. "New Evidence of the Genetics of the Palomino Horse," pp. 61-64.
Journal of Heredity. Vol. 42 November-December 1951
   Castle, W.E. "Genetics of the Color Varieties of Horses," pp. 297-99.
Journal of Heredity. Vol. 51 May-June 1960
   Castle, W.E. and W.R. Singleton. "Genetics of the 'Brown' Horse," pp. 127-30.
Journal of Heredity. Vol. 51 November-December 1960
   Castle, W.E. "Fashion in the Color of Shetland Ponies and Its Genetic Basis," pp. 247-48.
Journal of Heredity. Vol 52 May-June 1961.
   Castle, W.E. "Genetics of the Claybank Dun Horse," pp. 121, 128.

Journal of Heredity. Vol. 57 May-June 1966
 Singleton, W.R. and Q.C. Bond. "A Allele Necessary for Dilute Coat Color in Horses," pp. 75-77.
Journal of Heredity. Vol. 60 March-April 1969
 Pulos, W.L. and F.B. Hutt. "Lethal Dominant White in Horses," pp. 59-63.
Journal of Heredity. Vol. 63 March-April 1972
 Smith, A.T. "Inheritance of Chin Spot Markings in Horses," p. 100.
Journal of Heredity. Vol 65 January-February 1971
 Adalsteinsson, S. "Inheritance of the Palomino Color in Icelandic Horses," pp. 15-20.
Journal of Heredity. Vol. 66 No. 6 November 1975
 Gardner, E.J., J.L. Shupe, N.C. Leone and A.E. Olson. "Hereditary Multiple Exostosis, A Comparative Genetic Evaluation in Man and Horses," pp. 318-22.
Journal of Reproduction and Fertility. Suppl. Vol. 23 1975
 Bouters, R., M. Vandeplassche and A. Moor. "An Intersex (Male Pseudohermaphrodite) Horse with 64 XX/XXY Mosaicism," pp. 375-76.
 Chandley, A.C., J. Fletcher, P.D. Rossdale, C.K. Peace, S.W. Ricketts, R.J. McEnery, J.P. Thorne, R.V. Short and W.R. Allen. "Chromosome Abnormalities as a Cause of Infertility in Mares," pp. 377-83.
 Hughes, J.P., K. Benirschke, P.C. Kennedy and A.T. Smith. "Gonadal Dysgenesis in the Mare," pp. 385-90.
 McFeely, R.A. "A Review of Cytogenetics in Equine Reproduction," pp. 371-74.
 Short, R.V. "The Evolution of the Horse," pp. 1-6.
 Rossdale, P.D. and D. Leadon. "Equine Neonatal Disease: A Review," pp. 685-90.
Journal of the South African Veterinary Association. Vol 45 No. 4 1974
 Osterhoff, D.R., M.A.J. Azzie and J. OP'T. Hof. "Biochemical Genetics and Performance Ability in Horses," pp. 311-16.
Kalmus, H. **Genetics.** Bungay, Suffolk: Richard Clay and Company, LTD., 1965.
Kays, D.J. and J.M. **The Horse.** New York: ARCO Publishing Co., 1977.
Kirg, R.C. ed. **Handbook of Genetics.** Vol. 4. New York: Plenum Press, 1975. pp. 337-50.
Lasley, J.F. **Genetics of Livestock Improvement.** Englewood Cliffs, N.J.: Prentice-Hall, 1972.
Lasley, J.F. **Genetic Principles in Horse Breeding.** Houston: Cordovan Corp., 1976.
Leicester, Sir Chas. **Bloodstock Breeding.** London: J.A. Allen & Co., LTD., 1969.
Lerner, D.J. "Dental Care and Cavaties — or, How Often Should You Brush Your Horse's Teeth?" **Equine Health.** University of Illinois, College of Veterinary Medicine, Extension.
Levitan, M. **Textbook of Human Genetics.** 2nd ed. New York: Oxford University Press, 1977.
Louisiana Veterinarian. Vol. 4 No. 2
 Roberts, S.J. "Infertility in Mares," pp. 4-9.
Lush, J.L. **Animal Breeding Plans.** 3rd ed. Ames, Iowa: Iowa State University, 1945.
Martin, T.W. (Personal Communication)
McDonald, L.E. **Veterinary Endocrinology and Reproduction.** Philadelphia: Lea and Febiger, 1975.
Merck Veterinary Manual. 4th ed. Rahway, N.J.: Merck and Company, Inc., 1973.
Miller, R.W. **Appaloosa Coat Color Inheritance.** Undated Study.
Modern Veterinary Practice. Vol. 44 No. 5 1963
 Manning, J.P. "Equine Hip Dysplasia - Osteroarthritis," p. 44-45.
Montgomery, E.S. **The Thoroughbred.** New York: ARCO Publishing Co., 1973.
Napier, M. **Blood Will Tell.** London: J.A. Allen & Co., LTD., 1977.
New Zealand Veterinary Journal. Vol. 17 No. 8 1967
 Bain, A.M. "Foetal Losses During Pregnancy in the Thoroughbred Mare."
Pirchner, F. **Population Genetics in Animal Breeding.** San Francisco: W.H. Freeman and Co., 1969.
Proceedings of the American Association of Equine Practitioners Symposium on Equine Hematology.
 W.O. Kester, ed. Golden, Co.: The American Association of Equine Practitioners, 1975.
 Johnson, J.H., H.E. Garner and D.P. Hutcheson. "Some Coagulation Aspects of Epistaxis in the Conditioned Thoroughbred," pp. 561-63.
 McGuire, T.C., K.L. Banks and M.J. Poppie. "Immunodeficiency Disorders of Foals," pp. 190-92.
 Smith, A.T., C. Stormont and Y. Suzuki. "Alloantibodies: Their Role in Equine Neonatal Isoerythrolysis," pp. 349-53.
 Suzuki, Y., C. Stormont and A.T. Smith. "Alloantibodies: The Blood Groups They Define," pp. 34-41.
 White, J.G., K.L. Matlack, D. Mundschenk and G.H.R. Rao. "Platelet Studies in Normal and a Bleeder Horse," pp. 209-21.
Proceedings of the 1962 Annual Convention of the American Association of Equine Practitioners. 1963.
 Churchill, E.A. "Predisposing Factors to Lameness in Standardbreds," pp. 191-94.
Proceeding of the 1967 Annual Convention of the American Association of Equine Practitioners. Vol. 13
 Schebitz, H. and E. Dahme. "Spinal Ataxia in the Horse," pp. 133-38.
 Sponseller, M.L. "Equine Cerebellar Hypoplasia and Regeneration," pp. 123-26.
Proceedings of the 1968 Annual Convention of the American Association of Equine Practitioners. Vol. 14
 Prickett, M.E. "Equine Spinal Ataxia," pp. 147-51.

Proceedings of the 18th Annual Convention of the American Association of Equine Practitioners. F.J. Milne, ed. 1972.

Stormont, C. "Current Status of Equine Blood Groups and Their Applications," pp. 401-410.

Proceedings of the 19th Annual Convention of the American Association of Equine Practitioners. F.J. Milne, ed. 1973.

Johnson, J.H., H.E. Garner, D.P. Hutcheson and J.G. Merrian. "Epistaxis," pp. 115-20.

Proceedings of the 21st Annual Convention of the American Association of Equine Practitioners. F.J. Milne, ed. 1975.

Hamlin, R.L. "Lasix and Epistaxis in Horses," pp. 277-80.

Rurohit, R.C., J.M. Humburg, P.A. Teer, G.L. Norwood and R.F. Nachreiner. "Evaluation of Hypertension Factors in the Horse," pp. 43-49.

Proceedings of the 22nd Annual Convention of the American Association of Equine Practitioners. F.J. Milne, ed. 1976.

Stannard, A.A. "Equine Dermatology," pp. 273-92.

Proceedings of the 1963 Convention of the American Veterinary Medical Association.

Mackay-Smith, M.P. and C.W. Raker. "Mechanical Defects of the Stifle," pp. 82-85.

Proceedings of the 6th Annual Horse Short Course. November 6-8, 1966. Texas A&M University, College Station, Texas.

Banks, W.C. "Study of Epiphyseal Maturation in the Horse," p. 83.

Bullard, T.L. "Bowed Tendons," p. 90.

Cartwright, T.C. "Inbreeding and Outbreeding," pp. 50-56.

Flynt, M and B. Welch. "The Importance to the Breeder of Developing an Outstanding Cutting Horse," pp. 9-11.

Pumphrey, J. and T. Wells. "The Importance to the Breeder of Developing an Outstanding Race Horse," pp. 12-16.

Wolff, W.A. "Spavin," pp. 88-89.

Proceedings of the 7th Annual Horse Short Course. 1967. Texas A&M University, College Station, Texas.

Kieffer, N.M. "Guidelines to Consider in Choosing a Mate for Your Mare."

Proceedings of the 10th Annual Horse Short Course. November 20-21, 1970. Texas A&M University, College Station, Texas.

Burns, S.J. "Why She Won't Breed," pp. 39-41.

Proceedings of the 11th Annual Horse Short Course. 1971. Texas A&M University, College Station, Texas.

Kieffer, N.M. "Heretical Views on Some Theories of Horse Breeding," pp. 52-61.

Proceedings of the 14th Annual Horse Short Couse. (Texas Animal Agricultural Conference) January 21-24, 1974. Texas A&M University, College Station, Texas.

Householder, D.D. "The Genetics of Equine Coat Color," pp. 99-116.

Kieffer, N.M. "Methods of Estimating the Breeding Values of Potential Sires and Dams," pp. 95-98.

Proceedings of the 16th Annual Horse Short Course. April 8-9, 1976. Texas A&M University, College Station, Texas.

Atkins, D.T. "The Mare's Reproductive Cycle," pp. 5-11.

Proceedings of the 11th European Conference on Animal Blood Groups and Biochemical Polymorphism. Warsaw: The Hague PWN-Polish Scientific Publishers. July 2-6, 1968.

Osterhoff, D.R., D.O. Schmid and I.S. Ward-Cox. "Blood Group and Serum Type Studies in Basuto Ponies." p. 453.

Proceedings of the First International Symposium on Equine Hematology. W.O. Kester, ed. Golden, Co.: The American Association of Equine Practitioners, 1975.

Gerber, H., P. Tschudi and R. Straub. "Review of Serum Enzyme Activities in Equine Diseases," p. 417.

Kitchen, H. D. Boreson, S. Malkin and Il Brett. "Horse Hemoglobin," p. 42.

Schmid, D.O., S. Cwik and M. Emerich. "Lymphocyte Antigens of the Horse," p. 120.

Suzuki, Y., C. Stormont and A. Trommershavsen-Smith. "Alloantibodies: The Groups They Define," p. 34.

Proceedings of the Fourth International Conference on Equine Infectious Diseases. J.T. Bryans and H. Gerber, eds. Princeton, N.J.: Veterinary Publications, Inc., 1978.

Scott, A.M., "Immunogenetic Analysis as a Means of Identification in Horses," p. 259.

Proceedings of the 1st National Horsemen's Seminar. 1976. Virginia Horse Council, Inc., Fredericksburg, Virginia.

Farrell, R.K. "Horse Identification," p. 148.

Beech, J. "The Equine Eye," pp. 186-88.

Proceedings of the 2nd National Horsemen's Seminar. 1977. Virginia Horse Council, Inc., Fredericksburg, Virginia.

Squires, E.L. "Reproductive Physiology of the Mare," pp. 66-73.

Progress in Equine Practice. Vol. 1. E.J. Catcott and J.F. Smithcors, eds. Wheaton, Illinois: American Veterinary Publicaions, 1966.

Mahaffey, L.W. "Bacterial Diseases of the Bronchi, Bronchioles and Lungs," p. 161.

Progress in Equine Practice. Vol. 2. E.J. Catcott and J.F. Smithcors, eds. Wheaton, Illinois: American Veterinary Publications, 1970.

Boucher, W.B., G.A. Elliot and B. Schmucker. "Epistaxis Due to Rupture of an Aneurysm," pp. 273-74.

Cook, W.R. "Eitology of Epistaxis and Pharyngeal Paralysis," pp. 362-63.

Cross, R.S.N. "Etiology of Periodic Opthalmia," pp. 394-95.

Ommert, W.D. "Upper Respiratory Tract Problems," p. 361.

Rubin, L.F. "Cysts of the Iris," pp. 395-96.

Rapidan River Farm Digest. J.C. Hagan, ed. Winter 1976. Hintz, H.F. and R.L. Hintz. "I Bet on the Bay," pp. 107-08.

Jones, W.E. "A Good Breeding Program Needs a System," p. 21.

Smith, A.T. "Lethal Genes in Horses — A Breeder's Dilemma," pp. 131-34.

Rice, V.A. **Breeding and Improvement of Farm Aniamls.** New York: McGraw-Hill Publications, 1970.

Roberts, S.J. **Veterinary Obstetrics and Genital Diseases.** Ann Arbor, Michigan: Edwards Brothers., 1971.

Rooney, J.R. **The Lame Horse — Causes, Symptoms and Treatment.** Hollywood: Wilshire Book Company, 1976.

Rossdale, P.D. and S.M. Wreford. **The Horse's Health From A to Z.** New York: ARCO Publishing Co., 1974.

Rossdale, P.D. and S.W. Ricketts. **The Practice of Equine Stud Medicine.** Baltimore: The William and Wilkins Co., 1974.

Schalm, O.W., N.C. Jain and E.J. Carroll. **Veterinary Hematology.** Philadelphia: Lea and Febiger, 1975.

Science. Vol. 148 April 16, 1965

Benirschke, K., N. Malouf and R.J. Low. "Chromosome Complement: Differences Between Equus Caballus and Equus Przewalski Poliakoff," pp. 382-83.

Science. Vol. 151 January 1966

Koulischer, L. and S. Frechkop. "Chromosome Complement: A Fertile Hybrid Between Equus Przewalski and Equus Caballus," pp. 93-95.

Searle, A.G. **Comparative Genetics of Coat Color in Mammals.** New York: Lagos Press, Acad. Prss, 1968.

Singleton, W.R. **Elementary Genetics.** D. Van Nostrand Co., Inc., Princeton, 1968.

Sinnott, E.W., L.C. Dunn and J. Dobzhansky. **Principles of Genetics.** 5th ed. New York: McGraw-Hill Book Co., 1958.

Smith, H.A., T.C. Jones and R.O. Hunt. **Veterinary Pathology.** 4th ed. Philadelphia: Lea and Febiger, 1972.

Smith, J.D. **Horse Markings and Coloration.** New York: A.S. Barnes and Co., 1977.

Smythe, R.H. **The Mind of the Horse.** Brattleboro, Vermont: The Stephen Greene Press, 1965.

Snyder, L.H. **The Principles of Heredity.** Dallas: D.C. Heath and Co., 1935.

Stansfield, W.D. **Genetics.** New York: McGraw-Hill, 1969.

Steel, R.G.D. and J.H. Torrie. **Principles and Procedures of Statistics.** New York: McGraw-Hill, 1960.

Stud Manager's Handbook. M.E. Ensminger, ed. Clovis, California: Agriservices Foundation, 1974. Vol. 10.

Butterfield, R.M. "The Anatomy of Sleeping Standing Up," p. 68.

Killian, M. "The Syndication of Stallions," p. 186.

Lasley, J.F. "Are Inbreeding and Linebreeding Always Detrimental?" pp. 42-46.

Lasley, J.F. "Interaction of Heredity and Nutrition in Animals," pp. 202-207.

Rossdale, P.D. "Equine Reproduction — Infertility," pp. 15-17.

Rossdale, P.D. "The Evolution and Anatomy of the Horse — An Introduction," p. 200.

Stewart, D.D. "Horse Psychology Applied to Training," pp. 113-115.

Stud Manager's Handbook. M.E. Ensminger, ed. Clovis, California: Agriservices Foundation, 1975. Vol. 11.

Butterfield, R.M. "Male Genitalia," pp. 29-33.

Butterfield, R.M. "Structure and Function of the Hoof." p. 24.

Fraser, C.R. "How a Horse Thinks, Sees and Moves," pp. 129-133.

Myers, V.S. "Important Equine Reproductive Anatomy," pp. 34-35.

Ricketts, S.W. "Conception and Gestation in the Mare," pp. 1-4.

Ricketts, S.W. "The Diagnosis of Infertility in the Mare," pp. 10-14.

Taysom, E.D. "Selection of Horses," pp. 15-19.

Stud Manager's Handbook. M.E. Ensminger, ed. Clovis, California: Agriservices Foundation, 1977. Vol. 13.

Jeffcott, L.B. "Abortion in the Mare — Causes and Prevention," pp. 18-22.

Poppie, M.J. "Congenital Abnormalities of the Foal," pp. 31-33.

Potter, G.D. and B.F. Yeates. "Behavior Modification and Training of Horses," pp. 182-190.

Taysom, E.D. "More Efficiency in Horse Breeding," pp. 1-4.

Stud Manager's Handbook. M.E. Ensminger, ed. Clovis, California: Agriservices Foundation, 1978. Vol. 14.

Kirk, M.D. "Broodmares and Reproduction," pp. 126-129.

Kirk, M.D. "Unsoundness Related to Conformation," pp. 156-164.

Mackay, A. "Conformation of the Limbs of a Horse," pp. 153-155.

Mackay, A. "Conformation and Faults of the Body of a Horse," pp. 114-117.

Swenson, M.J. **Duke's Physiology of Domestic Animals.** 8th ed. Ithaca: Comstock Publishing Associates, 1970.

Tesio, F. **Breeding the Race Horse.** London: J.A. Allen and Co., 1958.

The Principles of Improved Horse Breeding. R. Albaugh. Berkeley, California: The University of California, USDA, 1977.

Thomas, H.S. **Horses: Their Breeding, Care and Training.** New York: A.S. Barnes and Co., 1974.

Thoroughbred of California. Vol. 33 No. 6 1961
 Britton, J.W. "Conformation and Lameness," p. 502.
Thoroughbred Record. Vol. 198 July 7, 1973.
 "Inheritance of Racing Ability in the Thoroughbred,"
Thoroughbred Record. Vol. 203 January 10, 1976.
 Couturie, J. "An Experiment in Inbreeding White Horses," pp. 127,180.
Thoroughbred Record. Vol. 204 November, 1976
 Zediker, J. "A Study in Grey," pp. 1534-1537.
Thoroughbred Record. October 26, 1977.
 Pratt, G.W. "Remarks on Lameness and Breakdown," pp. 1486-1489.
Thoroughbred Record. Vol. 206 No. 19 November 9, 1977.
 Kingsbury, E.T. "The Inheritance and Importance of Blood Transferrin Types in the Thoroughbred," pp. 1702-1705.
Veterinary Clinics of North America. (Symposium on Reproductive Problems) Vol. 7 No. 4. Philadelphia: W.B. Sanders Company, 1977.
 Rhoades, J.D. and C.W. Foley. "Cryptorchidism and Intersexuality," pp. 789-794.
Veterinary Medicine. Vol. 56 No. 6 1961
 "Bowed Tendons in the Horse," pp. 251-252.
Veterinary Record. Vol. 16 No. 8 1936
 Buckingham, J. "Hermaphrodite Horse," p. 218.
Veterinary Record. Vol. 53 No. 3 January 18, 1941
 Runciman, B. "Roaring and Whistling in Thoroughbred Horses," pp. 37-38.
Veterinary Record. Vol. 69 No. 49
 Mahaffey, L.W. and P.D. Rossdale. "Convulsive and Allied Syndromes in Newborn Foals," pp. 1277-1289.
Veterinary Record. Vol. 72 No. 31 July 30, 1960.
 Palmer, A.C. and J. Hickman. "Ataxia in a Horse Due to an Angioma of the Spinal Cord," pp. 611-613.
Veterinary Record. Vol. 73 No. 14 April 18, 1961
 Archer, R.K. "True Haemophilia (Haemophilia A) in a Thoroughbred Foal," pp. 338-340.
Veterinary Record. Vol. 74 No. 14 1962
 Jogi, P. and E. Norberg. "Malformation of the Hip Joint in a Standardbred Horse," pp. 421-422.
Veterinary Record. Vol. 78 No. 1 January 1, 1966.
 Cross, R.S.N. "Equine Periodic Ophthalmia," pp. 8-13.
Veterinary Record. Vol. 78 No. 18 April 30, 1966
 Fraser, H. "Two Different Types of Cerebellar Disorder in the Horse," pp. 608-612.
Veterinary Record. August 1969.
 White, D.J. and D.A. Farebrother. "A Case of Intersex in a Horse," p. 203.
Veterinary Record. December 23, 1972
 Archer, R.K. and B.V. Allen. "True Haemophilia in Horses," pp. 655-656.
Veterinary Record. Vol. 88 No. 7.
 Cook, W.R. and A.C. Palmer. "Arab Cerebellar Disease," p. 200.
Veterinary Reference Service Update 1977. Large Animal Edition.
 J.F. Smithcors, ed. Santa Barbara: American Veterinary Publications, 1977.
Veterinary Scope. Vol. 7 No. 1 1962
 Adams, O.R. "Lamenesses of Rodeo Horses," pp. 2-16.
Wagoner, D.M. ed. **Veterinary Encyclopedia for Horsemen.** Equine Research Publications, 1975.
Wagoner, D.M. ed. **Veterinary Treatments and Medications for Horsemen.** Dallas: Equine Research Publications, 1977.
Wall, J.F. **Breeding Thoroughbreds.** New York: Charles Scribner's Sons, 1946.
Wall, J.F. **A Horseman's Handbook.** Washington, D.C.: American Remount Association, 1945.
Way, R.F. and D.G. Lee. **The Anatomy of the Horse.** Philadelphia: J.B. Lippincott Co., 1965.
Willett, P. **An Introduction to the Thoroughbred.** London, England: Stanley Paul and Co., LTD., 1975.
Willis, L.C. **The Horse Breeding Farm.** New York: A.S. Barnes and Co., 1973.
Willoughby, D.P. **The Empire of Equus.** New York: A.S. Barnes and Co., 1974.
Wriedt, C. **Heredity in Livestock.** London: Macmillan and Co., LTD., 1930.
Zivocisna Vyroba. Vol. 10 1965.
 Dusek, J. "The Heritability of Some Characters in the Horse," pp. 449-456.

APPENDIX

1. Paternity Tests
2. Relationhips Between Blood Types and Performance Ability
3. Comparative Analysis of Plasma Protein in Various Breeds
4. Congenital Defects in Foals
5. Comparison Between Horse and Ass
6. Points of the Horse
7. Skeleton of the Horse

APPENDIX 1
PATERNITY TESTS

Paternity tests are used to solve questions of true parentage. Cases usually involve two or more possible sires for one foal (e.g., mare serviced by more than one stallion, or wrong stallion or mare brought to breeding shed) but may also include accidental exchange of foals between mares. Also, errors in breeding or registration records, which confuse a foal's true ancestry, can sometimes be detected through paternity testing.

Although parentage tests can involve eliminating possible parents on the basis of unrelated color patterns, careful analysis of inherited blood characters is far more efficient and accurate. Because of the many complicating factors involved in color inheritance, color tests introduce a greater possibility of error (i.e., a greater chance of excluding the true parent). By process of elimination, blood typing can solve at least 70% of all paternity cases involving two possible sires. In either blood typing or color testing, the principle of *genetic exclusion* must be used. Genetic exclusion simply means that test results cannot prove a horse to be the true parent; they can only show which horse could not possibly be the sire or dam in question.

COLOR TESTING

The complex nature of coat color inheritance (refer to **"Coat Color and Texture"**) explains why the expression of color may reflect very little of the horse's actual genotype. In some instances, it is obvious that a foal could not be the offspring of a certain color horse. In other cases, coat color differences may be misleading.

Without exception, the dominant white horse must have at least one dominant white parent. Because of its highly epistatic and dominant nature, the allele for white cannot be present in the horse's genotype without being expressed. (Its presence in the homozygous state is lethal.) Similarly, every horse with the epistatic-dominant G allele at their grey locus will be grey, regardless of any other genetic influences (except dominant white). For this reason, at least one of the parents of a greying foal must be grey (or, in some cases, may be a dominant white carrying the grey gene). The above statement is also true of a roan foal; at least one parent will be roan or, in rare cases, could be dominant white carrying the roan gene.

When coat color is used as a determinant, extensive investigation may be required. All possible complicating factors should be considered. For example, the multiple allelic restriction gene, A, permits a bay horse to produce bay, brown, or black offspring. The palomino foal's sire or dam must have contributed the dilution allele c^{cr}, but the parent may have expressed this allele as palomino, cremello, perlino, buckskin or dun. The expression may have also been hidden by an epistatic gene (e.g., grey, dominant white, etc.). These interactions between different loci, and contradicting terminology for coat colors, make it difficult to exclude one horse as a possible parent strictly on the basis of coat color. Although a thorough understanding of coat color inheritance lessens the possibility of error, color tests are not as accurate as blood tests in solving disputed parentage cases.

BLOOD TYPING

Blood typing is the identification of inherited blood characters which distinguish one individual from the next. (Only identical twins could have identical blood types.) As a paternity test, blood typing has been supported for several important reasons:

1. The blood types are present at birth and do not change throughout the life of the individual. They are not affected by environmental influences as are many other inherited characteristics.

2. The genes which determine blood type often follow simple patterns of inheritance. In many cases, the blood type alleles are codominant. Codominance allows the phenotypic expression (blood type identified through laboratory tests) to indicate the exact controlling alleles. In a four allelic system (A,B,C,D), each codominant combination of alleles would be expressed as a unique blood type (AA,AB,AC,AD,BB,BC,BD, CC,CD,DD). In these cases, tests which identify the phenotype also indicate the corresponding genotype.

3. The multitude of known alleles for several blood type loci decreases the possibility of excluding the true parent. Researchers have suggested that there could be over one million possible blood type *genotypes* among the various breeds.

4. The ability to identify each character through objective laboratory analysis increases the accuracy of the tests. Because the tests are subject to human error, however, qualified technicians and systematic procedures should be employed. (Disputed parentage cases are often handled by state veterinary diagnostic labs or indirectly through breed registries.)

Composition of Blood

Keeping the complex physical balances intact, protecting the body from invading organisms, and stopping the flow of blood from an injured area are only a few of the vital roles played by the components of blood. Hematology (the study of blood) and the technical aspects of identifying inherited blood factors are beyond the scope of this discussion. (The interested student should refer to related hematology or biochemical genetics texts.) A brief description of the basic components of blood provides sufficient background information for an analysis of inherited equine blood types.

1. Red blood cells (erythrocytes) carry the genetically determined protein,

hemoglobin, which is responsible for transporting oxygen to, and carbon dioxide from, the cells. (The presence of oxygen in hemoglobin is responsible for the deep red color of oxygenated blood.) These blood cells also carry inherited antigenic factors ("blood factors") which trigger the formation of detectable antibodies when injected into certain test solutions.

2. White blood cells (leucocytes) function primarily as infection fighters. These cells are less numerous than the red blood cells, except in cases of disease or cancer. Research indicates that inherited antigenic factors may also be located on the white blood cell surfaces.

3. Platelets (thrombocytes) play an important role in the clotting mechanism. (Refer to the discussion on hemophilia under **"Inherited Abnormalities."**) Studies also show that inherited antigenic factors may be located on the white blood cell surfaces.

4. Plasma is the suspension fluid for the red blood cells, white blood cells, platelets, blood proteins, and enzymes. The various plasma protein and enzyme types are also genetically controlled.

Blood Factors

Inherited antigenic factors located on the surface of the red blood cell are responsible for the hemolytic reaction in foals with neonatal isoerythrolysis. (Refer to "Neonatal Isoerythrolysis" under **"Inherited Abnormalities."**) These red blood cell antigenic factors were the first inherited blood characters to be identified. Eight different systems (loci) are involved, and the controlling alleles for each have been studied extensively. The interaction between corresponding alleles seems to be simple dominance, so that the presence of one allele may override the expression of another. For this reason, detection of the red cell antigens should be only one part of a complete blood typing parentage test. Several factors have been identified in the horse:

RED BLOOD CELL ANTIGENS*

System (Locus)	Controlling Alleles**	Genotypes (Possible allelic combinations)	Commonly Used Blood Typing Factors (phenotypes)
A	$A^a, A^b, A^c, A^{bc}, A^-$	15	Aa, Ab, Ac
C	C^a, C^-	3	Ca
D	$D^{bc}, D^c, D^{ce}, D^{cef}, D^d, D^{ad}, D^{de}, D^{df}$	36	Da, Db, Dc, Dd, De, Df
K	K^a, K^-	3	Ka
P	P^a, P^b, P^-	6	Pa, Pb
Q	Q^a, Q^R, Q^{RS}, Q^S	10	Qa, R, S
U	U^a	3	Ua

*According to A.M. Scott
**Although many red cell groups have been reported, only 15 are commonly used in blood tests.

Plasma Proteins and Enzymes

Plasma proteins are responsible for a complex balance of life-sustaining functions within the body. For example, transferrins and albumins transport important minerals, gamma globulins provide antibodies, and special enzymes control the coagulation mechanism. Other plasma proteins contribute to the structure of body tissue (e.g., muscle), while others influence the stability (pressure balance) of blood cells. Studies have revealed that there are many types of plasma proteins. By virtue of codominance, most of these forms indicate the exact identity of their two controlling alleles. If test results show that a horse's albumin is type FS, technicians know that he carries the alleles Alb^F and Alb^S at his albumin locus. Known characters for the protein typing complex* are listed:

Protein/Enzyme	Locus	Alleles	Phenotypes
Serum			
Albumin	A1	3	6
Transferrin	Tf	10	55
Prealbumin	Pr	9	45
Prealbumin	Xk	3	6
Postalbumin	Pa	2	3
Ceruloplasmin	Cp	2	3
Esterase	Es	7	22
Cholinesterase	Ch	4	9
Red-Cells			
Hemoglobin	Hb	2	3
Carbonic Anhydrase	CA	5	15
6 Phosphogluconate Dehydrogenase	PGD	3	6
Phosphoglucomutase	PGM	3	6
Phosphohexose Isomerase	PHI	3	6
Catalase	Cat	2	3
Acid Phosphatase	AP	2	3
NAD Diaphorase	Dia	2	3

*According to A.M. Scott

Continuing Studies

Although research suggests that blood platelets, leucocytes, gamma globulins, sperm and body tissue cells may also carry inherited factors, the identification of these factors (in horses) is not yet complete. Therefore, these characters are not yet applied to routine parentage tests.

APPENDIX 2
RELATIONSHIPS BETWEEN BLOOD TYPES AND PERFORMANCE ABILITY

Studies in biochemical genetics have recently suggested that superior racing performance may be associated with certain blood types. Further research may provide the horseman with an objective selection technique - selecting breeding stock on the basis of desirable blood types.

Preliminary studies* with Thoroughbreds have indicated that the following blood types show a higher frequency in the genotypes of outstanding race horses:

Transferrin	FF
Albumin	FS
Prealbumin	LL
Postalbumin	SS
Esterase	II
Phosphoglucomatase	FS
6-phosphogluconate dehydrogenase	SS

*Kingsbury, Osterhoff, Azzie and Op't Hof.

APPENDIX 3
COMPARATIVE ANALYSIS OF PLASMA PROTEIN FREQUENCIES IN VARIOUS BREEDS

The frequencies of inherited blood types are similar in horses of the same breed. Although breeds can be characterized by their blood types, genetic variation restricts categorizing individual horses into certain breeds strictly on this basis.

Blood type differences reflect the origin and history of the various breeds. Isolation, migration, interbreeding and indirect selection pressures can affect the frequency of certain blood types within one breed as compared to another. Breeds with a common heritage may show similar blood type frequencies. Thoroughbreds, for example, originated from crosses between British broodmare stock and Arabian or Barb stallions. This common ancestry is illustrated by average transferrin and esterase frequencies:

TRANSFERRIN ALLELES*
(average occurrence in %)

	D	F	H	O	R
Thoroughbreds	.30	.49	.04	.10	.06
Arabians	.30	.41	.17	.12	.00

*According to Kingsbury

ESTERASE ALLELES*
(average occurrence in %)

	Es^F	Es^I	Es^S	Es^{X_1}	Es^{X_2}	Es^0
Thorough-breds	0	0	.987	.022	0	0
Arabians	.037	0	.926	0	.037	0

*According to Osterhoff, Schmid and Ward-Cox

Selection for speed-related conformation traits in the Thoroughbred has caused distinct differences in the physical appearance between the two breeds. Because blood types have never been selected directly, similarities have not been significantly altered. (Slight changes may have been caused by selection for speed, since blood types and racing performance are believed to be correlated.)

Breeds which have been isolated for centuries (such as the Icelandic Pony) show independent blood type frequencies. The development of the same breed in separate countries also results in distinct blood type differences. Average transferrin frequencies for various breeds are listed in the table below:

BREED	TRANSFERRIN ALLELE* (frequency/breed in %)					
	D	F	H	M	O	R
Thoroughbred (U.S.A.)	.27	.56	.03	0	.09	.05
Thoroughbred (South Africa)	.26	.55	.04	0	.07	.07
Thoroughbred (Netherlands)	.30	.44	.04	0	.13	.09
Thoroughbred (Italy)	.41	.39	.03	0	.16	.01
Thoroughbred (Hungary)	.31	.47	.02	0	.10	.10
Thoroughbred (Belgium)	.31	.49	.07	0	.09	.04
Lippizaner (Austria)	.17	.23	.10	0	.44	.06
Lippizaner (Hungary)	.07	.51	.26	0	.15	.01
Shetland Pony	.17	.46	.03	.03	.11	.20
Icelandic Pony	.20	.27	.07	.01	.25	.20
Tarpan	.21	.62	.04	0	.07	.06

*According to Kingsbury, Bengtsson, Gahne, Rendel, Tomaszewska-Guszkiewicz and Zurkowski

APPENDIX 4
CONGENITAL DEFECTS IN FOALS

Congenital defects (abnormalities that are present at birth) may be caused by genetic and/or environmental factors. Inborn abnormalities that are thought to be inherited have been discussed in preceding chapters. Because some congenital defects are rare, research on their causing factors has been limited. These abnormalities have not been examined within the text, but are included in the following list of known congenital defects of the foal.

DIGESTIVE SYSTEM

parrot mouth
bulldog mouth
periodontitis
supernumerary teeth
cleft palate
distended guttural pouch
esophageal defects
atresia ani
atresia recti, atresia coli
white foal syndrome

CIRCULATORY SYSTEM

neonatal isoerythrolysis
hemophilia A (factor VIII deficiency)
interventricular septal defect
patent ductus arteriosus
patent foramen ovale
persistent right aortic arch
tetralogy of Fallot

RESPIRATORY SYSTEM

pharyngeal cysts

IMMUNE SYSTEM

combined immunodeficiency (CID)
B-lymphocyte deficiency
IgM deficiency

URINARY SYSTEM

patent urachus
one kidney missing

SKIN

dentigerous cyst
warts
malignant melanoma
epitheliogenesis imperfecta
hepatogenous photosensitization

EYE

nuclear cataracts
Y-type cataracts
persistent hyaloid vessel
aniridia
iridial heterochromia
coloboma iridis
iris cysts
hyperplasia of the corpora nigra
detached retina
optic nerve hypoplasia
night blindness
microcornea
corneal opacities (congenital keratopathy, corneal dermoids)
melanosis of the cornea
anophthalmia
microphthalmia
atresia of the nasolacrimal duct
entropion
ectropion
glaucoma

ENDOCRINE SYSTEM

goiter

NERVOUS SYSTEM
wobbler syndrome
occipito-atlanto-axial-malformation
cerebellar hypoplasia
cerebellar ataxia
Oldenburg ataxia
internal hydrocephalus
idiopathic epilepsy

REPRODUCTIVE SYSTEM
tipped vulva (pneumovagina)
cryptorchidism
intersex:
  male pseudohermaphroditism
  testicular feminization
63 XO aneuploidy

MUSCULOSKELETAL SYSTEM
contracted foal
contracted heel
contracted digital flexor tendons
weak digital flexor tendons
carpal joint deviations
upward fixation of the patella
lateral fixation of the patella
fibular enlargement
polydactyly
abrachia
multiple exostosis
hip dysplasia
osteochondritis dissecans
lordosis
scoliosis
kyphosis
torticollis
Mohammed's Thumbprint
Shistosoma reflexum

APPENDIX 5
COMPARISON BETWEEN HORSE AND ASS

STRUCTURE	HORSE	ASS
Chestnuts	usually four	usually on forelegs only; thin
Ergots	usually four	usually on forelegs only
Tail	full from dock	tufted (like ox)
Mane	full, pendant	erect, no forelock
Lumbar vertebrae	usually six	usually five
Lacrimal duct opening	inner wing of nostril	outer wing of nostril
False nostril	less extensive	extends higher up
Teats in male	absent	two rudimentary
Voice	neighing	braying, involves abdominal muscles, accessory structure on epiglottis
White markings	frequent	rare
Coat color	many variations	conservative, wild type
Croup	usually lower than withers	usually higher than withers
Dock	thick, strong, short	thinner, weaker, longer
Feathering	occurs at fetlocks	none
Gestation period	11 months	12 months
Hooves	rounded, wide	oblong, narrower
Teeth	incisors wide, angle becomes more acute with age	narrow, upright incisors
Ears	short	long

Although the domestic ass is a close relative of the modern horse, the two species differ in many ways. Differences can be attributed to evolutionary isolation and natural selection.

APPENDIX 6 POINTS OF THE HORSE

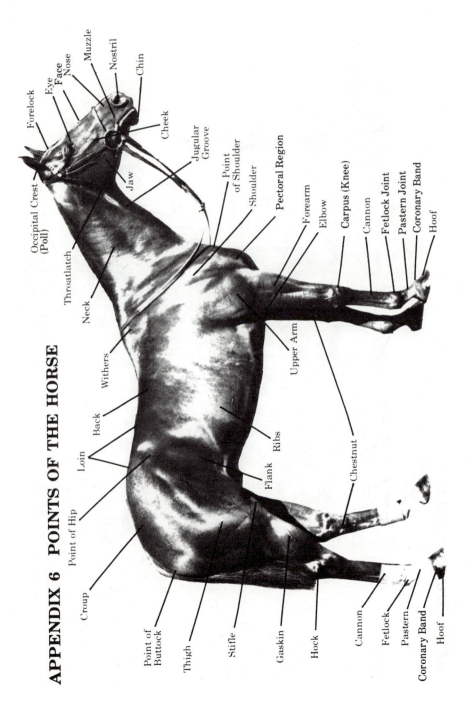

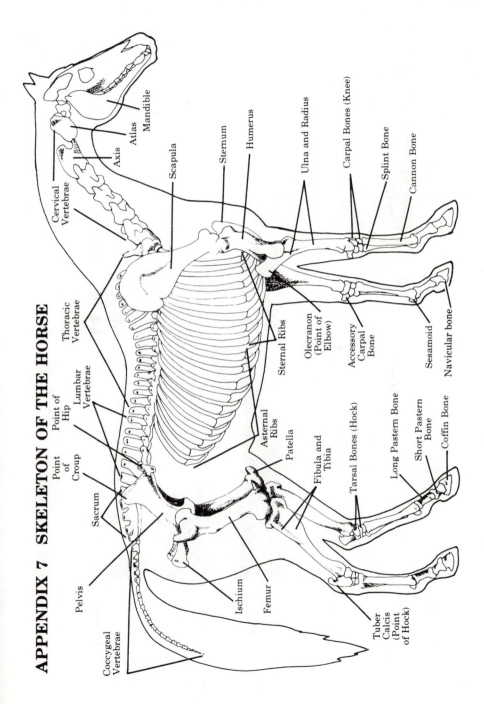

APPENDIX 7   SKELETON OF THE HORSE

Mandible
Atlas
Axis
Cervical Vertebrae
Scapula
Sternum
Humerus
Ulna and Radius
Carpal Bones (Knee)
Splint Bone
Cannon Bone
Olecranon (Point of Elbow)
Accessory Carpal Bone
Sesamoid
Navicular bone
Sternal Ribs
Asternal Ribs
Patella
Fibula and Tibia
Tarsal Bones (Hock)
Long Pastern Bone
Short Pastern Bone
Coffin Bone
Tuber Calcis (Point of Hock)
Femur
Ischium
Pelvis
Coccygeal Vertebrae
Sacrum
Point of Croup
Point of Hip
Lumbar Vertebrae
Thoracic Vertebrae

GLOSSARY

ABIOTROPHY: A loss of vitality or degeneration of certain cells or tissue that is present, but not evident, at birth. Abiotrophy is a term that is usually applied to inherited degenerative disorders.

ABRACHIA: A congenital defect involving the absence of arms or forelegs.

ACQUIRED TRAIT: An alteration in function or form resulting from a response to the environment. Acquired traits are not inherited.

ADAPTATION: Adjustment to changes in the environment.

ADDITIVE: A genetic term that refers to the contributing effects of many genes (polygenes) upon a single trait. Each gene has a small effect; each effect influences the same character so that the overall effect is cumulative.

ADRENALIN (EPINEPHRINE): A hormone produced by the adrenal gland. Adrenalin increases blood pressure, accelerates heart rate and slows the digestive system.

ALBINO: An individual characterized by absence of pigment in the skin, hair and eyes. The skin and eyes may appear pink due to the presence of blood vessels. Albinism is believed to be caused by a recessive allele at the C locus. This allele does not occur in horses.

ALLELE: Mutation of a gene to another form. There may be several possible alleles for a particular locus (corresponding points upon two homologous chromosomes). For example, the alleles A, A^t and a, which cause various color patterns in the horse, are mutations of the original "wild type" gene, A^+.

ALVEOLAR CAPILLARIES: The small blood vessels associated with the air sacs of the lungs. The alveolar capillaries receive oxygen from and return carbon dioxide to the lungs.

AMELANOTIC: Absence of melanin or pigment.

AMINO ACID: Organic building blocks for protein. Each protein molecule can be broken down into its amino acids. Each amino acid plays an important metabolic role.

AMNION: The innermost membrane around the fetus. The amnion is a fluid filled sac that protects the fetus by cushioning any sudden movement.

AMPLITUDE: Largeness or wideness; the extreme value of a fluctuating quality. The amplitude of a bell-shaped curve is the high point on the curve.

ANAPHYLAXIS: An exaggerated allergic reaction to a foreign substance.

ANCESTOR: An individual from which another individual descends (e.g., parent, grandparent, great-grandparent, etc.).

ANEMIA: A blood disorder characterized by abnormally low red blood cell count, hemoglobin concentration, or packed cell volume. Anemia is manifested by pale skin and mucous membranes, shortness of breath, and loss of energy.

ANEUPLOIDY: A chromosome number which is not an exact multiple of the haploid number, i.e., 2N+1 or 2N-1, where N is the haploid number. For example, 63XO aneuploidy is an abnormal chromosomal condition involving the absence of an X chromosome in the mare (2N-1).

ANHIDROSIS: The inability to sweat.

ANIRIDIA: Absence of the iris. The condition is hereditary and usually occurs in both eyes.

ANOPHTHALMIA: Absence of one or both eyes. The condition may be hereditary.

ANTIBODY: A protein substance which is evoked by the presence of an antigen. The antibody is usually antigen specific, meaning that it reacts against one particular antigen or group of antigens. Natural (inherited) antibodies also occur. Their identification has been helpful in solving cases of true parentage and has provided an objective means of identifying individuals.

ANTIGEN: A protein substance that the body recognizes as foreign. When an antigen is present, the body's immune system reacts by producing a specific antibody. Antigens include toxins, foreign proteins, bacterium and foreign tissue cells.

ARTIFICIAL SELECTION: Culling imposed by man to purposefully improve or eliminate a certain characteristic (as opposed to natural selection).

ASSOCIATION: The occurrence of a particular allele in a population more often than can be accounted for by chance. A positive association refers to an increase in the occurrence of one trait, as another is selected. A negative association is a decrease in the occurrence of one trait, as the occurrence of another increases.

ASSORTIVE MATING: Nonrandom mating with preferential selection for a particular genotype.

ATAXIA: An inability to coordinate voluntary muscular movements; incoordination.

ATRESIA: Congenital absence or closure of a normal body opening.

ATROPHY: A decrease in size or wasting away of a body part or tissue.

AUTOSOMAL RECESSIVE: Recessive allele located on a non-sex chromosome (i.e., on any chromosome other than X or Y). The expression of any recessive allele can be overridden by the presence of a corresponding (homologous) dominant allele.

AUTOSOME: Chromosome other than the sex chromosomes.

AVERAGE: The mean value of a population, found by adding together all values of the population, then dividing the sum by the number of values in the population.

AVERAGE EARNINGS INDEX: An index used to compare racing stallions, in which the average earnings of the progeny of the stallion under consideration are divided by the total average earnings of all starters. An average stallion would have an AEI of one, a below average stallion's AEI would be less than one, and a superior stallion's AEI would be greater than one.

BACKCROSS: The cross of a progeny with one of the parents, or with another individual of similar genotype.

BACKGROUND RADIATION: The natural radiation that all organisms are exposed to daily; this includes radiation from the sun, radioactive minerals, etc.

BASE NARROW: A conformation fault in which there is greater distance between the horse's legs at the top than at the bottom. The base narrow deviation is caused by an improper positioning at the elbow.

BASE WIDE: A conformation fault in which there is greater distance between the horse's legs at the bottom than at the top, caused by an improper positioning at the elbow.

BENIGN: Not malignant or recurrable; with a favorable outlook for recovery.

BILATERAL: Having two sides or pertaining to two sides.

BINOCULAR VISION: The ability to view a particular area with both eyes (as compared to monocular vision which pertains to viewing separate areas with each eye).

BLEEDER: A horse that suffers from epistaxis or nosebleed.

BLEMISHES: Minor faults caused by injury or occurring congenitally. Blemishes are considered unattractive but do not interfere with the horse's soundness.

BOWLEGS: A conformation fault in which the hocks are set too far apart. This deviation frequently causes intereference between the hind feet as they pass each other in travel. Bowlegs is also referred to as bandy legs.

BREED: A related group of animals showing certain inherited characteristics that separate them from other members of the same species.

BREEDING VALUE: A value assigned to a horse for a certain trait, calculated by comparing the considered horse to the population average. A high breeding value is given to a horse with an outstanding phenotype for a highly heritable trait.

BROKEN WIND (HEAVES): A respiratory disease in horses that involves rupture of the alveolar walls. Signs include chronic cough, lack of stamina, and the formation of a ridge (heave line) due to prolonged contractions of the abdominal muscles.

BUCK KNEES: Sprung over at the knees; forward deviation of the carpal joint.

BULLDOG MOUTH (UNDERSHOT JAW): A condition in which the lower jaw protrudes past the upper jaw.

CALF KNEES: A conformation fault in which the affected horse's legs bend back at the knees.

CAPILLARY: One of the minute blood vessels that form a network throughout the body, connecting the arteries and veins.

CARCINOMA: A malignant growth that tends to spread to the surrounding tissues.

CARRIER: An individual that carries the recessive allele for a trait without showing the trait.

CARTILAGE: A specialized type of fibrous connective tissue; precursor of bone.

CATARACT: An opacity (lack of transparency) in the lens of the eye.

CELL: Smallest unit within an organism that is capable of carrying on the essential life processes. Basically, the cell consists of a nucleus (which contains the chromosomes), a fluid-like cytoplasm, functional cytoplasmic bodies (organelles), and an enclosing cell membrane.

CENTRIC FUSION: The process by which two acrocentric chromosomes (chromosome with the centromere located toward one end) join to form a metacentric or submetacentric chromosome (chromosome with the centromere located at, or close to, the middle area).

CENTROMERE: The constriction of a chromosome which is the point of attachment for the spindle fibers and is concerned with chromosome movement during cell division.

CEREBELLUM: The part of the brain that is associated with coordination of movements.

CEREBRO-SPINAL FLUID: The fluid contained within the brain ventricles which is produced by the choroid plexus and the brain parenchyma. The cerebro-spinal fluid circulates through the ventricles and is absorbed into the venous system.

CHESTNUTS: The horny growths found on the inner side of horses' legs. They are located above the knee on the front leg, and below the hock on the hind leg. Chestnuts have no function in the modern horse except identification.

CHIMERISM: The presence of genetically and immunologically different tissues within an individual. Twins with two different types of erythrocytes may become chimeric if an exchange of blood occurs in utero. Somatic mutations may also cause the presence of genetically different tissues within one individual.

CHROMATIDS: Two strands formed by the duplication of a chromosome. The chromatids are attached by an area called the centromere. The centromere divides during one phase of cell division, and the two chromatids become separate chromosomes.

CHROMOSOMAL ABERRATIONS: Changes in chromosomal structure or number which affect the individual's genotype. Gross aberrations may cause apparent physical abnormalities.

CHROMOSOME: Thread-like deep-staining bodies within the cell nucleus that are composed of DNA. The chromosomes carry genetic information.

CLITORIS: The female structure which is homologous to the penis in the male.

CLOSEBREEDING: A breeding system which uses extreme inbreeding, e.g., sibling to sibling, parent to offspring.

CODOMINANCE: When heterozygous alleles are expressed by the individual, each contributing equally to the phenotype.

CODON: A triplet, or group of three bases, that codes for the production of a specific amino acid.

COLLATERAL RELATIVES: Relatives that share common ancestors, such as cousins, full and half-siblings, etc.

COLLECTION: Compression of the frame to form a short base, with weight moving to the hindquarters from the lightened forehand. Both forehand and hindquarters remain flexible, with hindlegs engaged under body. Poll should be flexed, and jaw, relaxed.

COLOSTRUM: The first milk given by a mammal. The mare's colostrum contains important antibodies which can be absorbed by the foal within 24-36 hours after birth. The antibodies provide protection against disease for the first few months of life.

COMBINED IMMUNODEFICIENCY DISEASE (CID): An inherited deficiency of infection fighting lymphocytes (B and T). CID has been reported in newborn Arabian foals and is believed to be caused by an autosomal recessive allele. The condition is always fatal.

COMPENSATORY GROWTH: The accelerated growth rate that a young, nutritionally deprived individual will experience when provided with an adequate ration.

CONGENITAL: Present at birth. Congenital defects may or may not be inherited.

CONJUNCTIVA: The membrane that lines the eyelids and covers the exposed surface of the sclera.

CONTINUOUS VARIATION: A constant spectrum of variation from one individual to another. When graphed, continuous variation for a measurable trait within a population will form a normal curve.

CONTRACTED HEELS: A condition involving an abnormally narrow heel, a shrunken frog, and parallel hoof bars. Excessive heat and/or lameness may result. Contracted heels may be inherited or may be caused by improper shoeing.

CORONARY BAND: Band of modified vascular tissue at the top of the hoof. This tissue is responsible for the growth of the hoof wall.

CORPUS LUTEUM: Yellow body in the ovary that secretes progesterone, an important reproductive hormone. The formation of the corpus luteum begins immediately after ovulation. The yellow mass forms at the site of the ruptured follicle; the cells which line the empty cavity make up the body. If pregnancy does not occur, the corpus luteum regresses. In the event of pregnancy, the corpus luteum may persist for several months.

CORTEX: The superficial grey matter of the cerebellum and the cerebrum.

CORTICAL: Pertaining to the cortex of the brain.

COUPLING: The section between the point of the hip and the last rib. A short-coupled horse ("short" is generally thought to measure less than the width of four fingers) is usually an easy keeper.

CRANIUM: The skeleton of the head.

CROSSBRED: An individual whose parents are of two different breeds.

CROSSBREEDING: The mating of horses from different breeds.

CROSSING OVER: The exchange of genetic material between homologous chromosomes.

CRYPTIC COLORATION: Coat color which tends to make the animal blend into his surroundings, such as that of the Przewalski horse.

CRYPTORCHIDISM: Failure of one or both testicles to descend to the scrotum after birth. The testes may be retained within the abdominal cavity (abdominal cryptorchidism) or within the inguinal canal (inguinal cryptorchidism).

CULLING: Elimination of undesirable animals from a breeding herd.

CUT OUT UNDER THE KNEES: A conformation fault in which there is an indentation just below the knee on the front of the cannon bone.

CUTANEOUS: Pertaining to the skin.

CYANOSIS: A bluish condition, especially of the skin and mucous membranes, that results from a lack of oxygen in the blood.

CYSTIC OVARIES: Ovaries which contain cysts in the area of the ovulation fossa. These cysts usually form after ovulation due to an infolding of the epithelial tissue in that area.

CYSTS: Any closed cavity, particularly one that contains a fluid.

CYTOGENETICS: Genetics of the cell. A branch of genetics that is concerned with cell division, chromosome structure and replication, functions of the various cytoplasmic organelles, and the identification of genetic material.

CYTOLOGIST: A scientist who studies the structure of cells.

CYTOPLASM: Fluid-like matter within the cell exclusive of the nucleus. The cytoplasm is surrounded by the cellular membrane and contains the various organelles of the cell (e.g., mitochondria, Golgi apparatus, etc.).

CYTOPLASMIC RETICULUM: A cell structure connected to the nucleus and bearing numerous ribosomes.

DAUGHTER CELLS: The multiple cells that result from the division of a single cell during mitosis and meiosis.

DEGENERATIVE: Pertaining to deterioration.

DELETION: A type of chromosomal aberration in which there is loss of part of a chromosome.

DERMOID: A growth or cyst which resembles skin.

DEVELOPMENTAL PROCESS: The process of growth and differentiation that occurs during gestation.

DIPLOID: The number of chromosomes normally found in the somatic cells. Each somatic cell contains both homologous chromosomes of each homologous pair.

DISRUPTIVE COLORATION: Coat color which tends to break up the outline of an animal, making the recognition of that animal at a distance difficult. Disruptive coloration is seen in the zebra.

DISTAL: Remote; removed from the point of attachment or any point of reference.

DNA: Deoxyribonucleic acid. A material that makes up the chromosomes of all cells. DNA carries the genetic information for every cell.

DOMINANT: A genetic term used to describe an allele that overrides the expression of a corresponding recessive allele.

DOUBLE HELIX: The paired spiral structure of the DNA molecule; the molecule resembles a twisted ladder.

DUCTUS ARTERIOSUS: A fetal blood vessel connecting the pulmonary artery directly with the descending aorta, thus allowing blood to bypass the lungs. Normally atrophies shortly after birth.

DUPLICATION: A type of chromosomal aberration in which part of a chromosome is duplicated.

ECTROPION: Outward drooping of the lower eyelid.

ELECTRON MICROSCOPE: An instrument which is used to obtain an enlarged view of an extremely small object (e.g., chromosome, Golgi apparatus, etc.). The electron microscope utilizes a beam of electrons instead of light.

ELECTRORETINOGRAPHY: The recording of the changes in electric potential in the retina after stimulation by light.

EMBRYO: An organism during the earliest stages of development. After the fertilized egg (zygote) divides, the organism (embryo) grows rapidly. The later stages of growth and differentiation are referred to as fetal.

EMPHYSEMA: See Heaves.

ENDOCRINE: Secreting internally. Anatomical classification that refers to glands and structures that secrete hormones into the blood or lymph. Endocrine hormones affect target organs in other parts of the body.

ENDOMETRITIS: Inflammation of the mucous membrane of the uterus.

ENTROPION: Inward turning of the lower eyelid.

ENVIRONMENT: Conditions and elements which make up the surroundings and influence the development of an individual.

ENZYME: A protein which is capable of accelerating or producing some change in another substance without itself being changed; enzymes act as catalysts in biological systems.

EPILEPSY: Disturbance of brain function which may result in unconsciousness, incoordination, convulsions, etc.

EPINEPHRINE: See Adrenalin.

EPIPHORA: Sudden flow of tears over the face.

EPIPHYSIS: Areas at both ends of a long bone that develop separately during the growth period. The epiphysis develops from an area of ossification (bone formation) and, until length-wise growth stops, is separated from the main portion of bone by a layer of cartilage.

EPISTATIC; EPISTASIS: An interaction between genes at different loci, in which a gene at one locus masks the presence of a gene at another locus. The masked gene is said to be hypostatic.

EPISTAXIS: Bleeding from the nose.

EPITHELIOGENESIS IMPERFECTA: Imperfect development of the covering of the internal and external surfaces of the body.

EQUUS: Designates several zoological classifications (subfamily, genus, and subgenus) within the animal kingdom. The subfamily and genus known as Equus include horses, zebras, asses and hemionids. The subgenus Equus includes the caballine (true horse) equidae.

ERGOT: A small, horny growth located on the back of the fetlock joint in most horses. Ergots have no function in the modern horse, but may be remnants of a foot pad found in early ancestors.

EUGENICS: The study of conditions that may improve the qualities of future generations.

EUMELANIN: The black/brown form of the pigment melanin which may color the coat, skin and eyes.

EUPLOIDY: Variations in chromosome number by whole sets or exact multiples of the haploid number; e.g., diploid, triploid.

EWE NECK: A neck in which the topline is concave, instead of straight or convex. a gene at one locus masks the presence of a gene at another locus. The masked gene is said to be hypostatic.

EXPRESSIVITY: The extent to which a heritable trait is shown by the individual carrying the gene or genes which determine it.

EXTENSION: In motion, refers to elongation of body and shifting of weight to forehand. A movement that brings a limb into a straight line.

FIXATION: Attainment of an allelic frequency of 1.0 within a herd or population. An allele is "fixed" within a population when it is the only allele for that particular locus.

FLAGELLUM: A mobile filament which serves as an organ of locomotion. A sperm has a flagellum which enables it to swim through the seminal fluid.

FOLLICLE: Cell that encases and nourishes the ovum as it develops within the ovary. When the ovum reaches maturity, the follicle bursts and expels the ovum into the oviduct.

FORAMEN OVALE: The opening in the wall of the fetal heart, between the atria, that normally closes shortly after birth.

FURLONG: One-eighth of a mile.

GAMETE: A mature sex cell, either a sperm or an egg, that contains the haploid number of chromosomes. In the horse, gametes normally contain 32 chromosomes.

GAMETOGENESIS: The development of the male and female gametes.

GENE: The determiner of a hereditary trait. The gene is a segment of a DNA molecule that instructs the cell to produce a certain protein.

GENE FREQUENCY: The proportion of one allele (in comparison to other alleles for the same locus) present within a designated population. Gene ("allelic") frequencies are calculated by dividing the number of loci at which an allele occurs by the number of loci at which it could occur. The resulting figure is between 0 and 1.

GENE POOL: The total of all genes in a population.

GENETIC CAPACITY: The potential of an individual as dictated by his genotype.

GENETIC CODE: The collection of base triplets of DNA and RNA which carry the genetic information for protein synthesis in the cell.

GENETIC DRIFT: Fluctuations in gene frequencies due to chance. Genetic drift tends to occur in small populations.

GENETIC EQUILIBRIUM: Constancy of a particular gene frequency through successive generations.

GENETIC LOAD: The average number of lethals per individual in a population.

GENITALIA: The reproductive organs.

GENOME: A haploid set of chromosomes inherited as a unit from one parent.

GENOTYPE: The gene types that an individual inherits from both parents.

GOLGI APPARATUS: A membranous cytoplasmic organelle consisting of several flattened vesicles that are closely stacked. Although little is known about the function of the Golgi apparatus, it is believed that the organelle maintains water balance during lipid (fat) metabolism.

GRADING: The outbreeding of a superior stallion to below average mares.

GROSS MUTATION: A change in a gene which involves an alteration in more than one nucleotide.

GROUND SUBSTANCE: The gel-like material in which connective tissue cells and fibers are imbedded.

GUT STRANGULATION: Death of a portion of the large intestine due to the loss of blood supply to that area.

HAPLOID: A set of one chromosome from each homologous pair. The haploid number of chromosomes is carried by the gametes (egg and sperm).

HEAVES: A respiratory ailment, characterized by forced expiration and dyspnea, resulting from the rupture of alveoli in the lungs, and caused by allergies, dust, etc. Similar to emphysema in man.

HEMIZYGOUS: The term used when describing the genotype of a male with regard to X-linked traits. Males have only one set of X-linked genes and, therefore, cannot be homozygous or heterozygous at any locus on the X chromosome.

HEMOPHILIA: A sex-linked recessive condition characterized by free bleeding which is due to a defect in the clotting mechanism.

HERITABILITY: The quality of being inherited; the extent to which a phenotype is influenced by the individual's genotype as opposed to environmental factors.

HERITABILITY ESTIMATE: An approximation, usually expressed as percentage, of the influence that heredity has on a trait, as opposed to environmental influences. For example, if the heritability of a trait is 20%, heredity accounts for 20% of the trait's expression, and environmental effects account for 80%.

HERMAPHRODITE: An individual exhibiting attributes of both sexes.

HETEROCHROMATIN: The chromosome material near the centromere constriction that contains very few genes. Heterochromatin stains darkly when treated with certain dyes.

HETEROCHROMIA: Differences in color between the irises of the eyes, or differences in color within one iris.

HETEROSIS: The beneficial effect of hybridization (crossing different species or types) on vigor, fertility, growth and overall physical quality.

HETEROZYGOTE: An individual that carries two different alleles at a certain locus (corresponding points on two homologous chromosomes).

HOLANDRIC GENE: A gene located only on the Y chromosome (in species that have the characteristic XY set of sex chromosomes in the male). Only males are affected by holandric genes. Holandric traits can only be passed from sire to son, etc.

HOMOGAMETIC SEX: The sex that possesses two identical sex chromosomes (e.g., XX in the mare).

HOMOLOGOUS CHROMOSOMES: The two chromosomes that contain identical loci.

HOMOZYGOTE: An individual that carries the same allele at corresponding points on homologous chromosomes.

HYBRID: The progeny of a cross between two organisms with very different genotypes, e.g., mare x jack = mule.

HYPERPLASIA: Overdevelopment of an organ or tissue due to an increase in the number of cells.

HYPOPLASIA: Underdevelopment of an organ or tissue due to a decrease in the number of cells.

HYPOSTATIC: Pertaining to a gene whose presence is masked by the presence of a gene (epistatic gene) at another locus.

IATROGENIC: Caused by man; man-made (artifical) as opposed to natural.

IDIOPATHIC: Of unknown cause.

IMPULSION: Forward, driving motion: the urge to advance.

INBREEDING: The mating of closely related individuals, or of individuals having similar genotypes.

INBREEDING COEFFICIENT: A measure of the number of homozygous gene pairs in a horse's genotype due to the common ancestry of his sire and dam.

INCOMPLETE DOMINANCE: The condition in heterozygotes where the phenotype is intermediate between the two homozygotes (e.g., the c^{cr} dilution).

INDUCED MUTATION: A change in gene structure caused by an external source other than background radiation.

INHERITED DISEASE: Any disease not caused wholly by environmental factors.

INTERFERENCE: Striking of the hoof of one leg against the fetlock of the opposite leg.

INTERSEX: An individual showing sex characteristics intermediate between male and female, or at least some characteristics of each sex.

INTERVENTRICULAR SEPTAL DEFECTS (IVSD): Failure of the interventricular partition to close after birth. This defect interferes with normal circulation of oxygenated and nonoxygenated blood and may cause heart failure.

INTRAUTERINE: Within the uterus.

INVAGINATION: The infolding of one part within another.

INVERSION: A chromosomal aberration caused by the section of chromosome between two breaks rotating before fusing back into place.

IRIDOCYCLITIS: Inflammation of the iris and the ciliary body.

KARYOTYPE: The complete set of chromosomes from the nucleus of a cell, usually arranged according to centromere position, from the largest to the smallest.

KERATIN: A protein which is the principal component of skin, hair, nails, etc.

KNOCK KNEES: A conformation fault in which the knees deviate toward each other.

LATERAL: Pertaining to a side or outer surface.

LETHAL GENE: A gene whose phenotypic effect is sufficiently drastic to kill the bearer. Lethal genes may be dominant, incompletely dominant, or recessive. They may result in the immediate death of the embryo, fetus or newborn foal (true lethal), they may cause death as the individual matures (delayed lethal), or they may result in death only when a certain environmental factor is present (partial lethal).

LIGAMENT: A band of fibrous tissue that connects bones and/or cartilage.

LINEBREEDING: A conservative program of inbreeding designed to concentrate the blood of a certain famous ancestor within the linebred offspring.

LINECROSSING: A type of outcrossing in which members from two distinct family lines are mated.

LINKAGE: Two genes at different loci on the same chromosome are said to be linked.

LINKAGE GROUP: All of the genes located on a given chromosome.

LOCUS: The specific site of a gene on a chromosome. The term also applies to corresponding positions on homologous chromosomes.

LYMPHOCYTES: A type of white blood cell, produced mainly by the lymph tissue. Lymphocytes contribute to cell-mediated immunity.

LYSOSOME: Organelle in the cytoplasm of the cell that is responsible for intracellular digestion. The lysosome material (accumulation of enzymes) is enclosed by a single membrane. The organelle may also serve as a storage site for insoluble material.

MAIN GENE: The principle gene responsible for the expression of an inherited trait, as opposed to modifying genes.

MALIGNANT: Tending to become progressively worse and end in death; in the case of neoplasms, refers to uncontrollable growths which tend to spread and recur after removal.

MEAN: The arithmetic average value of a population.

MEDIAN: The middle value of a series of readings arranged serially according to magnitude.

MEIOSIS: The method of cell division during the production of gametes. Meiosis results in sperm or ova that contain the haploid number of chromosomes (e.g., 32 chromosomes in the horse).

MELANIN: The basic pigment of the skin, coat and eyes.

MELANOBLASTS: Cells that originate from the embryonic neural crest and develop into pigment-producing melanocytes.

MELANOCYTES: The cells responsible for the production of melanin.

MELANOMA: A tumor composed of melanin-pigmented cells.

MELANOSIS: A disorder characterized by abnormal pigment deposits.

MELANOTIC: Containing melanin.

METABOLIC: Referring to the process by which life is produced and maintained and to the transformation of energy into a form usable by an organism.

METASTASIZE: To spread disease from one part or organ of the body to another.

MICROPHTHALMIA: Abnormally small eyes.

MIS-SENSE CODON: A new triplet, or sequence of three bases, that is caused by a change in the DNA molecule. This mis-sense codon codes for the production of a different amino acid.

MITOCHONDRIA: Structures within a cell which break down foodstuffs and generate energy for cellular processes.

MITOSIS: The type of cell division which occurs in somatic cells. Mitosis results in the production of genetically identical daughter cells.

MODE: The value within a population that occurs with the greatest frequency.

MODE OF INHERITANCE: The manner in which the gene, or genes, for a specific trait are expressed. The mode of inheritance of a characteristic may be recessive, dominant, codominant, etc.

MOLECULES: A combination of two or more atoms to form a specific chemical substance.

MOSAIC: An individual that possesses cells that have different genotypes. Normally, all cells within an organism carry identical genotypes (except for the haploid sex cells).

MULTIFACTORIAL: Trait that is controlled by many genes. Each gene has a small additive effect (polygenetic effect). Multifactorial traits are often influenced heavily by the environment.

MULTIPLE ALLELES: The existence of more than two alleles for a particular locus in a population.

MUTAGEN: Any agent that causes a mutation (e.g., radiation).

MUTATION: A sudden and permanent change in an individual's genotype. If the mutation occurs in a sex cell, the mutant gene may be passed to future offspring.

MUTATION RATE: The number of mutations of any one particular locus which occurs per gamete per generation.

NATURAL MUTATION: A permanent change in an individual's genotype that occurs spontaneously, due to background radiation or another natural cause.

NATURAL SELECTION: Culling process based on "survival of the fittest." Animals that cannot adapt to changes in the environment are removed, while adapted animals survive to reproduce. The traits which enhanced survival are transmitted to the next generation. Therefore, biological fitness influences the frequency of a trait in a population.

NAVICULAR DISEASE: A degenerative condition involving the navicular bone, navicular bursa and the deep flexor tendon. Inflammation of the bursa results in excess synovial fluid production; the navicular bone becomes eroded, and small fibers of the tendon are torn by the roughened bone. The tendon may eventually become attached to the bone.

NEONATAL ISOERYTHROLYSIS: Severe anemia in the newborn foal caused by the hemolysis of its red blood cells by isoantibodies from the dam's colostrum. The mare's isoantibodies are produced when incompatible blood (blood that contains an inherited factor that the dam does not have) from the foal enters her blood stream. These isoantibodies accumulate in the colostrum and, if the foal ingests the colostrum within 24-32 hours after birth, the antibodies may be absorbed by the foal's intestine, enter his blood stream and cause the hemolytic reaction.

NEURAL CREST: Early embryonic cellular mass that forms an elongated ridge along the anterior-posterior axis of the embryo. This tissue eventually forms the brain, spinal cord and complex nervous system.

NEUROMUSCULAR: Pertaining to muscles and nerves.

NICK: A lucky combination of genes that results in superior offspring. In horses, crosses between certain families are thought to produce nicks.

NITROGEN RETENTION: The retaining of nitrogen in the body for use in the production of muscle tissue. Nitrogen retention is increased by testesterone.

NONDISJUNCTION: Failure of two homologous chromosomes to separate and migrate to different daughter cells during the first meiotic division, or failure of the two chromatids to separate and migrate to different daughter cells during mitosis or the second meiotic division. As a result, one daughter cell has two chromosomes or two chromatids, while the other is missing a chromosome.

NON-HOMOLOGOUS: Refers to chromosomes that do not have corresponding loci.

NON-SENSE CODON: A new triplet, or sequence of three bases, that cannot direct the production of any amino acid. Non-sense codons result from a change in the normal sequence of bases (e.g., a deletion or an addition of a base).

NUCLEIC ACID: A chemical substance that contains sugar, phosphorus and several bases. Nucleic acids carry the genetic information (codons) for each cell. There are two types of nucleic acid: DNA and RNA.

NUCLEOTIDE: A component of nucleic acid that is composed of a base and a phosphate group.

NUCLEUS: The structure within the cell which contains the chromosomes.

OCULAR: Pertaining to the eye.

OFFSET KNEES: A conformation fault in which the leg does not follow a straight line through the radius and cannon bones.

OOCYTE: The diploid germ cell that will undergo meiosis (oogenesis) to form an egg.

OPTIC PAPILLAE (DISCUS NERVI OPITICI): A circular area on the retina that has no light receptors. At this point, the optic nerve fibers converge to leave the eye.

OUTBREEDING: The mating of unrelated individuals.

OUTCROSSING: The mating of unrelated animals within the same breed.

OVARIECTOMY: The removal of an ovary or ovaries.

OVARY: The female gonad; the organ that produces the female gametes (ova) and reproductive hormones.

OVERDOMINANCE: The ability of heterozygous alleles to produce superior results compared to what each allele produces in its homozygous state.

OVERLAP TIME: Occurs when two or more feet are on the ground at the same time. Shorter overlap time means a horse travels faster, because he spends less time on the ground and more time moving forward through the air.

OVERO: The recessive type of white spotting, usually characterized by white face markings and ventral body markings. The white areas typically have ragged edges.

OVUM: The female reproductive cell (pl. ova).

PACER: A horse whose natural gait is the pace, a lateral two-beat gait in which fore and hind legs on each side of the body move forward and backward together.

PADDLING: Faulty action in which the foreleg below the knee is thrown outward as the horse moves forward.

PALPATION: The act of feeling with the hand.

PARROT MOUTH (OVERSHOT JAW): A congenital defect involving imperfectly meshed teeth. The condition is similar to bucked teeth in humans.

PARTI-COLORED: Having more than one color; e.g., Appaloosa, Pinto, etc.

PATENT: Open.

PEDIGREE: The record of an individual's ancestral history.

PENETRANCE: The frequency with which a heritable trait is shown by individuals who carry the gene, or genes, for the trait.

PERIPHERAL VISION: The ability to view a circular field surrounding the body. The wide-set eye placement of many herbivores allows them to view their surroundings with just a slight movement of the head.

PHAEOMELANIN: The red/yellow form of melanin pigment.

PHARYNGITIS: Inflammation of the pharynx.

PHENOCOPY: A physical condition which is due to environmental factors but resembles an inherited characteristic.

PHENOTYPE: The visible and measurable appearance of an individual which results from the interaction of his genotype and the environment.

PHOTODYNAMIC: Term that describes the force exerted by light. Photodynamic is also used in reference to ingested substances that cause a skin reaction in sunlight.

PHOTOPIC: Pertaining to vision in light.

PHOTOSENSITIVITY: A condition involving the skin's abnormal reaction to sunlight.

PHYSIOLOGIC: Refers to the vital physical processes of any normal body system (e.g., circulatory, respiratory, nervous, etc.), as opposed to pathologic which denotes an abnormal physical process.

PITUITARY: Pertaining to the pituitary gland which is located at the base of the brain. This gland secretes several vital hormones and is important to growth, maturation and reproduction.

PLASMA PROTEIN: Protein molecules within the non-cellular fluid portion of the circulating blood, including antibodies and blood clotting proteins.

PLEIOTROPIC: Refers to the ability of one gene to affect several traits. Pleiotropic genes often control the production of an enzyme that controls several biological reactions.

POINT MUTATION: A change in one nucleotide (nucleic acid, sugar, phosphate group) of a codon. Point mutations may involve the loss of a nucleotide, the addition of a nucleotide, or the substitution of one nucleotide for another.

POLAR BODY: Cell produced during oogenesis (formation of the egg) that contains a complete set of chromosomes and very little cytoplasm. When a primitive egg cell divides, one daughter cell (the future ovum) receives most of the cytoplasm, and the other becomes a polar body.

POLYDACTYLY: The presence of more than one digit on either hand or foot. In horses, polydactyly involves the presence of two hooves and two pastern bones on one limb. The condition is thought to be hereditary.

POLYGENES: A group of genes (i.e., at different loci) that influence the same trait. Each polygene has a small cumulative effect upon the trait.

POLYPLOID: Having more than two full sets of homologous chromosomes (e.g., 3N, 4N, etc.).

POPULATION: A group of individuals that comprise a certain category or inhabit a certain location.

PREPOTENCY: The ability to consistently transmit hereditary characteristics to offspring. Prepotency is directly related to the number of homozygous alleles in the individual's genotype.

PROBABILITY: The likelihood that a certain event will occur; $0 =$ certainty that the event will not occur, $1 =$ certainty that the event will occur.

PROGENY: Offspring; descendants.

PSEUDODOMINANCE: The expression of a recessive allele at a locus due to the absence of any corresponding allele on the homologous chromosome.

PSEUDOHERMAPHRODITE: An individual that possesses either testes or ovaries, but shows one or more characteristics of the other sex.

PUREBRED: Belonging to a breed with recognized characteristics maintained through generations of unmixed descent.

PURE LINE: A strain of individuals that breed true for certain characteristics. This prepotency is caused by a "fixed" homozygous condition for the controlling alleles of each characteristic.

QUALITATIVE: Pertaining to quality. The term is also used to describe a trait that has discontinuous phenotypes. Eye color, for example, is expressed in specific ways — brown, amber, blue, etc. Qualitative traits have no continuous array of phenotypes.

QUANTITATIVE: Pertaining to quantity. The term is also used to describe traits that show a continuous distribution of phenotypes. Height, for example, is seen as a continuous spectrum of measurements.

RANDOM MATING: Selection of a mate without regard to genotype.

RANDOM SELECTION: Choosing individuals from a population without a selection formula. With random selection, every individual has an equal chance of being selected.

READING FRAME-SHIFT: A mutation that involves a deletion or addition of a base from a DNA molecule. As a result, every codon after the change is shifted, and the instructions from the DNA molecule are altered.

RECESSIVE: A trait that cannot be expressed when its controlling allele is in the homozygous state. The expression of the recessive allele is overridden by the presence of a corresponding dominant allele.

RECIPROCAL CROSS: A second cross of the same genotypes in which the sexes of the parental generation are reversed.

RECOMBINATION: A new association of genes in an individual derived from crossing over (exchange of homologous material) in the parent's genotype.

RED CELL ANTIGEN FACTORS: Protein substances carried on the surface of an individual's red blood cells that, when injected into another individual or into a test solution, may cause the formation of antibodies.

REFINEMENT: Delicacy and elegance of build. A refined horse has a graceful appearance — a well-chiseled head, a slender neck, a smooth coat, etc.

REGRESSION: A genetic term used to describe the movement of the offspring of superior parents back to the level of the population. Offspring of superior parents frequently "regress" to the population average, rather than maintain their parents' quality.

REPLICATE: Synthesis of new DNA from preexisting DNA during nuclear division (separation of genetic material during cell division).

RIBOSOME: A cytoplasmic organelle that functions as the site of protein synthesis. The ribosomes usually adhere to the cytoplasmic reticulum.

RNA (RIBONUCLEIC ACID): One of the two nucleic acids involved in the transmission of genetic material. RNA carries the genetic messages from the nucleus to the ribosomes where the messages are decoded and the designated proteins are produced.

ROARING: A loud whistling or roaring noise that is made by a horse with laryngeal hemiplegia during exercise. Roaring is caused by paralysis of certain laryngeal muscles.

ROBERTSONIAN TRANSLOCATION: See: centric fusion.

SCOTOPIC: Referring to the lowest light to which the eye is dark-adapted.

SEGREGATION: The separation of homologous chromosomes during meiosis, so that each gamete contains only one member of each homologous pair (i.e., the haploid number of chromosomes).

SELECTION DIFFERENTIAL: With respect to a specific measurable trait, the difference between a selected individual and the population from which he was selected.

SELECTION INTENSITY: The stringency used in choosing individuals. With a high selection intensity, only a few superior individuals are chosen; with a low intensity, many are acceptable.

SELECTION PRESSURE: See: selection intensity.

SENSE CODON: A triplet, or sequence of three bases in the DNA molecule, which has been changed in some way, but still codes for the original amino acid.

SEPTUM: A wall dividing two cavities.

SEX CHROMOSOMES: The chromosomes responsible for sex (XX in the mare, XY in the stallion).

SEX INFLUENCED: An inherited trait that is expressed more frequently in one sex than in the other. Sex influenced traits are influenced, to a great extent, by reproductive hormones. If a trait is only expressed in one sex, it is known as a sex limited trait.

SEX LINKED: A term used to describe genes that are carried on the sex chromosomes. Sex linked is usually used in reference to X-linked genes.

SHIVERING: A nervous disorder in the horse that interferes with locomotion. Shivering is characterized by overflexion of the hocks, an inability to lower the foot and tremors.

SIBLING: Brother or sister.

SICKLE HOCK: Deviation in the angle of the hock as seen from the side. The cannon slopes forward due to excessive angulation of the hock.

SISTER CELLS: Daughter cells resulting from division of the same parent cell.

SLAB-SIDED: A conformation fault in which the barrel, in cross-section, is not adequately round. The ribs lack sufficient curvature and the horse is frequently narrow-chested.

SOMATIC MUTATION: A change in the genetic information within any body cell except the germ cells of the ovary or the mature gametes (sperm and ova).

SOUNDNESS: The quality of physical fitness.

SPECIES HYBRIDIZATION: Crossing members of two separate species e.g., jack x mare = mule.

SPERMATID: A cell that is produced by the second meiotic cell division of spermatogenesis. The spermatid eventually matures into the motile spermatozoa which carries the sire's genetic contribution to his offspring.

SPERMATOCYTE: Reproductive cell that develops from the testicular germ cell (spermatogonium). The primary spermatocyte is produced by a mitotic cell division (spermatogonium to primary spermatocyte), while the secondary spermatocyte is produced by the first meiotic division (primary spermatocyte to secondary spermatocyte).

SPERMATOGENESIS: Development of sperm.

SPERMATOZOA: Male gamete or sex cell which is composed of a head and tail. The head contains the genetic information to be transmitted during fertilization. The cell is propelled by the tail (flagellum) for a limited duration.

SPLINTS: New bone growth that results from irritation to the interosseous ligament. Splints may cause arthritis or, if located near the carpal joint, may cause lameness. Usually, splints constitute a blemish. Poor conformation, such as bench knees, may predispose a horse to this condition.

SPRINTER: A horse that performs best at speed over short distances.

SQUAMOUS CELL: A flat, scale-like cell.

STANDARD DEVIATION: The average difference between members of a population and the mean value (population average).

STAYER: A horse whose best race length is a longer distance.

STENOSIS: A stricture or narrowing of a canal.

STRINGHALT: Involuntary, greater than normal flexion of the hock while the horse is in motion.

STRUCTURAL MUTATION: A change in genetic information caused by a change in the physical structure of the chromosome.

SUBCUTANEOUS: Beneath the skin.

SWING TIME: The time that each leg of a galloping horse spends in the non-weight-bearing (airborne) phase of his stride.

SYNAPSE: Attraction between two homologous chromosomes during one phase of meiotic cell division (prohase I). This attraction allows the homologous chromosomes to pair and later migrate to opposite ends of a dividing cell. The end result is the production of two cells that contain one chromosome from each homologous pair.

SYNECHIAE: An adhesion or union between two surfaces. Synechiae often refers to the adhesion of the iris to the cornea.

TENDON: A fibrous cord of connective tissue which attaches muscle to bone or other structures.

TERATOGEN: An agent that causes congenital abnormalities (toxins, drugs, etc.).

TESTCROSS: A cross between an individual with a dominant phenotype and an indiviual with a recessive phenotype. Testcrosses are generally used to determine whether the individual with the dominant phenotype is homozygous dominant or heterozygous, or to determine the degree of linkage between two traits.

TERATOGENESIS: The production of an abnormal fetus. The growth process may be disturbed by a nutritional deficiency, toxicity, a genetic abnormality, radiation, etc.

TETRAD: A genetic term used to describe the four-stranded chromosome structure formed during one phase of meiosis. The structure consists of synapsed homologous chromosomes in their duplicated form (i.e., two chromatids attached at their centromere, paired with two homologous chromatids attached at their centromere).

THYROID GLAND: A gland in front and along the sides of the upper part of the trachea.

TIED IN BEHIND THE KNEE: Pertaining to a deviation in leg set where the flexor tendons appear to be too close to the cannon bone just below the knee.

TOE IN: A conformation fault in which the feet are pointed in; pigeon-toed.

TOE OUT: A conformation fault in which the feet are pointed out; splay-footed.

TORTICOLLIS: Wry or crooked neck.

TRAINABILITY: The physical and mental characteristics which affect the horse's ability to be trained for a particular purpose.

TRANSCRIPTION: The process whereby genetic information is transmitted from DNA in the nucleus to messenger-RNA.

TRANSFERRIN: A blood serum protein that controls iron levels in the blood.

TRANSLATION: The process whereby genetic information from messenger-RNA is decoded for protein synthesis.

TRANSLOCATION: Exchange of genetic material between two non-homologous chromosomes. The switching of segments is a result of breakage and refusion of chromosome arms.

TRIPLET: A codon, or group of three bases, that codes for the production of a specific amino acid.

UNIFACTORIAL: Trait that is controlled by a single gene (pair of alleles).

UNILATERAL: On one side.

VARIANCE: The state of being different.

VARIATION: Deviation from the standard type.

VARIEGATED: Diversified; having different colors.

VERTEBRAE: The bones which form the spinal column.

VITREOUS FLUID: Clear, glass-like fluid.

WALL EYE: Blue, white, or colorless eye.

WHITE FOAL SYNDROME: An inherited disorder involving limited pigmentation and atresia coli (closure of the intestine). The condition is associated with the overo color pattern and is especially predominant when two horses with strong modifiers for white are crossed. It is believed that the defect is controlled by a recessive allele at the overo locus.

WHORLS: An area of hair growing in a twisted manner. Whorls are caused by a change in the direction of hair growth.

WILD TYPE: The original, environmentally-adapted form of a species or trait before artificial selection has taken place.

WOBBLES: A disease of horses characterized by incoordination, most evident in the hind legs. It is the result of compression of the spinal cord and may be inherited.

ZEBROID: Offspring resulting from zebra - horse crosses. Although breeders once hoped to combine the tractability of the horse with the disease resistance of the zebra, hybridization between the two species has not been commercially successful.

ZEONKEY: Hybrid offspring resulting from crosses between the ass and the zebra. Zeonkies commonly have shoulder stripes, long ears, faded zebra striping over body, and distinctive striping on the legs and ears.

ZYGOTE: Diploid cell that results from the union of sperm and ovum; a fertilized egg cell.

INDEX

A

A locus, 243, 245
abnormalities
  abnormal leg set, 334
  congenital, 327
  inherited, 325
abrachia, 343
absence of the retina, 357
acrocentric chromosome, 417
additive genes, 481
adenine, 419, 422
age, 68
agouti, 245
Al Hattab, 251
albinismus partialis, 350
albinismus totalis, 350
albino, 255
Allan F-1, 53
alleles, 142, 416, 458
  multiple, 549
amble, 303
American Fox Trotting Horse, 54
American Paint Horse Association, 49
American Saddlebred, 52, 224
amino acid, 418
anaphase, 428, 432, 435
Andalusian Breed, 20
aneuploidy (63 XO), 444
anophthalmia, 359
anhidrosis, 365
aniridia, 350
antigen, 402
  red blood cell, 503
Ap gene, 270
Appaloosa, 47, 268, 301
  color pattern, 268
apron face, 292
Arabian, 38
ass vs. horse, 510
association of traits, 488
  positive, 203, 492
  negative, 203
astral bodies, 432
ataxia, 382
  spinal, 382
  cerebellar, 382
  equine spinal, 383
atheroma of the false nostril, 367
atresia ani, 377
atresia coli, 267, 377
atresia dental, 333

atresia of the nasolacrimal duct, 360
atresia recti, 377
autosomes, 142, 417
Average Earnings Index, 151

B

B locus, 244
backcrossing, 218
backing, 314
Barb, 20, 30, 38
Bartlett's Childers, 40, 81
base, 418
*Bask+, 39, 71
Baskir Horse, 299
bay, 245, 501
behavior, 198
Belgian, 56, 134
bell-shaped curve, 210, 485
Big head, 333
black, 244, 247
black chestnut, 247
blanket, 48, 270
blaze, 292
blemishes, 65
blood, 401
  combined immunodeficiency, 34
  152, 183, 328, 404
  composition, 502
  factors, 503
  flow, 396
  types & performance, 505
  typing, 502
bloody shoulder markings, 251
Bold Ruler, 71, 81
bone spavin, 335
brachygnathia, 330
brain, 381, 389
  brain hypoxia, 390
breeder
  market, 147
  non-market, 147
breeding, 147
  backcrossing, 218
  breed improvement, 170
  closebreedng, 171, 175
  crossbreeding, 169, 171, 480
  inbreeding, 169, 171, 480
  linebreeding, 171, 177
  linecrossing, 188
  outbreeding, 169, 186

outcrossing, 187
selective, 25
systems, 169
true breeding, 171, 295
value, 208, 470
breeds, 35, 64
American Fox Trotting Horse, 54
American Paint Horse, 49
American Saddlebred, 52, 224
Andalusian Breed, 20
Appaloosa, 47, 268, 301
Arabian, 38
Barb, 20, 30, 38
Baskir Horse, 299
Belgian, 56, 134
Canadian Pacer, 52
Celtic Pony, 17, 36, 56, 59
Cleveland Bay, 37, 66
Clydesdale, 56, 134
development, 170
Exmoor Pony, 59
Fjord Pony, 27, 36
Fox Trotting Horses, 54, 316
Galiceno Pony, 59
Gotland Pony, 27, 36
Holstein, 122
improvement, 170
Isabella, 261
Lippizaner, 66, 478
Miniature horse, 59
Morocco Spotted Horse, 49
Norwegian Dun, 51
Paint, 49, 266
Palomino Horse Association, 50
Palomino Horse Breeders of America, 50
Paso Fino, 55
Percheron, 35, 56, 134
Peruvian Paso, 55
Pinto Horse Association, 49
plasma protein frequencies, 505
Pony of the Americas, 59
Quarter Horse, 42
Saddlebred, 52, 224
Shetland Pony, 35, 59, 264
Shire, 56, 134
Sorraia, 51
Suffolk Punch, 37, 56, 66, 134
Trakehner, 122
Turkmene, 28
Welch Cob, 59
Welch Pony, 59
broken wind, 394
bronchitis, 394
broodmare, 155
maternal influence, 162
performance vs. pedigree, 163

selection of, 155
broodmare sires, 154
brown, seal, 245
Bruce Lowe System, 237
buck-knee, 111, 118
buckskin, 51, 259, 262
Bull Rock, 41
bulldog mouth, 331
Byerley Turk, 40

C

C locus, 256, 262, 451
calf-knee, 111, 118, 195, 335
Canadian Pacer, 52
canter, 310
Caslick's operation, 375
cataracts, 353
complete, mature cortical, 354
nuclear, 354
y-type, 353
cell, 413
structure, 414
Celtic Pony, 17, 36, 56, 59
Cenozoic Era, 2, 6
centric fusion, 463
centriole, 427
centromere, 416, 428
cerebellar ataxia, 388
cerebellar hypoplasia, 387
chestnut, 247
chestnuts, 365
chromatids, 416, 428
chromosome, 1, 141, 415
aberration, 462
abnormalities, sex, 444
acrocentric, 417
banding, 445, 462
chromatids, 416, 428
deletion, 465
diploid, 430
double helix, 420
homologous, 143, 415
rearrangement, 464
ring chromosomes, 465
sex, 141, 439
X, 417, 141, 439
Y, 141, 417, 439
CID, 34, 152, 183, 328, 404, 477
circulatory system, 395
interventricular septal defects, 398
patent ductus arteriosus, 397
patent foramen ovale, 398
persistent right aortic arch, 399

Citation, 73
claybank dun, 246, 260
cleft palate, 376
Cleveland Bay, 37, 66
close ancestors, 76
closebreeding, 171, 175
Clydesdale, 56, 134
CNS, 380
coat
  color, 239
  texture, 296
codominance, 451
codon, 418, 461
coefficient of inbreeding, 181
 breeds, 183
collateral relatives, 77, 180
coloboma iridis, 351
color, 66
  bay, 245, 501
  black, 244, 247
  black chestnut, 247
  blanket, 48, 270
  bloody shoulder markings, 251
  breeds, 47
  brown, seal, 245
  buckskin, 51, 259, 262
  chestnut, 247
  claybank dun, 246, 260
  coat, 239
  cremello, 239, 253, 255, 257, 472
  cryptic coloration, 240
  dapple
    grey, 251
    silver, 263
  diluted, 256
  diminishing contrast, 273
  disruptive coloration, 240
  dominant white, 249, 255, 502
  dun, 51, 259, 262
  eye color, 242
  grey, 142, 249, 250, 455, 472, 501
  grulla, 51, 261
  Isabella, 261
  Leopard Appaloosa, 48, 271
  liver, 227, 244
  overo, 49, 224, 243, 266, 267, 480
  parentage, 501
  parti-colored, 47, 266, 457
  perlino, 260
  piebald, 50, 266
  pinto, 49, 266
  pseudoalbino, 255
  red dun, 246, 260
  roan, 249, 252, 501, 253
  seal brown, 245, 246, 259

  skewbald, 50, 266
  sorrel, 247
  tobiano, 219, 266
  white, 240
  white blanket, 270
combined immunodeficiency (CID), 34,
152, 183, 328, 404, 477
combined training, 125
competitive trail horse, 130
complete dominance, 450
condition, 69
conformation, 65, 82, 329
  body, 89, 344
  head and neck, 85, 330
  legs, 93, 132, 334
  race horse, 109
congenital defects, 327, 507
  congenital night blindness, 357
  congenital keratopathy, 352
  congenital occipito-atlanto-axial
    malformation, 386
  congenital hydrocephalus, 389
contracted digital flexor tendons, 337
contracted heels, 339
contrast, 273
convulsive syndrome, 390
coronet, 293
cornea, 352
corpus luteum, 437
cow hock, 335
cow sense, 127
cremello, 239, 253, 255, 257, 472
cresty neck, 334
cribbing, 33
crossbreeding, 189
crossing over, 464
cryptic coloration, 240
cryptorchidism, 149, 165, 369
  unilateral, 369
  bilateral, 369
  abdominal, 369
  inguinal, 369
culling, 169, 194
  minimun level, 203
curb bit, 31
curly hair, 298
cutting horse, 127
cyanosis, 397
cysts
  of the eye, 352
  pharyngeal, 395
cytogenetics, 414
cytoplasmic genetics, 163
cytoplasmic reticulum, 414
cytosine, 419, 422

D

D locus, 256, 260, 262, 451
Damara zebra, 15, 365
Dancing Dervish, 251
dapple
  grey, 251
  silver, 263
Darley Arabian, 40
daughter cell, 414
Dawn Horse, 2
defects
  inherited, 325
  congenital, 327
deletion, chromosome, 465
Denmark, 52
dental atresia, 333
deoxyribonucleic acid, 418, 419
dermoids, 353
Diamond, 46
diet, 33
digestive system, 7, 33, 376
  atresia ani, 377
  atresia coli, 377
  atresia recti, 377
  cleft palate, 376
  esophageal defects, 379
  shistosoma reflexum, 379

diluted coat colors, 256
diploid, 430
disposition, 32, 67, 159, 381
disruptive coloration, 240
distal leg spot, 294
DNA, 418, 419, 460
domestication, 23
dominance, 450
  complete, 450
  codominance, 450
  incomplete, 451
  overdominance, 451
dominant trait, 145, 219, 441
dominant white, 249, 255, 502
donkey, 425, 511
dorsal stripe, 51, 259
dosages, standard, 235
double helix, 420
draft horse, 55, 133
  development, 56
dressage, 30, 126
dry coat, 365
dun, 51, 259, 262

E

E locus, 243, 247, 452
earnings index, 108
Eclipse, 40, 44, 81
ectropion, 360
egg, 141, 437
electroretinograms, 359
electroretinography, 357
embryo, 414
emphysema, 394
endurance horse, 130
entropion, 360
environment, 146, 165, 193
environmental variation, 170, 194, 470
enzyme, 460, 504
Eohippus, 3
epilepsy, idiopathic, 390
epiphysitis, 336
epistatic, 145, 249, 451, 502
epistaxis, 395
epitheliogenesis imperfecta, 367
equine osteomalacia, 333
Equus
  asinus, 423, 425
  burchelli boehmi, 423, 426
  caballus, 423, 425
  grevyi, 423, 426
  hemionus onager, 423, 425
  caballus, przewalksi, 423, 425
  zebra hartmannae, 423, 426
ergot, 365
erythrocytes, 133
esophageal defects, 379
  dilations, 379
  strictures, 379
  megaesophagus, 379
eumelanin, 241, 243, 261
evolution, 1, 459
  Cenozoic Era, 2, 6
  dawn horse, 2
  Eohippus, 3
  forest horse, 4, 18
  Merychippus, 4
  Mesohippus, 4
  Mesozoic Era, 2
  migration, 11, 29
  Miocene epoch, 6
  Miohippus, 4
  Oligocene epoch, 4
  Paleozoic Era, 2
  plains horse, 6
  Pleistocene epoch, 14
  Pliocene epoch, 11
  Pliohippus, 11, 17
  Precambrian Era, 2
  Theriodant, 2, 457
ewe neck, 334

Exmoor Pony, 59
expressivity, 227, 452
eye, 348
  cornea, 352
    congenital keratopathy, 352
    dermoids, 353
    melanosis, 353
    microcornea, 352
  iris, 349
    albinismus partialis, 350
    albinismus totalis, 350
    aniridia, 350
    coloboma iridis, 351
    glass eye, 350
    hyperplasia of the corpora nigra, 352
    wall eye, 350
  lens, 353
    cataracts, 353
    lens luxation, 355
    periodic ophthalmia, 355
    persistent hyaloid vessel, 356
  retina, 356
    absence of the retina, 357
    congenital night blindness, 357
    glaucoma, 357
    optic nerve hypoplasia, 358

F

F locus, 265
facilities, 70
Factor VIII Deficiency, 408
families, 155
feathering, 295
feet, 10, 101, 119, 132
fertility, 34, 164
fertilization, 439
fetal heart, 396
fetlock, 293
fetus, 157
fibular enlargement, 342
five-gaited, 53
Fjord Pony, 27, 36
flat feet, 340
flaxen mane and tail, 265
Flying Childers, 40, 44
foals, congenital defects, 507
foramen ovale, 398
forest horse, 4, 18
fox-trot, 54, 316
Fox Trotting Horses, 54, 316

G

G locus, 250
gaited horses, 52
gaits, 303
  amble, 303
  backing, 314
  canter, 310
  fox trot, 54, 316
  gallop, 111
  rack, 322
  running walk, 316, 318
  slow gait, 53
  trot, 307
  walk, 304
Galiceno Pony, 59
gallop
  concussion, 111
  suspension, 111
Galton's law, 231
gamete, 141, 440, 429
gametogenesis, 191, 439
gelding, 66
gene, 1, 142, 415, 458
  detrimental, 327
  frequency, 475
  interactions, 450
  modifying, 450
  recessive, 218
General Stud Book, 41
genetic
  exclusion, 501
  potential, 196
  variability, 469
  variation, 170, 194, 470, 549
genotype, 142, 196, 414
genotypic frequency, 480
glass eye, 49, 350
glaucoma, 357
Godolphin Barb, 40
Golgi apparatus, 415
Gotland Pony, 27, 36
grading, 189
Grant's zebra, 429
Great Horse, 57, 133
Grevy's zebra, 15, 429
grey, 142, 249, 250, 455, 472, 501
  dappled, 251
  flea-bitten, 251
gross mutation, 462
growth rate, 198
grulla, 51, 261
guanine, 419, 422
gut strangulation, 344

H

hair follicle, 242, 243
half-breed, 227
Hambletonian, 10, 44
Hardy-Weinberg law, 476
harness race horse, 116
Hartmann's Mountain zebra, 14
heading horse, 130
health, 69
heart, 396, 400
    circulatory system, 395
heartscore, 112
heaves, 394
heavy horse, 17
heeling horse, 130
height, 198, 212, 486
hematoma, 407
hemolysis, 403
hemolytic foal, 403
hemophilia, 406, 442
heredity, 146
heritability estimate, 74, 76, 207
    table, 208
hernia, 344
    umbilical, 344
    scrotal, 371
    inguinal, 371
Herod, 44
heterochromatin, 463
heterochromia, 349
heterosis, 172
heterozygous, 142, 416
hinny, 190
hip dysplasia, 341
hobbles, 117
Hoist The Flag, 81
holandric genes, 443
Holstein, 122
homologous chromosome, 143, 415
homozygous, 142, 171, 174, 416
hoof, 10, 101, 119, 132
hormones, 165
horse vs. ass, 510
horseshoe, 34
hunter, 120
    pony, 122
hybrid, 428
hybrid vigor, 187
hybridization, 190
hydrocephalus, 389
hyperplasia of the copora nigra, 352
hypostatic, 145
hypothyroidism, 372

idiopathic epilepsy, 390
inbreeding, 169, 171, 480
    advantages, 174
    coefficient, 181
    disadvantages, 171
    relationships, 186
incomplete dominance, 451
independent assortment, 452
infertility, 197
inguinal hernia, 371
inheritance, 415, 549
    mode of, 329
intelligence, 68, 102
interphase, 420, 427
intersex, 159, 445
interventricular septal defects, 398
inversions, 465
iridial heterochromia, 349
iris, 349
Isabella, 261
isoantibodies, 402
isoimmunization, 401
IVSD, 398

J

jack, 190
jenny, 190
Jockey Club, 40
jumper, 120
Justin Morgan, 46

K

keratitis sicca, 360
keratopathy, 352
karyotypes, 422
kyphosis, 348

L

lacrimation, 355
laryngeal hemiplegia, 391
laryngeal paralysis, 391, 492
larynx, 391
lateral curvature of spine, 348
lateral luxation of the patella, 339
leg set, abnormalities, 334
lens, 353
    luxation, 355
leopard pattern, 48, 271

I

lethal genes, 166, 253, 255
  carriers, 173
  delayed lethal, 327
  partial lethal, 327
  true lethal, 166, 327
libido, 150
light horse, 38
linear keratosis, 367
linebreeding, 171, 177
linecrossing, 188
linkage, 452
Lippizaner, 66, 478
liver, 227, 244
locus, 142, 415
  A, 243, 245
  Ap, 270
  B, 244
  C, 256, 262
  D, 256, 260, 262
  E, 243, 247
  F, 265
  G, 250
  O, 267
  R, 252
  S, 263
  Sp, 271
  T, 266
  W, 255
  Wb, 270
lordosis, 345

M

male pseudohermaphorditism, 445
mallenders, 367
Mambrino Chief, 52
management, 165
manners, 150, 159
markings, 274
  apron face, 292
  blaze, 292
  snip, 291
  sock, 293
  star, 291
  stocking, 293
  strip, 291
Matchem, 40, 44
maternal influence, 163
  ability, 159
  instinct, 162
mature size, 198
meiosis, 421, 429
melanin, 241
melanocytes, 241
melanoma, 250, 363

melanosis, 353
Merychippus, 6
Mesohippus, 4
Mesozoic Era, 2
Messenger, 44, 53
metacentric chromosome, 416
metaphase, 428, 432, 434
microcornea, 352
microphthalmia, 359
migration, 11, 29
milking ability, 162
Miniature horse, 59
minimum culling level, 203
Miocene epoch, 6
Miohippus, 4
"mis-sense" codon, 461
Missouri Fox Trotting Horse, 54
mitochondria, 414
mitosis, 426
modifying genes, 450
Mohammed's thumbprint, 367
monorchidism, 370
moon blindness, 355
Morgan, 46
  horse club, 46
Morocco Spotted Horse, 49
mosaicism, 463
mothering ability, 159
mule, 190
multiple alleles, 549
multiple exostosis, 348
mutation, 2, 146, 194, 327, 416, 457
  structural, 458
  point, 460
  gross, 462

N

Narragansett Pacer, 52
navicular disease, 335
negative association, 490
neonatal isoerythrolysis, 401
population genetics, 469
natural selection, 1
Native Dancer, 81
nervous system, 380
  ataxia, 382
  cerebellar hypoplasia, 387
  congenital hydrocephalus, 389
  congenital occipito-atlanto-axial
    malformation, 386
  convulsive syndrome, 390
  idiopathic epilepsy, 390
  shivering, 390

temperament, 381
wobbler syndrome, 382
Nez Perce, 48
NI, 401
nicks, 154, 188
night blindness, 357
non-disjunction, 463
"non-sense" codon, 461
normal curve, 485
Norwegian Dun, 51
Numidian, 20, 30, 38
nymphomania, 159, 373
  pseudonymphomania, 373
  true nymphomania, 373

O

O locus, 267
Oligocene epoch, 4
onager, 25, 425
oocyte, 436
oogenesis, 436, 439
optic nerve hypoplasia, 358
organelles, 414
osteochondritis dissecans, 341
osteomalacia, 333
outbreeding, 169, 186
outcrossing, 187
ovary, 157
overdominance, 172, 451
overlap time, 113, 115
overo, 49, 224, 243, 266, 267, 480

P

pace, 317
pacer, 44, 117
Paint, 49, 266
Paleolithic period, 17
Paleozoic Era, 2
Palomino, 50, 204, 243, 253, 257, 471, 501
  Palomino Horse Breeder's of America, 50
  Palomino Horse Association, 50
parentage, 501
parrot mouth, 330
parti-colored coat, 47, 266, 457
Paso
  gaits, 217
  horses, 54
  Fino, 55
  Peruvian, 55

pastern, 293
patent ductus arteriosus, 397
patent foramen ovale, 398
paternity tests, 501
pedigree, 25, 64, 71, 106
  annotated, 72
  balance, 107
  mare, 161
  stallion, 153
penetrance, 227, 452
Percheron, 35, 56, 134
performance, 64, 78, 106
  rate, 109
  records, mare, 161
  records, stallion, 151
periodic ophthalmia, 355
  moon blindness, 355
  recurrent uveitis, 355
periodontitis, predisposition to, 332
perlino, 260
persistent hyaloid vessel, 356
persistent right aortic arch, 399
Peruvian Paso, 55
Persian onager, 16
phaeomelanin, 241, 243, 257, 261
pharyngeal cysts, 395
pharyngitis, 395
phenocopies, 325
phenotype, 142, 196, 414
phosphate group, 419
photophobia, 355
photosensitization, 361
piebald, 50, 266
pigment, 240
pinky syndrome, 366
Pinto, 49
Pinto Horse Association, 49
pituitary tumor, 299
placentation, 402
plains horse, 6
plasma, 503
plasma proteins, 504
platelets, 503
pleiotrophic, 243, 454
Pleistocene epoch, 14
Pliocene epoch, 11
Pliohippus, 11, 17
pneumovagina, 374
point mutation, 460
points, 245
points of horse, 512
polar body, 438
polo, 42
polydactyly, 342
polygenes, 212, 549
pony breeds, 58

Pony Of The Americas (POA), 59
population mean, 482
positive association, 492
posterior synechiae, 355
Precambrian Era, 2
prepotency, 174, 454
price, 70
primary oocyte, 436
Princequillo, 71
probability, 471
progeny records, 77
progeny test, 152, 175
prognathia, 330
prophase, 427, 431, 434
protein, 418
Przewalski's horse, 17, 19, 32,
67, 245, 425
pseudoalbino, 255
pseudohermaphroditism, 445

Q

qualitative
  inheritance, 213
  traits, 145, 195, 214
quantitative,
  inheritance, 218, 481
  traits, 145, 188
  Quarter Horse, 42

R

R locus, 252
race horse, 106
racing, heritability, 108
rack, 53, 316, 322
Raffles, 71
rafter hips, 336
recessive
  carrier, 218
  trait, 145, 218, 224, 478
records
  objective, 79
  subjective, 79
rectal palpation, 158
recurrent uveitis, 355
red blood cells, 502
red dun, 246, 260
regression, 483
relationships, 179, 186
replication, 416, 460
reproductive system, 67, 80, 81, 368

anatomy
  mare, 156
  stallion, 148
exam
  mare, 158
  stallion, 149
defects
  cryptorchidism, 369
  inguinal hernia, 371
  hypothyroidism, 372
  monorchidism, 370
  nymphomania, 373
  pneumovagina, 374
  scrotal hernia, 371
records
  mare, 160
  stallion, 150
respiratory system, 391
  bronchitis, 394
  emphysema, 394
  epistaxis, 395
  heaves, 394
  pharyngeal cysts, 395
  roaring, 391, 492
retina, 356, 382
ribonucleic acid, 422
ribosomes, 414
rickets, 334
ring chromosomes, 465
Riva Ridge, 114
RNA, 422
roach back, 348
roan, 249, 252, 501
  blue, red, strawberry, 253
  Appaloosa, 271
Roan Allan F-38, 53
roaring, 391, 492
Robertsonian translocation, 463
roping horse, 128
running walk, 316, 318

S

S locus, 263
Saddlebred, 52, 224
sallenders, 367
scoliosis, 348
scotopic vision, 357
scrotal hernia, 165, 371
seal brown, 245
  light, 246
  dark, 246
  dilute, 259
secondary oocyte, 437

Secretariat, 81, 113
Selection, 63, 105
  against a trait, 227
  artificial, 31, 32
  broodmare, 162, 163
  controversial theories
    Galton's law, 231
    Stamina Index, 233
    stamina, progeny, 234
    Vuillier dosage system, 234
    Bruce Lowe system, 237
    telegony, 238
  differential, 470
  index, 205
  intensity, 469
  methods, 200
    number of traits, 200
    tandem method, 201
    minimum, 203
    selection index, 205
    selection pressure, 207
  natural, 1, 194
  performance, 105
  pressure, 207
  progress, 201, 202, 206
  recessive, 478
  stallion, 147
  tandem method, 201
  theories, 154
    broodmare sires 154
    controversial, 231
    families, 155
    nicks, 154
  trait, 217
semen examination, 149
seminiferous tubules, 430
"sense" codon, 461
sex, 66
  chromosomes, 141, 439
  determination, 157, 439
  influenced, 443
  limited, 443
  linked, 441
sheer mouth, 333
Shetland Pony, 35, 59, 264
Shire, 56, 134
shistosoma reflexum, 379
shivering, 390
show horse, 137
sickle hock, 335
sidebone, 335
silver dapple, 263
siring records, 151
size, 66, 190
  at maturity, 198
skewbald, 50, 266

skin, 361
  anhidrosis, 365
  atheroma of the false nostril, 367
  epitheliogenesis imperfecta, 367
  chestnuts, 365
  ergot, 365
  linear keratosis, 367
  mallenders, 367
  sallenders, 367
  melanoma, 363
  Mohammed's thumbprint, 367
  photosensitization, 361
  pinky syndrome, 366
  squamous cell carcinoma, 362
  subcutaneous hypoplasia, 367
  variegated leukotrichia, 366
slow gait, 53
snaffle, 31
snip, 291
sock, 293
Somali Wild Ass, 16
somatic cells 141
Sorraia breed, 51
sorrel, 247
soundness, 65, 69, 111
Sp gene, 271
species, 423
species hybridization, 190
speculum exam, 158
speed, 10, 113
  blood types, 506
  horse, 107
sperm, 141, 430
spermatogenesis, 430, 439
spermatogonium, 430
spindle, 427
splint bones, 11
splints, 335
sprinter, 107
squamous cell carcinoma, 362
stance time, 115
standard deviation, 483, 487
stallion, 147
  reproductive tract, 148
  exam, 149
  reproductive records, 150
  performance records, 151
  siring records, 151
Stamina Index, 233
  progeny, 234
Standardbred, 44
  stud book, 45
star, 291
stayer, 107
steeplechaser, 119
sterility, 444, 445, 447

steroids, 164
stiff joints, 342
stock horse, 126
stocking, 293
stride, 113, 488
strip, 291
structural mutation, 458
stud fee, 153
subcutaneous hypoplasia, 367
submetacentric chromosome, 416
sugar group, 419
Suffolk Punch, 37, 56, 66, 134
supernumerary teeth, 333
suspensory apparatus, 9
sway back, 345
swing time, 115
synapse, 431

T

T locus, 266
tandem method, 201
tapir, 3
Tarpan, 21
teasing, 159
teeth, 6
  canine, 6
  wolf, 6
telegony, 238
telocentric chromosome, 417
telophase, 429, 433, 435
temperament, 102, 381
Tennessee Walking Horse, 53, 264
test cross, 218
testes, descent, 368
  testes, fetal, 370
testicular feminization, 447
texture, coat, 296
theories, selection, 231
  Galton's law, 231
  Stamina Index, 233
  stamina, progeny, 234
  Vuillier dosage system, 234
  Bruce Lowe system, 237
  telegony, 238
Theriodant, 2, 457
three-gaited, 53
thymine, 419
tipped vulva, 165, 374
tobiano, 219, 266
TPR, 131
trainability, 32, 68, 103
training, 68
traits, 200

associated, 203, 488
combination of, 455
culling, 169
genes affecting, 212
introducing, 219
number considered, 200
selecting for, 217
selection against, 227
Trakehner, 122
Transcaspian kulan, 16
transferrin, 112
  alleles, 505
transcription, 460
translocation, 464
triplet, 418
trotter, 44, 117
trichoglyph, 300
trot, 307
true-breeding, 171, 295
True Briton, 46
Turkmene breed, 28
twinning, 166

U

umbilical hernia, 344
undershot jaw, 332
upward fixation of the patella, 338
uracil, 422
uterine biopsy, 160
uterus, 167

V

variation
  genetic, 170, 194, 549
  environmental, 170, 194
variance, 483
variegated leukotrichia, 366
vices, 33
vision, 5, 8
Vuillier dosage system, 234
  standard dosages, 235

W

W locus, 255
wall eye, 350
walk, 304
wave mouth, 333
wavy mane and tail, 300

Wb gene, 270
weight, 210
Welch Cob, 59
Welch Pony, 59
Whirlaway, 81
windsucker, 33
white, 240
  dominant, 249
white blanket, 270
white blood cells, 503
"white foal" syndrome, 243, 378, 267
white markings
  apron face, 274, 292, 457
  blaze, 292
  snip, 291
  sock, 293
  star, 291
  stocking, 293
  strip, 291
whorl, 300
Wild Horse of Central Asia, 17, 20
wild type, 248, 459
Wise Exchange, 251
wobbler syndrome, 382
Wright's formula, 184

XYZ

X chromosome, 417, 141, 439
X-linked
  dominant, 441
  recessive, 442
  traits, 155, 408, 452
XY female, 462
Y chromosome, 417, 141, 439
Y-linked trait, 153, 443
zebra, 15
  Damara, 15, 365
  Grant's, 429
  Grevy's, 15, 429
  Hartmann's Mountain, 14
  markings, 51, 259
zeonkey, 190
zygote, 414, 429

NOTES

NOTES

NOTES

NOTES